YARDSTICK NATION

YARDSTICK NATION

The Metric System in America

HECTOR VERA

VANDERBILT UNIVERSITY PRESS
NASHVILLE, TENNESSEE

Copyright 2025 by Vanderbilt University Press.
First printing 2025.
All rights reserved.

Library of Congress Cataloging-in-Publication Data on file

Names: Vera, Héctor (Vera Martínez) author
Title: Yardstick nation : the metric system in America / Hector Vera.
Description: Nashville, Tennessee : Vanderbilt University Press, [2025] |
 Includes bibliographical references and index.
Identifiers: LCCN 2024044570 (print) | LCCN 2024044571 (ebook) | ISBN
 9780826507839 paperback | ISBN 9780826507846 hardcover | ISBN
 9780826507853 epub | ISBN 9780826507860 pdf
Subjects: LCSH: Metric system--United States--History
Classification: LCC QC92.U54 V47 2025 (print) | LCC QC92.U54 (ebook) |
 DDC 389/.160973--dc23/eng/20250422
LC record available at https://lccn.loc.gov/2024044570
LC ebook record available at https://lccn.loc.gov/2024044571

Front cover image: Vector artwork of rulers by YZ vector

To Claudia Tania and Rufina
To Witold Kula (†)

Contents

Acknowledgments ix

INTRODUCTION. A World Too Small for Two Systems:
Or, A Luxury Only the Rich Can Afford   1

1. An Irresistible Force Meets an Immovable Object:
The Global Expansion of the Metric System and the United States   29

2. Leviathan's Foot: State Capacity and
the Establishment of Measurement Systems in America   69

3. A Troubled Lingua Franca: Scientists, Engineers,
and Educators in the Battle of the Standards   115

4. Searching for a Perfect Language for Commerce:
Measurement and Economy   159

CONCLUSION. Why Is There No Metric System in the United States?:
Five Reasons   207

APPENDIX. Adoption of the Decimal Metric System of Weights
and Measures by Country   221

Chronology of the Metric System in the United States 229
Notes 237
Archives 281
Bibliography 283
Index 317

Acknowledgments

> There are few activities more cooperative than the writing of history. The author puts his name brashly on the title-page . . . but none knows better than he how much his whole enterprise depends on the preceding labours of others.
> —CHRISTOPHER HILL, *The World Turned Upside Down*

My initial interest in the systems of measurement came from a fascination with unsuccessful reforms by the French and Soviet revolutions to replace the Gregorian calendar. A casual chat with Luis Fernando (†) and Tomás Granados directed me, unexpectedly, toward the history of weights and measures. Luis Fernando, a historian, told me, "You know, the French revolutionaries failed to change the calendar, but they invented the metric system." Intrigued, I asked if they knew if something had been written about the history of the metric system. Tomás, a statistician, said, "There is a book titled *Measures and Men*, which, I think, deals with that." It most certainly did! It was no other than Witold Kula's masterpiece on historical metrology. I read it with feverish enthusiasm and, since then, I have dedicated my scholarly life to understanding how people invent, appropriate, and give meaning to measurement systems and how different forms of quantification shape and transform the life of social groups.

Writing a book is an exciting but time-consuming task. As years pass, one acquires multiple personal, intellectual, and institutional debts. Thus, there are personal acknowledgments to be made.

Thinking is an inner conversation, but one that is nurtured by readings and face-to-face exchanges. The outlook developed in *Yardstick Nation* was greatly influenced by Jorge Bartolucci, Eiko Ikegami, and Eviatar Zerubavel. Their works on the history of astronomy, historical sociology, and the sociology of time taught me a great deal about how to understand the articulation of meaningful human actions, sociocultural change, and the making of measurement and reckoning methods.

My conception of how to do social research that is both historical and theoretically informed comes from my experience with a stimulating community at The New School for Social Research, among them, Oz Frankel, Carlos Forment, Michael Donnelly (†), José Casanova, Sarah Daynes, Eli Zaretsky, Federico Finchelstein, Amy Stuart, Monica Brannon, Yoav Mehozay, Abdul Quader, Kimberly Spring, Gema Santamaría, Peter Marina, Keerati Chenpitayaton, Yifat Gutman, Aysel Madra, Dan Sherwood, Lindsey Freeman, Jeff Zimmerman, Ritchie Savage, and Samuel Tobin. Marisol López Menéndez gave me encouragement and advice on countless occasions.

In Mexico, I have enjoyed a prolonged sociological dialogue with Jorge Galindo, Olga Sabido, Priscila Cedillo, and other partners from the Sociedad Mexicana de Sociología (SMS). Álvaro Morcillo provided me with precious advice to find a suitable publisher.

Esteemed friends and colleagues listened to ideas and read drafts, proposals, and chapters that culminated in this book: Rubén Flores, Camila González, Andrei Lecona, Carla Escobar, Gustavo Rojas, Miruna Achim, Yuval Feinstein, Gil Eyal, Diane Vaughan, Jean-Claude Hocquet, Virginia García Acosta, and Aashish Velkar.

I want to thank Steven Rodriguez, editor at Vanderbilt University Press, for his unflagging enthusiasm; finding a knowledgeable and supporting editor is a rare fortune in academic life.

Cristóbal Henestrosa greatly improved the maps displayed in the book. Braulio Güémez helped me to find a rare brochure at Duke University's library. Amber Paranick, reference librarian in the Serial and Government Publications Division at the Library of Congress, provided vital clues to identify the creator and source of the cartoon used in the Introduction. The physical and digital collections at Columbia University's Butler Library and the New York Public Library were essential for this work.

I am deeply indebted to UNAM's Instituto de Investigaciones sobre la Universidad y la Educación for providing a splendid space for conducting research. Financial support from the Consejo Nacional de Humanidades, Ciencias y Tecnologías (Conahcyt) was crucial to developing and culminating this book. The writing of Chapter 3 was possible thanks to the support of UNAM's Programa de Apoyos para la Superación del Personal Académico (PASPA-DGAPA). Their backing gave me the "precious boon of time" necessary for detailed and thoughtful work.

Cultural products, like books, owe their existence not only to the efforts of their nominal creators but also to the "anonymous toil of their contemporaries." This book would have been impossible without the work of librarians, archivists, reviewers, discussants, research assistants, students, editors, copyeditors, designers, secretaries, and many other members of the indispensable "support personnel" that sustain the academic world.

My family—Luciela, Jorge (†), Graciano, Claudia, and Jorge Luis—have been a constant source of pride and support.

This book is dedicated, with love, to Claudia Tania (who listened and commented on many ideas and eased multiple worries) and to Rufina, our greatest co-authorship.

Copyright Acknowledgments

Some paragraphs in Chapters 1 and 3 appeared in "Melvil Dewey, Metric Apostle," published in *Metric Today* 45, no. 4 (July–Aug. 2010): 1, 4–6 (reproduced with permission of the US Metric Association) and "Decimal Time: Misadventures of a Revolutionary Idea, 1793–2008," published in *KronoScope* 9, no. 1–2 (2009): 33–37 (reproduced with permission of Brill Academic Publishers). Sections of Chapter 4 were previously published in "Breaking Global Standards: The Anti-Metric Crusade of American Engineers," in *Technology and Globalisation: Networks of Experts in World History*, edited by David Pretel and Lino Camprubí, 189–215 (Palgrave Macmillan, 2018; reproduced with permission of Palgrave Macmillan).

Metric Menu

	Grams
Cape Cod Oysters, Mignonette	120
Potage Mongole	150
Olives	20
Radiches	80
Rolls	12
Butter	8
Broiled English Lamb Chop with bacon	330
Potatoes sautes	70
Green Peas a l'Etuvee	50
Salad, Jurassienne	40
Glace Napolitaine	60
Coffee	60
Total	1000

INTRODUCTION

A World Too Small for Two Systems

Or, A Luxury Only the Rich Can Afford

> Now after World War II, new interest in the metric system has arisen and we think it will be accepted universally. Our world is too small for two systems.
> —JOHN T. JOHNSON, *Morning World-Herald* (1946)

"Just Because We Never Did It"

On April 7, 1923, the luxurious Hotel Astor in New York was the location of a "Metric Luncheon," held under the auspices of the American Metric Association. The menu was carefully crafted to match the motive of the occasion. The invitation promised that the five-course meal would weigh exactly one kilogram.[1] (See opposite page) The luncheon was organized to raise funds for the Association. The charge for this "balanced menu" was $2.00; a precise rate, the organizers boasted, of "5 grams for one cent." Metric Luncheons were repeated several times every year for the rest of the decade. The 1920s—as the 1860s and the 1900s before, and the 1970s afterward—was a period of great hopes and effervescence for metric devotees. The advent of a metric era in America seemed imminent. The toil of participating in political lobbying, educational campaigns, international meetings, scientific discussions, and public controversies was about to pay off, or so they thought.

There have been other periods when the idea of a metric America did not appear so imminent; on the contrary, it seemed as a fact of nature or providence—as a self-evident truth—that the United States is not, and never will be a metric country. In the first two decades of the twentieth century, metric hopefuls were active and confident; but in the 1940s they were apathetic. The tone of metric exhortations became less wholehearted and more whiny and self-pitied. An editorial page in the *New York Daily News*, a few months after World War II, conveyed that sentiment. The metric system, it said, "is standard throughout most of the civilized world and some of the uncivilized, except in the United States and the British Empire. We and our British friends persist in limping along with the ancient inches, feet, miles, pounds, short tons, long tons, gills, pints, gallons, magnums, jeroboams, cubic feet, cubic yards, etc."[2] The editorial was accompanied by a cartoon by the syndicated illustrator Leo O'Mealia (fig. 0.1), who captured the aura of taken-for-grantedness surrounding the notion of a non-metric America. It shows a couple in their daily routine, the wife, reading the newspapers, asks "Why on Earth don't we adopt the metric system if it's so much better?"; bothered, the husband replies, "Why, just because we never did, my dear."[3]

"Just because we never did" expressed neatly a widespread attitude; but it did not answer the question. Why? Why were the expectations of metric enthusiasts not fulfilled? Why did the whole world embrace the metric system, but not America?

Yardstick Nation is a book about the metric system in the United States, which is to say, a history of something that did not happen—but not a history about nothing.[4] It answers the question why is there no metric system in the United States? A question that metric supporters of different generations have asked so many times. To address it, we need to understand the history of the United States, the global history of metrication, and the social nature of measurement.

A Yardstick Nation

It is a widely popular assumption that the United States is an example for the world and its history is (or should be) a model that the rest of the world ought to emulate. This is mixed with the assumption that our whole era is inexorably fashioned by the cultural influence of the United States.[5]

Figure 0.1. Cartoon by Leo O'Mealia. *New York Daily News*, November 15, 1945.

This image of America as a yardstick, a standard against which all other countries compare themselves, is a self-aggrandizing myth. The history of weights and measures serves as an antidote to this fable. More than that, it shows that the United States is a yardstick nation in another sense, it is the country of the yard in a world where the meterstick is the ruler.

The United States is the only country in the world that could—but decided not to—adopt the metric system. Why is it that a nation known for its openness to the future, its scientific innovations, and its preference for practicality has not embraced the most practical, scientific, and innovative system of measurement? *Yardstick Nation* explains why there is no metric system in the United States and how things got to where they are now. To do so, it analyzes the political, economic, and intellectual factors that determined the trajectory of the United States as a self-excluded nation from the most successful measurement language in the world.

Measurement—How Almost Everybody Does It

Measurement is "The act of measuring or calculating a length, quantity, value, etc."[6] So defined, it looks like an uncomplicated action. However, on this simple act rests most economic calculations, the possibility to communicate scientific results, the political administration of sensitive issues like land tenure and taxes, and the cognitive apprehension of the physical dimensions of all objects.

Measurement helps people to make information about their physical environment transmissible and comparable; as such it is ubiquitous.[7] Weights and measures are present in an infinity of ordinary activities.[8] As John Quincy Adams pointed out in 1821:

> Weights and measures may be ranked among the necessaries of life to every individual of human society. They enter into economical arrangement and daily concerns of every family. They are necessary to every occupation of human industry, to the distribution and security of every species of property, to every transaction of trade in commerce, to the labors of the husbandman, to the ingenuity of the artificer, to the studies of the philosopher, to the reaches of the antiquarian, to the navigation of the mariner and the marches of the soldier; to all the exchanges of peace and all the operations of war. The knowledge of them, as an established use, is among the first elements of education and is often learnt by those who learn nothing else, not even to read and write. This knowledge is riveted in the memory by the habitual application of it to the employments of men throughout life.[9]

Measurement is an activity set within cultural and social practices that give it meaning and purpose even in its most elementary expressions.[10] It is historically variable, as it is produced on each occasion by the technical capabilities of an epoch, the notions of precision, the moral economies that regulate "fair game," the shared understandings of justice, and the multiple forms of political authority in any given society.

If measures work properly, they are mostly imperceptible. The deep cultural significance of measures can be appreciated in the moments when they are taken for granted and become invisible.[11] To do their job in the best possible way, measurement standards need to operate as shared assumptions, like a native language. People speak the idiom of measurement when they exchange information and trade objects. This allows them to strike agreements, make distinctions, organize their thoughts, and quantify the world. How social groups use the idiom of measurement expresses their sense of order and fairness. The methods of measurement define who people are and what they value.[12]

Measures, however, do not always work properly. When that happens, they become opaque, troublesome, and disorienting. See, for example, what happened in Brooklyn in 1902, when city surveyors reported that they recognized as legal four different units of measurement called *foot*, all of different lengths: the Williamsburg foot, the Bushwick foot, the foot of the 26th Ward, and the United States foot. Besides the expected confusion about which foot was used to appraise what piece of land, metrological homonymy created serious fiscal problems. Some areas of the borough's real estate were untaxable because different surveys, using different feet, became legally nonexistent![13]

Up to the nineteenth and twentieth centuries, mix-ups like that were the norm around the world. In Russia, roads were measured with different *versts*; in Japan, the amount of rice necessary to feed a person for a single year was calculated using different *kokus*; in Turkey, weighing was done with different *okkas*; in Mexico, land areas were estimated with different *fanegas*. In those countries, the solution to the predicaments caused by the heterogeneity and overabundance of units of measurement was to adopt the decimal metric system. Meter, liter, and kilogram replaced a myriad of units to provide governments, merchants, scientists, and regular people with a stable, simple, and standardized measurement system. And the same happened almost everywhere else; but the *almost* should be explained.

The United States is one of the five countries that have not fully adopted the metric system. Among those, it is the only one that does not have the excuse of a chronic lack of science and technology funding to explain that conspicuous absence. The United States has not transitioned to the metric system despite being the country with the biggest national spending on research and development—or perhaps it has not done it for reasons related to having the biggest R&D budget. The rest of the non-metric countries (Liberia, Palau, Micronesia, and the Marshall Islands) spend so few resources on technological infrastructure that they hardly gather any data about it.

The other countries that have not made their transition to the metric system are in that position due to a lack of technical and economic resources. Take the case of Samoa. For decades it had an outdated legislation on weights and measures that existed before its independence from New Zealand in 1962. That legislation only allowed imperial measures (even though in practice they accepted metric units). In 2010 a spokesperson of the Samoa's Ministry of Commerce explained to me that they "have yet to identify financial assistance to assist us with drafting a new bill on measures as there is a lack of expertise around our islands in this field."[14] Adding to this, Samoa needed, for lack of independent regulations, to adhere to testing recommendations set in New Zealand and Britain for the calibration of petrol pumps and trader scales. It was not until 2015 that Samoa was able to pass a Metrology Act that made the use of metric units mandatory and exclusive in commerce—even though they still need to verify and authenticate its primary standards at national measurement institutes of countries that are signatories to the Metre Convention, limiting thus its technical self-determination.

Of course, this is not the case in the United States. This is the only country that has had the economic and technical means to switch to the metric system but has decided not to. In this book we shall see why.

The Social History of Measurement

Unthankful only in appearance, metrological studies, in the hands of an intelligent researcher, became an instrument capable of revealing the great streams of civilization.
— MARC BLOCH, *"Le témoignage des mesures agraires"* (1934)

Yardstick Nation belongs to a tradition of sociological and historical studies that highlights the importance of the social relations intertwined in all forms of quantification. The significance of this scholarship cannot be stressed enough. In an age obsessed with measurement, we have inquired too little into how the most basic instruments and conventions we use to measure came to be in the first place. Systems of measurement are the hidden infrastructure that makes possible the social acts of weighing, gauging, computing, reckoning, and quantifying.

Despite a lack of new research about the history of weights and measures in the United States—aside from a few journalistic or panoramic works—in recent years a series of studies have focused on the consequences of measurement and quantification in modern society.[15] A few monographs analyze how technologies of quantification, like censuses and accounting, have penetrated the fabric of American life; also, researchers have explored the history of systems of measurement in Europe.[16]

The forefather of the social studies of measurement was Polish historian Witold Kula. For him, in a social history of weights and measures the technical details of measuring are not as important as the broader activities and institutions related to weighing and measuring. Historical metrology, Kula insisted, should take into account all the "elements associated with measuring: systems of counting, instruments of counting, methods of using these instruments, . . . the different methods of measuring in different social situations, and finally, the entire associated complex of interlinked, varied, and often conflicting social interests."[17] All these elements form part of an articulated structure. The objective of historical metrology is thus to locate this structure within the society that produced it.

In America, sociologist Otis Dudley Duncan was one of the first to advocate for this kind of research.[18] A sociology of measurement—or a "sociological metrology" as he called it—is necessary for understanding the role of quantification in society.[19] Physical dimensions and techniques for measuring them are social constructs that were invented to solve social problems. Systems of physical units evolve through complex social processes that invite investigation in the fields of social change, class conflict, social movements, bureaucratization, and the sociology of knowledge. As Duncan put it, "The obligate symbiosis of sovereignty and mensuration, the bureaucratization of science consequent upon its opting for measurement standardization, the resilience of custom, the role of social upheaval in cracking custom's cake—would seem to be grist for the mill of a sociology of measurement."[20]

The Metric System as a System of Social Relations

Renamed in 1960 as the International System of Units (SI), the decimal metric system of weights and measures was created during the French Revolution. A reform in weights and measures had been demanded for a long time in France, but it was with the revolution that social conditions made the metrological overhaul possible.[21] The general design of the new system culminated in 1795 with the creation of an elegant scheme with three interrelated basic units—meter, liter, and kilogram—that replaced hundreds of local measures that coexisted, in disorganized fashion, all around France. In that country alone there were approximately 700 different units with distinct names, plus their local variations (altogether totaling approximately 250,000 measures).[22]

The metric system is a language: an interrelated vocabulary, at once verbal and mathematical. It involves a series of standards (both as socially agreed definitions and objects in which those definitions materialize), and a set of socially sanctioned practices about who, when, and how the units, standards, and instruments of measurement should be used.

Like other scientific and technical languages, the metric system appears to be an innocuous device, a dry standard devoid of meaning and restricted to the practical purposes of gauging length, volume, and mass. However, as Karl Marx's table, which at first sight seems an "extremely obvious and trivial thing" but when it enters in relation to other commodities "stands on its head, and evolves out of its wooden brain grotesque ideas, far more wonderful than if it were to begin dancing of its own free will," so once the metric system enters social life it is transformed into an entity entrenched in deep and mutable chains of meaning and interests.[23] In social practice the metric system becomes an instrument whose mere measuring capabilities are relatively unimportant, while its economic and political potencies are forceful and menacing; it turns into a language judged not just for its scientific prowess but for its nationalistic, ideological, and religious connotations.

Weights and measures do not have a history of their own, or an existence independent of the social structures, personal interests, and collective beliefs in which they operate. Systems of measurement are systems of social relations.

Three Principles: Centralization, Regeneration, and the Geometric Spirit

The metric system bears the stamp of three political and ideological features of the French Revolution: regeneration, centralization, and the "geometric spirit." Centralization was the way to organize the dealings of the state in accordance with the dictums of a single authority—homogeneity of practices and state regulation of even minute activities were objectives of the new government. Regeneration was the ideological outlook that saw the revolution as an opportunity to create a new France, a new history, and a new "man." The geometric spirit was a way to conceive and organize thinking, art, people, territory, and government according to symmetry, regularity, and exactness.

One of the main consequences of the revolution was to augment the power and the rights of public authority. It replaced the structure of the feudal society with a social and political organization marked by uniformity and simplicity. As Alexis de Tocqueville noted, "what suddenly emerged from the entrails of a nation that had just overthrown the monarchy was a power more extensive, more minute, and more absolute than our kings had ever exercised."[24] The revolution heightened the power of a unique central organ of government, with its intendants and professional administrators dispatched to rule the provinces. Administrative courts and claims bureaus trespassed on traditional local tribunals. It forced even the apparently most independent provinces to the same rules and administrative patterns as the rest of the country.[25] The idea of public authority vindicated state intervention and the idea of public service "justified its exorbitant character."[26]

With the abolition of feudal rights, the revolutionaries destroyed the social foundation of the old metrological regime. The privileges of the lords to regulate the weights and measures used in their seigniory existed no more. The necessary conditions for the metrification of France began there.

Centralization was married to the idea of homogenization. The language, laws, administration, festivities, and measures of the nation had to either resemble the center of authority or follow the plans designed there. People should no longer be Alsatians, Bretons, Corsicans, Normands,

Occitans, but Frenchmen."[27] All vernaculars, provincial customs, and local idiosyncrasies had to be purged. The different patois and local idioms were to be replaced with *français national* (national French). All local units of measurement were to be supplanted in favor of the metric system.[28]

The revolutionary state did a great deal to materialize the abstract plans of the metric system into actual practices of the citizens. It financed the scientific research needed to base the meter on "nature," including the years-long arc measurement done by Jean-Baptiste Delambre and Pierre Méchain (research necessary to satisfy the original definition of the meter as "one ten-millionth of the distance from the equator to the North Pole").[29] More expensive and troubling were other imperative tasks: confiscate old instruments of measurement, produce and distribute new ones, verify and seal measuring instruments, create tables of equivalences between old and new units, police the markets to prevent wrongdoings during the transition period, and teach the population how to use the new measures and make calculations with decimals.[30] An abundance of conviction, stubbornness, and singlemindedness were required to push through this slow and painful process.

Regeneration was a central political idea in the French Revolution. The concept conveyed a political, moral, and social program aimed at the creation of a "new people." The ultimate goal was to "reform everything." Images of rebirth and renewal were part of the revolution's symbolic rupture with the past. As the revolution became more radical, however, "regeneration" lost its association with the hope of bringing back an idyllic society. Nothing was to be expected from previous eras, the past was seen as a source of corruption; there was nothing worth saving from history.[31] Let us get back to Tocqueville, "in 1789 the French tried harder than any other people has ever done to sever their past from their future, as it were, and hollow out an abyss between what they had been and what they wished to become."[32] The Revolution inspired the idea that they were not merely changing France's social system but regenerating the human race.[33]

The metric system reflected this ideal. Reforming weights and measures was one of the most recurrent issues raised in the lists of grievances (*cahiers de doléances*) in which the three estates expressed their complaints to the King in 1789.[34] When the revolutionaries tackled the problem, they were not satisfied with just perfecting the old units of measurement. They did not try to simply set the measures of Paris as national standards.[35] Instead, they opted for the creation of a completely new system that would not carry anything from an imperfect and corrupt past. A whole system of measurement was invented afresh, all the nuclear elements of a measurement system were newly made: the magnitude of the units, their names and prefixes, and the arithmetic logic of grouping and dividing. The history of measurement—as the history of humanity—had to be restarted. If the republican calendar signaled the first day of an emerging era of liberty, the metric system marked a new length, a new volume, and a new weight for a nascent age of equality.

The geometric spirit was a philosophical idea that characterized many eighteenth-century thinkers. It involved an aspiration to think clearly, rationally, and scientifically to liberate people from the tyranny of superstition, ignorance, and theology.[36] In aesthetic terms, it aimed to replace the stylistic exaggerations of the Baroque and the irregularity of rococo with elementary geometrical forms. Writers imagined utopian cities based on a rigid geometry, with quadrangular or circular shapes, divided into equal juxtaposed parts or symmetrical rings arranged around a clearly defined center. The great ideas of equality by nature and equality before the law were given spatial expression by means of rule and compass.[37] The geometric spirit was a "quest for simplicity."[38]

The Revolution gave sway to the people inspired by these ideas, and they put them into practice. When radicalism was in the ascendant, intellectuals acquired their maximum of authority and power of intervention. Their geometric concepts became an explosive substance to shatter the edifice of tradition.[39] Draconian measures were introduced to reshape the rules of the government and the mores of people according to the new spiritual grid. As they say, "geometric reason is as hard and sharp as steel."[40] The marriage of state power and the geometric spirit framed a veritable fearful symmetry.

The geometric spirit, combined with the energy of the revolution, gave birth to memorable inventions. The ingenious optical telegraph, the flawless metric system, and the terrible guillotine.[41] Of those, especially the guillotine captured the collective imagination and has come to represent those tumultuous years. Victor Hugo described it as an artifact of wood, iron, and cords with a "sombre initiative," a mechanism that "eats flesh and drinks blood."[42] Alejo Carpentier called it "the Machine" (with a capital M), an instrument that stunned people with its "implacable geometry" and its "necessary precision."[43] It was the perfect symbol of terror because it standardized executions. It was equalitarian, and methodical; and it was used quantitatively—the revolutionaries were "oddly preoccupied with counting the guillotined."[44] Poets and novelists never wrote inspired lines about the metric system as they did about the guillotine, but their affinities were unmistakable. The metric system was created to conduct life orderly and systematically, the guillotine to end life orderly and systematically.

Critics of the revolution, like Edmund Burke, were horrified by the destructive potential of a draconian application of the quest for simplicity. He despised the annihilation of the old corporations and tribunals, the confusion of orders and ranks, and the "equalizing principle" behind it. He was acutely aware of how the geometric spirit was removing traditions and local specificities. When the revolutionary government was reorganizing France's political geography by dividing the national territory into geometrically equivalent departments (subsequently subdivided into communes and cantons), Burke expressed that "nothing more than an accurate land surveyor, with his chain, sight, and theodolite, is requisite for such a plan.... When these state surveyors came to take a view of their work of measurement, they soon found that in politics the most fallacious of all things was geometrical demonstration."[45] Burke condemned the Republic's efforts for national uniformity and homogenization of space, time, language, and weights and measures. The Revolution aspired to restart the world and to remeasure everything: it tried to reform people and to redefine space and time. The departments succeeded the provinces, the republican calendar succeeded the Gregorian calendar, and the metric system succeeded the measures of the lords.[46]

The metric system was the perfect embodiment of the geometric spirit. When Condorcet defended the idea of a reform of weights and measures based on a "natural, uniform, and unchangeable standard," he insisted that "ideas of uniformity and regularity, please all minds, and particularly sound minds."[47] The metric system fulfilled that vision. First, it was an interconnected system: the meter was the basic unit, the liter was equal to one cubic decimeter, and the kilogram was equal to the mass of a liter of water at the temperature of its maximum density. Second, the revolutionaries aimed for "the decimalization of everything measured or metered."[48] Their plans to thoroughly restructure all forms of measurement included weights and measures, the circumference, currency, and time reckoning, all of which were fit into a decimal grid.[49] The election of decimals was a particularly radical option because regular folk were not familiar with decimal fractions.[50] As an early critic of decimal measures put it, "every girl and every unlettered tailor knows what half a quarter-ell stands for; but we would lay a hundred to one that many professional accountants would be unable to assure you that half a quarter-ell is equal to one hundred and twenty-five thousands."[51] But this did not stop the revolutionary fervor for decimalization, for its potential to facilitate calculations, and for its arithmetical simplicity.[52]

Summing up, in revolutionary France (and later in other countries) the geometric spirit provided a vision of the future in which the metric system had a central place. The centralized state provided the muscle to set in place the building blocks of the new order by regulating the lives of citizens even in their most mundane aspects. And regeneration provided the reason to justify the destruction of old mores and institutions.

Three Counter Principles: Laissez-Faire, Tradition, and Cultural Specificity

Were the principles of centralization, regeneration, and geometric spirit present in the United States? Not all of them. Regeneration and the geometric spirit were widely followed ideas in the American republic; but the most crucial element, centralization, was absent. That is one of the missing pieces in the American metric puzzle. Additionally, critics of metrication in America articulated their opposition around three counter-principles: laissez-faire instead of centralization, tradition instead of regeneration, and cultural specificity instead of the geometric spirit.

Americans were no strangers to the geometric spirit. Their territory, cities, and currency were products of the grid mentality that orders the world with straight lines and right angles. The United States was the first country to implement a fully decimal currency, in 1792.[53] American librarians invented the first systems of book classification—the Dewey decimal classification and the Library of Congress classification—that organize knowledge following a clear and systematic logic. American cities and towns are arranged following a rectilinear design (quite different from typical medieval cities) and were pioneers in developing a method to systematically number every single house in each street.[54] Many states—from Pennsylvania to Oregon and from North Dakota to New Mexico—have the distinctive rectangular shape that characterizes American political geography, a shape that confirms that since its beginnings (with the Land Ordinance of 1785) the American republic "was not underdeveloped in measurement activities."[55]

American metric advocates adored the simplicity of decimal weights and measures. One of their favorite slogans was that the entire metric system could be described with a single sentence: "Measure all lengths in meters, all capacities in litters, all weights in grams, using decimal fractions only, and saying deci for tenth, centi for hundredth, milli for thousandth, deka for ten, hekto for hundred, kilo for thousand, and myria for ten thousand."[56] Simplicity was a plus.

Regeneration was also a staple in the creation of the country, a new republic self-conscious of not carrying the weight of a medieval past, a nation that imagined itself as starting its history with a clean slate. In the words of Thomas Paine, "We have it in our power to begin the world over again."[57] *Novus ordo seclorum*, the United States motto, was aptly used to signify the beginning of the "new American era." Rejecting the metric system cannot be attributed to lack of affinities with an ideology of regeneration.

The point where the social conditions that shaped the metric system in France and the political and cultural circumstances in America diverged significantly was in the importance of administrative centralization. This is not to say that the state apparatus in the United States has been a stranger to centralized decisions and resources. The American state had effectively achieved centralization in many areas. The federal government was active in creating a single, homogenous national currency and in dividing the territory with a coherent system of zip codes, to mention just a couple of examples. However, the kind of efforts made to eliminate privately issued monies and subdue counterfeiters to secure the circulation of a single national currency were not replicated to secure a single national measurement system. In the Department of the Treasury, the

United States has a single authority and technical overseer in monetary matters. In the National Institute of Standards and Technology (NIST), the country has a single technical supervisor of weights and measures, but it *does not have a single, nationwide metrological authority*.

Why is centralization so important? For the metric system to be socially operational, a critical mass of people and institutions must use it synchronically. Piecemeal transitions, with uncoordinated actions by a few local authorities, schools, and occupational groups bring messy, unsatisfactory results. Take for example what happened in Ohio in 1973, when the State Department of Transportation placed two pairs of road signs along Interstate Highway 71 that showed the distance to Columbus, Cleveland, and Cincinnati in kilometers and miles.[58] It was not part of an overall plan to change signs on all national roads, not even in the whole state. The rest of the signs remained indicating only miles and speedometers in cars kept indicating miles per hour. Those signs were a couple of metric droplets in the flow of a customary river. They amounted to nothing.

The scheme of laissez-faire, non-compulsory standardization that has been used for more than two centuries in the United States has functioned to a limited degree. Although there have been intense periods of unforced adoption of technical conventions, when "simplification" was a battleground in government and industry, many of those processes ended up truncated. In the 1920s, for instance, the Secretary of Commerce carried out multiple standardization projects aiming to eliminate needless sizes and styles among everyday commodities like bedsteads, bed springs, and blankets. Simplification was seen as a necessity to improve industrial and commercial efficiency. Having too many options was inefficient. The Secretary pointed out that there were more than half a million different styles of shoes on the market and one kind of wood that was sold under twenty different names. In his plan, the government should coordinate the *voluntary* adoption of a reduced number of standards in the industry. However, it had limited success. Today, for example, the number of mattress sizes is smaller than a century ago, but alongside the regular king size (76 × 79.5 inches) still exist a variety of local sizes, like California king (72 × 83.5), Texas king (80 × 98), Wyoming king (84 × 84), Alberta king (96 × 96), and Alaskan king (108 × 108). The policy of non-compulsory adoption of weights and measures had similar results, with two measurement systems cohabiting in the same country, making the redundancies of a dual economy inevitable.

Another hindrance to metrication has been the fear that a change in weights and measures will destroy part of the American tradition and eliminate some of the "uniqueness" and "exceptionality" of the country. The instigation of moral panics has been a fixture in anti-metric rhetoric. Metrication has been recurrently depicted as part of international complots against the United States. A constant among those who oppose the metric system is the notion that the introduction of the meter is an attack on Americanness, on the values and traditions that define national identity. The metric system has been associated with whatever political enemy of America—real or imaginary—was more menacing at the time. In the 1880s railway engineer Charles Latimer chastened the "followers of Darwin and the infidel" because they "will deny the inspiration of our weights and measures and will hail the appearance of the new French unit, . . . abandoning the Sabbath and burning the Bible."[59] In the 1920s, textile industrialist Samuel Dale revealed an alleged German plot according to which a victory over England in World War I would have meant "the immediate enforcement of the metric system" throughout the English-speaking world.[60] During the Cold War, Dean Krakel, director of the National Cowboy Hall of Fame and Western Heritage Center, in Oklahoma, asked Congress to repeal the Metric Conversion Act of 1975 and presented an appeal to the Supreme Court to impede the federal government from spending $2.5 million to promote metrication. Krakel argued that the metric system was part of a communist plot inspired by a "Marxist doctrine of one world, one monetary system, one language, one educational system, . . . one measurement system."[61] In recent times of anti-globalist populism, conservative TV host Tucker Carlson warned in a program that "almost every nation on Earth has fallen under the yoke of tyranny: the metric system."[62] Carlson then interviewed James Panero, an art critic and anti-metric activist, who proclaimed that metric was the "original system of global revolution and new world orders." Carlson finished the segment by declaring "I'd accept the kilometer when we accept the euro. Never!"[63]

Ironically, metric advocates have welcomed this kind of rhetoric.[64] They usually try to portray metric opponents as persons with a weak grasp of reality—and people like Latimer, Krakel, and Carlson seem to fit that stereotype perfectly. However, metric enthusiasts usually miss that even if those "backward" individuals present the issues in a paranoic and disproportionate way, there are elements of their discourse that have found an echo in larger social circles. Rejecting the metric system did not save America from an atheist-nazi-communist-globalist invasion; nevertheless, metric opponents are not irrational brutes. The non-adoption of the metric system in the United States is not the result of irrationality, stupidity, or shortsightedness—as many want to believe. Certainly, several anti-metric personalities are guided by nationalistic and even chauvinist attitudes, but many others have been sophisticated thinkers interested in rational and practical solutions to social and technical problems—and some others were at once rational and paranoic, like Harold Urey, winner of the Nobel Prize in chemistry, who called the metric system the "communist's secret weapon."[65] Most of the decisions that culminated in the metrological marginalization of America were rational and intelligible choices—but as frequently happens, a chain of rational actions may produce an irrational outcome.

Other accounts of the absence of the metric system in the United States emphasize either "American exceptionalism" or the influence of anti-metric movements as the main reasons for the repeated failures to introduce the meter, liter, and kilogram into American life.[66] These narratives fail to provide a convincing argument for why the United States follows a great number of international measurement codes—like the Gregorian calendar, Greenwich mean time, daylight saving time, decimal currency, and Global Positioning System (GPS)—but not the metric system. And they do not clarify why anti-metric movements (usually formed by market protectionists, manufacturers, and mechanical engineers) were successful despite being confronted by pro-metric campaigns (organized by exporters, scientists, and educators) that were usually better funded, superiorly organized, and backed by more influential organizations—organizations that promoted the adoption of the metric system, as the American Metrological Society, actually had great success with others endeavors, like the International Meridian Conference of 1884, where Greenwich time was established.[67]

Considering that after more than two centuries of metric attempts the United States is still part of the list of not fully metricated countries, it is tempting to think that the anti-metric movements have made a significant difference. However, their particular contribution to the final outcome (a non-metric America) should not be blown out of proportion. Maybe in *some* historical junctures organized opposition was a necessary condition to impede metrication, but it was never a sufficient condition. After all, there were occasions when securing compulsory metric legislation failed even in the absence of visible opposition. Prior to the 1870s there had not been organized movements to uphold the retention of customary measures, and still, a mandatory metric policy never came about. American anti-metric groups were successful because they fought a battle in which structural conditions tipped the scales in their favor every single time.

Some have argued that "it was not the powers of state that brought about" standardization of weights and measures in modern times; the change was possible, they argued, because the "responsibility for standards passed from the hands of rulers into those of scientists."[68] This is a misunderstanding or a very narrow view of metrological matters—scientists tend to overestimate their influence on society. Experts may control the technical nuances of measurement, but the collective operation of a system of measurement is well beyond the scope of influence of engineers and scientists. People who assume the science-centric perspective tend to forget that to promote the broad adoption of a measurement system experts need to create networks of professional societies and require public funding, assistance from international diplomacy, and help from domestic legislators; not to mention a robust intervention by the state in all matters related to measurement.[69]

The question of how a few professional groups (physicians, chemists, cartographers, drug dealers, etc.) adopted the metric system is quite different from the question of how full metrication is possible. Full metrication means that most people execute most of their measurement activities using the metric system. This only happens when a majority of the population in a country has a practical command of the metric system. We call it "full metrication" and not "total metrication" because in all countries you can find some people who employ, at least occasionally, non-metric measures. Getting to full metrication requires a lot of time. In Mexico, for instance, it took more than forty years of an active metric policy for the metric system to start reaching important portions of the rural population; in Canada, after fifty years since the original adoption of the metric system, it is still possible to find multiple activities performed with English units; not even in its native France metric measures had displaced entirely the medieval units more than a century after the transition had begun.[70] To explain how full metrication is possible we must pay special attention to the role of the state and its administrative capacities. As we shall see in Chapter 2, only states (and not economic actors or scientific societies alone) can guarantee two essential conditions to achieve full metrication: policing the employment of metric units and providing populations with the intellectual and material means to learn the metric language.

Double Handicap: State Capacity and Symbolic Power

Two of the main reasons why there is no metric system in the United States are a limited state capacity and a reduced accumulation of symbolic power. In other words, centralization is precarious and the authority of state agents to intervene in everyday matters has never been taken for granted.

State capacity refers to the "degree of control state agents exercise over persons, activities, and resources within their government's territorial jurisdiction."[71] An increase in state capacity involves, among other things, growth in the means of carrying out intended policies, including the standardization of state practices. State capacities require a substantial amount of "infrastructural power," the development of logistical techniques that aid the effective penetration of the state in social life.[72] Symbolic power, on the other hand, designates new domains of administrative activity that people accept as legitimate state practices—for example when people see as obvious that a state issues birth certificates, then that state is displaying symbolic power. More than the mere recognition of authority as valid, symbolic power means the naturalization of state power. It is the ability to make appear as natural, inevitable, and apolitical, practices that are the product of contingent historical struggles. Thus, along with the recognition of legitimate authority comes a misrecognition through the appearance that no power is being exercised whatsoever.[73]

In America, the state apparatus suffers a double handicap. First, there is a loose form of centralization that limits the state administrative capacities and reduces its ability to control several aspects of social life that in other countries are regulated more strictly (including weights and measures). This is why in America the regulation of standards is not done by nationwide, mandatory programs, as happens elsewhere. It is done, partly, by suggesting "model laws" (that may or may not be adopted by local legislatures) and voluntary standards (which are usually defined by the industry). Public authorities play only an advisory role. The creation of the National Bureau of Standards, in 1901, was supposed to regulate the establishment of standards, but it did not receive the necessary resources to transform American processes of standardization into one controlled by the federal government.[74] Multiple and varied forms of regulating weights and measures coexist at the city, county, and state levels.[75] Using Robert Jenks's apt description, in America standards are regulated by "governing in the absence of government."[76]

Second, the state lacks the necessary symbolic power (legitimacy) to execute firm and comprehensive plans that could interfere with the interests of the industry and the everyday routines of people. This is why the controversies and animosity surrounding the idea of the compulsory use of the metric system is not an isolated case of public suspicion against government intrusion into citizens' businesses. Much less consequential issues have been the object of intense hostility, like the ban on single-use plastic bags or the prohibition of incandescent lightbulbs.[77] If plastic bags and light bulbs cause trouble to government officials, they know that going all in for the metric system and banning yards and pounds would be nightmarish—a costly electoral proposition with no political gains in the short term.

If one compares the metric debates in the United States with those in other countries, a single but striking difference becomes visible. In America, the arguments in favor of metrication (global coordination, easiness in calculations, etc.) and against it (retooling costs, popular resistance, and so forth) were almost identical to those in other countries. What was markedly different were the arguments over who should decide what system of measurement is better and who should be responsible for directing standardization. In other countries, the switch to the metric system created intense disputes around its pertinence and technical qualities, but it rarely raised questions over who should oversee the implementation and who was, in the last instance, responsible for deciding whether to start the transition or not. It was taken for granted that the government was the proper vehicle to address those issues. Debates were about what the government should decide, not if the government should be in charge of it. In the United States, the opposite happened. The main issue was who should call the shots—and even many metric advocates doubted if a government mandate was the right path to metrication. The legitimacy of the state in metrological matters was never assumed. Regarding weights and measures, the American state had an insufficient accumulation of symbolic power; it was not seen by all actors as the legitimate director to orchestrate the process.

When American functionaries have not faced the suspicions and antagonism of their own citizens, they have been more open to pushing through the whole metrication process via state mandates. That is why the United States was one of the original signatories of the 1875 Metre Convention, the conference that cemented the leading position of the metric system internationally.[78] It also supported the Pan-American Conferences of the 1890s and their plans for a fully metric continent. And more telling is that the United States occupation governments introduced the metric system in the Philippines after the 1898 Spanish-American war, and in Japan after World War II.

If there is a feature that has shaped the destiny of the metric system in America, it is not nationalism, traditionalism, or the practical advantages of customary measures. Those aspects have colored the discussions but have not been determining factors. The one factor with lasting consequences has been a loosely centralized state and the aversion to compulsion.

Path-Dependence

In the early years of the American republic, between 1791 and 1826, nine House and Senate committees were appointed to draft and approve a weights and measures legislation that would secure the use of reliable standards, first among public officials (like custom officers, public land surveyors, and postmasters) and then among the wider public. Members of these committees acknowledged the need for action but disallowed all plans that involved general enforcement.[79] This was the beginning of a pattern that continues to this day: the rejection of mandatory policies and the hope that homogenization will be achieved through the voluntary adoption of standards.

Every missed opportunity to adopt the metric system in the nineteenth century opened the door for the customary system to get more entrenched in the social and economic life of the country, and reduced, in consequence, the possibilities of making the transition in the future. Pragmatist philosopher Charles S. Peirce, who worked for decades as metrologist in the United States Coast Survey, described the problem like this in 1890:

> The whole country having been measured and parcelled in quarter sections, acres, and house-lots, it would be most inconvenient to change the numerical measures of the pieces. Then we have to consider the immense treasures of machinery with which the country is filled, every piece of which is liable to break or wear out, and must be replaced by another of the same gauge almost to a thousandth of an inch. Every measure in all this apparatus, every diameter of a roll or wheel, every bearing, every screw-thread, is some multiple or aliquot part of an English inch, and this must hold that inch with us, at least until the Socialists, in the course of another century or two, shall, perhaps, have given us a strong-handed government.[80]

Even if Peirce's point on socialism was off the mark (metrication is not exclusive to socialists or authoritarian governments; most liberal democracies have voluntarily transitioned to the metric system), he was right on the increasing complications of changing systems when a process of industrialization has already taken place.[81]

Measurement standardization is a path-dependent process. Early decisions can have a disproportionally big influence on future decisions. Once a particular standard has gained an initial lead it becomes increasingly difficult to reverse direction. The process locks in a particular standard and tends to lock out other potential options. This does not mean that the optimal solution will prevail: if an inferior standard gains an initial lead it may succeed over more suitable alternatives.[82]

Since the end of the eighteenth century, there have been multiple negatives in the United States to act on the metric system, but they did not have identical consequences. Each time authorities opted for inaction or delayed the decision, an incremental effect followed in the direction of non-metric-adoption. Every missed opportunity meant that pushing for transition would be more complicated the next time around. Nevertheless, even if the range of options narrows down with time and the transition becomes more costly, the switch is never impossible. Dozens of countries went metric in the second half of the twentieth century, like the United Kingdom and Canada, despite having a large industrial infrastructure already built around English units of measurement.

A Luxury Only the Rich Can Afford

The size of the American economy gives the country the ability to impose its own measurement standards on other countries. This advantage remedies many of the economic drawbacks that come with being one of the few remaining non-metric countries in the world. No other nation could afford such a luxury. Even England had to accelerate its metrication process to smooth its integration into the European market, while the United States had no need to do that to sign its commercial agreements with Mexico and Canada, for example. Daniel Immerwahr calls it "the stupefying privilege" the United States enjoys in the realm of standards; it could force other countries to adopt its standards, but it is "never bound by those imperatives itself."[83]

The technological ascendancy of the United States lessens the aches of maintaining a measurement system that everyone else has abandoned. By designing new technologies—or updating old ones—using customary units, the United States maintains those units operative. Take, for instance, the pica, an old unit used in typographic design that corresponds to one-sixth of an inch.[84] Or computer components, like chip leads, which are spaced one-tenth of an inch apart. If metric countries decide to make chips spacing the leads with a round number of millimeters, those parts would not be compatible in the international market. That happened in Russia. In the 1990s they discovered that Soviet assembly equipment could not be used to produce Western-type chips, and the potential exports of microelectronics were lost.[85] You either play along with American standards or you are out of the game.

In 1946, John T. Johnson, president of the American Metric Association, prognosticated that the world had become "too small for two systems." English measures had to die and let the metric system reign supreme.

His prediction nearly came true. Since the end of World War II more than seventy countries joined dozens of other nations that had already embraced metric measures—but the world, for the meantime, has proven big enough for two systems.

Structure of the Book

Yardstick Nation traces the history of the metric system in America, but it does not follow a strict chronological order. Linear stories are easier to follow, but they can be misleading; important associations and affinities between events in different periods could be lost in rectilinear narratives. The chronology at the end of the book shows the temporal sequence of how the main events unfolded, but the chapters are focused on four different dimensions of the social life of measures: the global circulation of standards, the consolidation (or lack thereof) of metrological state capacities, the mobilization of expert knowledge, and the conflicting economic interests around measurement.

Reading a book with a nonlinear structure is similar to watching *Citizen Kane*, where viewers go through a story that zigzags in and out of various points in the protagonist's life. Here, instead of following the boisterous life of Charles Foster Kane as remembered by his friends and colleagues, we will follow the tumultuous trajectory of the metric system in America as seen from the point of view of 1) the place of the United States in a metric world; 2) the resources of the American state in its attempts to establish a sole system of measurement; 3) the experts who fought to convince the public regarding which measures were better suited for their lives; and 4) the incompatible economic interests that wanted to protect American markets maintaining customary units or to participate in a global economy switching to the metric system. Each standpoint (developed in a corresponding chapter) offers a key to figuring out why America is not metric.

The standpoint of global history (Chapter 1) is necessary because contrasting the metric trajectory of the United States with other countries clarifies what is missing in this atypical case. If we want to understand the absence of the metric system in the United States, the first thing to do is to place the country in a larger framework. The planetary conquest of the metric system followed well-defined patterns. Distinct waves of metrication followed recognizable geographical, economic, and geopolitical paths. In some regions, like Africa, the metric system was, in good part, the product of colonialism. In others, like Latin America and the Commonwealth, it came as a result of independence from colonial rule. Regions like the Caribbean went metric to foster international cooperation; others, like England and Ireland, to get inserted in larger economic markets. Some, like Italy and Turkey, went metric to complete processes of national unification or modernization; others to deepen the reach of revolutionary regimes, like France, Russia, and China. What was common to all instances, is that countries opted to go metric in periods of significant social change. Economic transformation, when countries wanted to "catch up" with other parts of the world; political transformation, when states went through ruptures in sovereignty. The negative case that does not fit in the model is the United States. Sometimes its historical timing was unfavorable. For once, America may be the only country in the world that has not gone through a major rupture in sovereignty since before the invention of the metric system.[86] When it gained independence from England, the metric system did not exist yet; so, when the new government wanted to reorganize its weights and measures, metrication was not an option. Later, when it entered a process of national unification, after the Civil War, the government did not want to go metric without the company of its main commercial partner, England. And when the United Kingdom finally switched, in the 1970s, the United States was already in such a commanding international position that going metric did not seem like a necessity anymore. In the 1860s America was so weak that it could not afford to go metric without England; in the 1970s it was so strong that it could remain non-metric without the United Kingdom. Whatever the case, all the propitious opportunities were missed. This chapter traces the patterns of global diffusion of the metric system from its inception during the French Revolution until its most recent adoption by Samoa in 2015. It underlines the significant influence of colonization in spreading the metric system outside Europe, stresses the pioneering role of Latin American countries in making the metric system a transcontinental standard, and highlights the American imprint on the remaining four countries that have not adopted the metric system. Also, the diffusion of the metric system is compared with the spread of Hindu-Arabic numerals, the international time zones system, and the abandoned plans of a decimal system of time reckoning.

From the standpoint of the state and political power (Chapter 2), the metric system is a scientific instrument that states use to make territories, resources, and populations "legible." Nation-states have been the single most important factor for the global consolidation of the metric system. States adopted, regulated, and enforced the metric system among populations at large. They guaranteed the necessary conditions to achieve full metrication: policing the employment of metric units and teaching people how to use them. This was part of the state's interest in establishing a monopoly on the legitimate means of measurement. Controlling weights and measures was a tool that allowed centralized governments to enhance their administrative control (improving the extraction of revenue, undermining the influence of local authorities, consolidating internal markets, and introducing homogenizing institutions that aid the creation of a common national experience). The chapter shows that modern nation-states need metrological unification (obtained usually through metrication), and that metrication needs the states. If we want to understand the history of the metric system in the United States, we need to see it through the eyes of the American state. Two moments were particularly important, the years after the War of Independence, and the period of Reconstruction (1865–1877). The central difference between America and most countries that successfully transitioned to the metric system is the active participation of a centralized state. The United States had to deal with a fragmented federal system where the states of the Union retained greater autonomy vis-à-vis the federal government.

From the standpoint of expertise (Chapter 3), the metric system is an instrument that the scientific community uses as a universal language that aids international and interdisciplinary communication among experts. This view was opposed by some experts who preferred their national standards and by certain occupational groups that invented or perfected their own measurement systems and saw the adoption of metric measures as costly and unnecessary. This opposition was at times vitriolic because units and techniques of measurement are banners of group identity among professions. The metric system was thus confronted with a considerable number of measurement systems created *ex professo* to replace it. The triumph of the metric system over its challengers in scientific circles contributed greatly to its ultimate global success. The chapter shows that in the United States, more than any other country, scientists and engineers were divided regarding metrication. The lack of a unified front of experts to push the government into the path of metrication hindered the chances to implement an aggressive metric policy. Scientists played a limited role in the government; their opinion on metrication was heard but it was not very influential. The chapter examines the leading pro- and anti-metric organizations and the role of their respective founders. There is an analysis of the momentous debate between Herbert Spencer and Lord Kelvin, two of the most famous scientists in the world at the time, which reshaped the content of the discussions over metrological reform on both sides of the Atlantic.

From an economic standpoint (Chapter 4), weighing and measuring are cornerstones among the cognitive processes involved in economic activities. Systems of measurement are forms of social knowledge entangled in practically all economic processes—from estimating land and industrial productivity to transaction costs and asymmetric information. Setting shared standards of weights and measures—like establishing a national currency—is crucial for a national economy. The metric system, additionally, became a valuable instrument as an internationally accepted standard in a nascent economy of global scope. In America, the interests of the economic elite were not homogeneous regarding measurement standardization. While exporters wished to be in harmony with the rest of the world, manufacturers fought to maintain the status quo. The chapter deals with how economic actors in the United States struggled to define what standards of weights and measures would benefit them the most. Looking at different treaties planned for the formation of free international markets, the chapter analyzes the battle between American manufacturers and exporters over the adoption of the metric system. Companies interested in international trade financed campaigns to secure the adoption of the metric system, with the aim of making American companies more competitive in Latin America and the Pacific. Manufacturers, on the other hand, who faced the financial burden of retooling and acquiring new equipment if metric legislation were to be passed, defended "American and Anglo-Saxon" measures, traditions, language, and culture that were—in their opinion—under attack by a foreign-minded, pro-metric elite. This opposition helped to deter metric legislation. Besides competing interests, these groups had different visions of how states and markets should relate to each other. Exporters wanted free markets, but also a state capable and willing to set appropriate international standards (like the metric system). Manufacturers wanted a state restrained from defining standards, but active in protecting national markets from international competition with tariffs and local standards (like customary weights and measures). These positions were in line with their respective ideological outlooks: cosmopolitanism, in one case; and nationalism, in the other.

The Conclusion recaps the argument of the book summarizing the five reasons why there is no metric system in the United States: loosely centralized state, divided experts, international economic predominance, aversion to compulsion, and unfavorable historical timing.

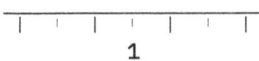

1

An Irresistible Force Meets an Immovable Object

The Global Expansion of the Metric System and the United States

> To challenge the metric system is like challenging the rising tide.
> —CHARLES SANDERS PEIRCE, *The Nation (1890)*

The decimal metric system of weights and measures is one of the most successful intellectual devices ever conceived. Few ideas, if any, have penetrated so deeply and widely into humanity's collective mind as this system of measurement. It can be said, without sarcasm, that the metric system is more popular than Jesus—and more popular than The Beatles. Millions of people who have never read a line from the Bible conduct their routine activities with meters, liters, and kilograms. You can mention any world religion, language, piece of popular culture, or icon of globalization and their achievements pale into insignificance compared with what the metric system has done in its roughly two centuries of existence.

Today eight billion people around the globe use the metric system, which is more than three times the number of Christians in the world (Christianity being the religion with the largest number of followers, with 2.4 billion). The language with the biggest number of native speakers is Chinese, with 1.3 billion (one-sixth of metric users); the language with the biggest total number of speakers is English, with 1.8 billion. Neither The Beatles nor Michael Jackson, the best-selling music artists in history, have sold more than half a billion records. I do not want to suggest that the metric system has been more influential or meaningful in world history than Christianity—or that people adore it as they loved Paul, John, George, and Ringo. It is simply much more widely spread.

Today 95 percent of the world population lives in countries where the metric system is exclusive legal system of measurement; the other five percent lives in the five countries where the metric system is only optional (i.e., it can be used legally, but it is not mandatory). This makes the metric system the only measurement system with which a commercial or civil contract can have validity in every single nation on the planet. The metric system has had an uncanny success not only in its geographical expansion but also in its penetration among social groups and classes; as it is used by landowners, microbiologists, peasants, machinists, butchers, and housekeepers alike. The immense majority of people in almost all countries use the metric system as their main instrument of calculation when it comes to apprehending their own selves and their environment in terms of quantities of length, volume, and mass.

The metric system has been naturalized to such a degree in most of the world that we lose sight of the astonishing fact that the same measurement system employed to calibrate pipettes in Canberra, to indicate the scale of maps of Siberia, and to weigh the chemical elements in a laboratory in Cairo, is also used by milkmen in Vienna, butchers in Ottawa, and shoemakers in Guanajuato. This amazing coordination was literally unimaginable at the beginning of the modern era. The article on "Measure" in Diderot's *Encyclopedia* claimed that "It is a well-known fact that people will never agree to use the same weights and the same measures."[1] It is a historical irony that the intellectual heirs of the encyclopedists proved them wrong and created a universal system of weights and measurement: the decimal metric system. How was this universal language possible?

My aim in this chapter is to show what social and historical conditions account for the improvable but colossal accomplishment of the metric system—and also elucidate the circumstances that have halted its development. What social conditions made it possible for the decimal metric system to be adopted by 95 percent of the world's population? Why did the metric system succeed in its quest for global expansion while similar projects, like the Republican calendar and the decimal system of time reckoning, failed to do so? What are the circumstances that eventually halted the expansion of the metric system and have impeded its complete global diffusion? Why is the United States the only Western country that has not adopted the metric system? Let us find out.

"For All Time, for All People"

Geometricians and algebraists were consulted upon a question which was of administrative jurisdiction. They thought that the unit of weights and measures should be deduced from a natural constant, so that it might be adopted by all nations. They thought it was not enough to provide for the advantage of forty million men; they wanted the whole universe to participate in it.

—NAPOLEON BONAPARTE, *Mémoires (1823)*

The efforts to design a rational system of measurement by the French revolutionaries culminated in 1799 with an international meeting of scientists held in Paris, the Congress on Definite Metric Standards—considered the first international scientific conference in history.[2] There French savants participated with colleagues of nine other nations of continental Europe to finish the calculations and verifications to determine the length of the basic unit of the decimal metric system, the meter.[3] The congress was a solid initial step for the internationalization of the system, which was one of the central aims of its creators.

The original proposal to create the metric system at the beginning of the Revolution considered a collaboration with Great Britain. However, the growing animosity between France and England in the 1790s derailed that part of the plan. Something similar happened with the United States when Thomas Jefferson distanced himself from the proposal of the National Assembly, after having shown some initial interest in the idea of an international measurement system. England's and America's reluctance to participate in the 1799 conference has had lasting consequences in the history of the metric system.[4]

In 1791, when describing the aspirations behind the creation of a new system of measurement designed by enlightened men, Condorcet declared that such a system should embrace all people and ages. That eventually turned out to be the motto of the metric system, "For all time, for all people." Condorcet's vision—so ambitious and optimistic—has come very close to reality.[5] The metric system has become the only universal language to express quantities of length, mass, and volume in all realms of human activity. From astrophysics and textile design to food recipes and pharmacy, we use meters, liters, and kilograms as gauging tools. In two centuries, the metric system has ousted thousands of local, regional, and national units of measurement, some of which had been around for more than a thousand years. How did this happen?

To answer this question we shall trace the path of the metric system from the narrow corridors of scientific societies and bureaucratic agencies in Paris at the end of the eighteenth century to the rest of the countries in the globe in the twenty-first century.

The "Spread of Understanding"

In 1922 noted science historian George Sarton drew attention to what he called "the spread of understanding." Through a series of historical vignettes, he tried to illustrate the complications in disseminating scientific knowledge. Some of those vignettes were dedicated to the practical applications of decimal notation and Hindu-Arabic numerals. These numerals, he said, were a time- and labor-saving invention of the first magnitude, and with it the Hindus made to mankind an inestimable gift: "No strings of any kind were attached to it, nor was the suggested improvement entangled with any sort of religious or philosophic ideas." Those who wanted to use the new numerals were not asked for anything in return, "they were asked simply to exchange a bad tool for a good one." Despite all its advantages, more than a millennium elapsed between its invention and its general acceptance. This tardiness shocked Sarton. Mountains, seas, and deserts, he said, "are smaller obstacles to the diffusion of ideas than the unreasonable obstinacy of man. The main barriers to overcome are not outside, but inside the brain."[6]

Sarton then moved to the figure of Simon Stevin—who developed a simple system to express decimal fractions (what we know now as the decimal point)—and noted that Stevin realized that one of the "logical consequences" of the introduction of a decimal system of numbers and fractions would lead to a decimal system of weights and measures and that the adoption of one was not truly complete without the adoption of the other, "to measure according to one system and to count according to another destroyed the economy of both."[7] Sarton described the invention of the metric system during the French Revolution as the embodiment of Stevin's vision. Assessing the diffusion of the metric system Sarton observed:

> During the last century [the metric system] spread all over the world, except, strangely enough in the Anglo-Saxon countries where it met— and still meets—with a resistance, which is stronger in that it is irrational.... [There] are still many English and American apostles, full or learning, who will prove to everybody who will listen that their incongruous sets of weights, measures and moneys are much more convenient than the metric system! How can they do it? I really don't know, but they do it with a fervor only equalled by the paradoxical absurdity of their plea.[8]

These comments and questions are relevant and intriguing. But Sarton's harsh statements on the "unreasonable obstinacy of man" and the "absurdity" of British and American metric opponents show little sensitivity to relevant issues that may help to explain the form that the spread of human understanding took under diverse social conditions. We should address the issues raised by Sarton from a more sympathetic and sophisticated perspective.

Global Diffusions: Numbers, Time, Measures

The global expansion of the metric system is a phenomenon of extraordinary proportions. Since we live today in a metric world it is difficult to grasp what a colossal enterprise that was. To better understand it, we need to contrast global metrication with other conventions that also operate at a planetary scale, like Hindu-Arabic numerals and International Time Zones.

It took more than one thousand years (from around 500 AD to the end of the sixteenth century) for Hindu-Arabic numerals to be known in Asia, Africa, Europe, and the Americas; and even at the beginning of the nineteenth century it could not be taken for granted that people of all walks of life would know how to use them (symptomatic of this is that introductory pages in some manuals of weights and measures in the early nineteenth century usually contained a brief explanation of the names, symbols, and meaning of those numerals). Their travel from India to Islamic Asia and Africa took around three centuries (Al-Khwarizmi's treatise on calculation with Hindu-Arabic numerals appeared around 825 AD) and started to penetrate Europe around the end of the tenth century (Fibonacci's *Liber Abaci*, a work crucial for their popularization, was not printed until 1202); but they were not widely known in the old continent until the fifteenth century.[9]

Two main difficulties made this diffusion process very slow—in contrast to the 200 years that the metric system needed to go from Paris to Quito and Sydney and everywhere in between. First, Hindu-Arabic numerals demanded to be learned by a critical mass of people in order to be socially operational; this was the barrier of a language lacking enough speakers at a time when communication and transportation were long-winded. Second, Hindu-Arabic numerals clashed with other numeral systems that were preferred—and protected—by authorities in many places. In fourteenth-century Florence, for example, money changers were obliged to use Roman numerals and could not use Hindu-Arabic figures.[10] As long as the eighteenth century, Cyrillic numerals were favored in Russia over their Asian counterparts; it was Peter the Great who ordered the use of Hindu-Arabic digits.

On the other hand, the spread of Hindu-Arabic numerals had the advantage of not needing much of a material or technological infrastructure to be employed; no special instruments are required to count and calculate using the symbols *0, 1, 2, 3, 4, 5, 6, 7, 9*—if something ink and paper were all it was required for most purposes. In the long run, the triumph of Hindu-Arabic numerals set the indispensable conditions for the future attainment of the *decimal* metric system; had a non-decimal numeral system conquered the world instead, the need to revolutionize the arithmetic base of customary measurement systems would not be so pressing.

A more recent example suitable for comparison with the expansion of the metric system is the international time zones system. The Meridian

Conference that brought about this arrangement was held in Washington, DC, in 1884—in part thanks to the lobbying efforts made by the American Metrological Society, which was one of the first and most energetic pro-metric organizations in America. The final agreement that put in place our familiar partitioning of the world into twenty-four different hour zones, with Greenwich serving as the prime meridian, came after fierce resistance from the French delegation—they suggested that if France was to accept the British meridian the English should correspond adopting the metric system. The main obstacle was not a cognitive barrier—like the one that halted the spread of Hindu-Arabic numerals—because setting international time zones did not require clock users to get familiar with a new and fancy time reckoning technique; neither did they need to change how their timepieces work or to replace their old watches with new models. For lay people, the international standard time only meant synchronizing their watches with a public clock—usually the one in train stations. What hampered the acceptance of this institution was surpassing political disagreements between rival nations and subduing the spirit of autonomy of some cities that did not want to set their time according to a British observatory.[11] Nevertheless, and despite this opposition, in less than fifty years international time zones prevailed globally—and became a notable instance of effective internationalism. Few challenges have come about to alter this chronological order. One of those exception came in 2008, when a group of Muslim clerics created "Mecca Time." It is based on the idea that the holy city of Islam in Saudi Arabia is the center of the world and, as such, should replace Greenwich as the prime meridian. The world's largest clock, sited atop the Royal Mecca Clock Tower, started ticking in August 2010 to mark this time—even if its use is not widely followed, it serves as a reminder of the negotiated and contingent character of international standards.[12]

The metric system was able to gain world acceptance much quicker than Hindu-Arabic numerals, but slower than international time zones. As the time reform, the metric system had the advantage of being created in a moment when global interconnectivity was growing rapidly; it was also helped by a vision of internationalism that pushed forward several worldwide initiatives—and these initiatives, in return, enhanced greatly the idea of internationalism by being a living proof of its benefits.[13] However, the metric reform shared some of the inherent limitations that impeded the march of Hindu-Arabic numerals and international time zones.

As Hindu-Arabic numerals, the metric system represented a cognitive barrier for most of its target users—which at the dawn of the nineteenth century included all humankind except for a few thousand individuals, the majority of them scientists and French public servants. Created afresh—negating history in all its specifications—the metric system was novel for users in its dimensions, names, and arithmetic, To work as a universal language for quantities, as its creators envisioned it, the metric system needed to be employed not only by an enlightened minority, but by peddlers, cooks, contractors, tailors, and regular folk. Again, a language—even a technical and scientific one—requires speakers, and the metric system, in its inception, had very few of them. Everything in the front of metric literacy had to be done. The importance of having speakers makes the different strategies of appropriation of the metric system crucial to explain its success; the problem of appropriation is then essential for any research on metrication that tries to understand how people started talking—or babbling—the metric language and how they eventually fully make it their own. Also, as the setting of the Greenwich meridian as the benchmark for international time zones, the metric system had to fight against all kinds of national traditions and needed to answer questions like "Why should we (Americans, Britons, Chinese…) have to accept a French system in favor of our customary weights and measures?"

A complication that neither Hindu-Arabic numerals nor universal chronological coordination had to endure, but one that heavily curbed the pace of global metrication was that metrication demands a massive renovation of tools and machinery—accompanied by the parallel task of destroying old instruments, which had to be burned, broken, melted, or thrown into the sea to avoid confusions and commercial wrongdoings. When compulsory metric laws entered into effect and customary units became obsolete by means of legislation—usually after a window of tolerance between one to ten years, depending on the country—the possession and employment of thousands of measuring instruments became unlawful and their owners subject to fines or prison. Calipers, gauge blocks, yardsticks, rods, measuring chains, surveyor's wheels, tape measures, rulers, odometers, bushels, barrels, pints, measuring coups, pipettes, water meters, balances, weights, steelyards, scales, and a long list of other measuring, gauging, and weighing appliances had to be adapted to metric units or confiscated and burned into ashes. Also, technical manuals, blueprints, cadastral maps, and the like had to be translated into the metric idiom. Brand new metric devices and technical literature had to be acquired or produced.

It is difficult to offer an approximation of how many metric instruments have been built due to the sudden demand produced by compulsory metrication around the world. Just to give a sense of the dimensions of this forced retooling, an estimation made in Mexico in 1886 of the number of instruments necessary to equip the 3,422 testing offices in districts and municipalities indicated that about 200,000 metric tools had to be built or bought to provide those offices with barely the essential equipment (wooden meter sticks, tinplate liters, brass kilograms, etc.).[14] This fifth of a million devices were only good to equip the inspectors and did not include all the devices used in commerce, agriculture, and domestic life that should be tested and calibrated annually against the official instruments.

An enormous operation of distribution of knowledge—to ensure that people could master the new measuring concepts and instruments—had to be accompanied by the industrial production of physical standards in which the system was materialized.[15] Both processes are necessary for the appropriation of the metric system. If someone has a meterstick and a kilogram scale but possesses no knowledge of the metric magnitudes, names, and decimal logic of grouping and subdivisions, those instruments are useless. And the same goes if a person understands the system conceptually but does not have the appropriate measuring gadgets. Both circumstances have occurred repeatedly in all corners of the world during the last two centuries and represented steep challenges for states, societies, and individuals. Considering all these obstacles, the transition to the metric system around the world is a towering achievement of humanity.

Metric Triumph, Calendar Failure

Parallel to the creation of the metric system, French revolutionaries designed and briefly implemented a new calendar (known as the

Republican calendar) and a decimal time-reckoning system.[16] The new calendar was conceived to signal the beginning of a new period in human history, the Republican era. Day one of this new epoch was September 22, 1792, the day when the French republic was proclaimed.

A brief look at the calendar's decimal architecture is instructive to see its similarities with the metric system.[17] Time reformers retained the year of twelve months, but they changed the irregular months of the Gregorian calendar (28, 30, and 31 days of duration) to have a more symmetrical division. All months were 30 days each. Since that only sums to 360 days, the five additional days needed to approximate the solar year were placed at the end of the year without being counted as part of any month. Months in the calendar were divided into three equal "weeks" of ten days each, called *decades*.[18] The day was divided into ten hours, and "so on up to the smallest measurable portion of duration."[19] The hundredth part of the hour was called "decimal minute," and the hundredth part of the minute "decimal second."

The similarities of this plan with the project of weights and measures were abundant, but the time reform was short-lived. The decimalized day, hour, and minute survived less than two years, and were barely used in practice. Some clocks were manufactured to display decimal time, but the whole plan was doomed from the beginning.[20] Authorities said that the project was suspended because the cost of replacing clocks and watches was too high, and because of popular confusion due to the novelty of the decimal units. These were not convincing arguments, considering that the metric system faced the same adversities and was pushed through nonetheless. More persuasive was the argument that counting hours was not a commercial activity susceptible of police regulation and the old practices would continue "due to the immense force of habit."[21] This was an acceptance that these reforms had to be introduced more by state regulation than by popular agreement.

The rest of the calendar, including the *décade*, lasted officially for twelve years. The new ten-day "week" was the most controversial element of the whole calendar. It represented a disruption of the rhythms of commerce, festivities, and labor, and a direct confrontation against religious practices.[22] The entire experiment produced mixed results. Some embraced enthusiastically the new calendar, but, in general, it created confusion and many people simply kept using the "old" calendar and week. At the end, Napoleon restored the Gregorian calendar and the seven-day week in 1805.[23]

A key to understand the metric success and the chronological failure is that the metric system substituted a *multitude of systems* of measurement, while the republican calendar aimed to replace *one system* of time reckoning. The metric system confronted a disarray of local and uncoordinated measures, while decimal time clashed against the Gregorian calendar, a firmly established convention. A reform of weights and measures was a popular request, and the metric system was an answer to that demand; a reform of the system of time reckoning was just the preoccupation of a tiny group of philosophers and politicians.

More importantly, metrication, in the long run, was sponsored by scientific societies and large-scale merchants, and financed and implemented by nation-states. The institutions and political authorities that sustained the existence of pre-metric units of measurement in Europe (feudal lords) were demolished by social revolutions. Something similar happened with decimal currencies—their diffusion was in concert with the spread of national money produced under government monopoly.[24] Decimal money replaced duodecimal and vigesimal monetary systems because it counted with permanent support from national states during the last two centuries. Decimal time, in contrast, was an institutional orphan.[25]

An Irresistible Force: Global Metrication

There is a considerable lack of research on the international propagation of the metric system. Most accounts pay a great deal of attention to the creation of the system and its technical novelties, but not so much to the details of how and why the meter spread throughout the world. Those who have traced the international diffusion of the metric system have done it with a strong emphasis on a handful of countries. What has happened outside Europe has been almost completely ignored. The role played by colonialism in spreading the meter has been overlooked; and the same goes for the role of Latin America, during the second half of the nineteenth century, to create the critical mass needed to generate momentum for the metric system outside the old continent. The information on international metrication has been deficient and outdated, and the analytical

Metric System in the World, 1800 TO THE PRESENT

FIG. 1.1 **1800**

ADOPTER: France.

FIG. 1.2 **1801–1850**

NEW ADOPTERS (in chronological order): United Kingdom of the Netherlands [today Belgium and Netherlands], Luxembourg, **Algeria**, **Senegal**, Spain.

VOLUNTARY ADOPTIONS: 3
COERCED ADOPTIONS: 2

> **Names in bold** indicate countries where adoption of the metric system was coerced due to countries being under the control of a colonial power at the time.

FIG. 1.3 **1851–1900**

NEW ADOPTERS (in chronological order): Portugal, Colombia, Monaco, Mexico, Venezuela, **Cuba**, Italy, Brazil, Peru, Uruguay, Romania, Chile, Ecuador, Dominican Republic, Bolivia, Germany, Turkey, **Suriname**, Austria [and today Croatia, Czech Republic, Liechtenstein, Montenegro, Slovenia, Slovakia], Serbia, Hungary, Sweden, Switzerland, **Mauritius**, Argentina, **Seychelles**, **Bosnia and Herzegovina**, Costa Rica, Norway, **French West Africa** [today Benin, Chad, Côte d'Ivoire, Mauritania], **Niger**, **Congo** [today Republic of the Congo], El Salvador, Finland, Bulgaria [including today Macedonia], **Sao Tome and Principe**, **Tunisia**, Nicaragua, Honduras, **Djibouti**, Paraguay, **Puerto Rico**, Iceland, **Equatorial Guinea**.

VOLUNTARY ADOPTIONS: 31
COERCED ADOPTIONS: 13

FIG. 1.4 **1901–1950**

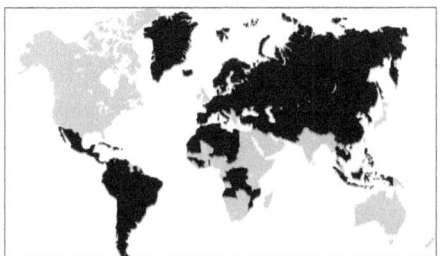

FIG. 1.5 **1951–2000**

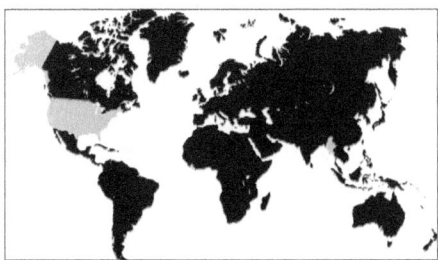

NEW ADOPTERS (in chronological order): **Guinea, Angola, Cape Verde, Guinea-Bissau, Mozambique, Philippines**, Denmark, San Marino, **Burundi, Congo** [today Democratic Republic of Congo], **Rwanda**, Guatemala [including today Belize], **Malta, Vietnam**, Thailand, China, **Comoros**, Panama, Mongolia, Russia/USSR [today Armenia, Azerbaijan, Belarus, Estonia, Georgia, Latvia, Lithuania, Moldova, Ukraine, Kazakhstan, Kyrgyzstan, Tajikistan, Turkmenistan, Uzbekistan], Poland, Haiti, **Morocco, Western Sahara, Cambodia, Libya, Indonesia**, Afghanistan, **Togo**, Iran, Iraq, Lebanon, **Syria**, Andorra, North Korea, South Korea, Israel, Albania.

VOLUNTARY ADOPTIONS: 19
COERCED ADOPTIONS: 19

NEW ADOPTERS (in chronological order): Japan, Egypt, Bhutan, **Taiwan**, Jordan, Sudan [including today South Sudan], India, Madagascar, **Macau, Timor-Leste**, Greece, Maldives, Burkina Faso, Central African Republic, Gabon, Mali, Somalia, Cameroon, Kuwait, United Arab Emirates, Nigeria, Ethiopia [and today Eritrea], Laos, Nepal, Saudi Arabia, United Kingdom, Kenya, Tanzania, Uganda, South Africa [and today Namibia], Pakistan, Ireland, Singapore, Botswana, Swaziland, Zimbabwe, the Bahamas, Dominica, Grenada, Saint Kitts and Nevis, Saint Vincent and the Grenadines, Bahrain, Australia, New Zealand, Lesotho, Zambia, Canada, Trinidad and Tobago, Sri Lanka, Papua New Guinea, Solomon Islands, Guyana, Malaysia, Ghana, Cyprus, Qatar, Fiji, Barbados, Jamaica, Nauru, Antigua and Barbuda, Oman, Tonga, the Gambia, Malawi, Sierra Leone, Tuvalu, Yemen, Bangladesh, Kiribati, Brunei, Vanuatu, Saint Lucia.

VOLUNTARY ADOPTIONS: 71
COERCED ADOPTIONS: 3

FIG. 1.6 **2001–2024**

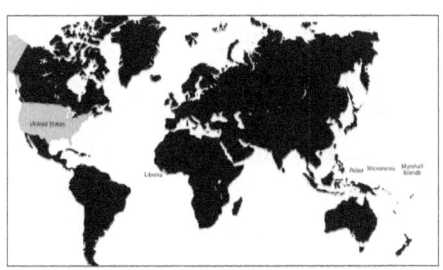

REMAINING NON-METRIC COUNTRIES:

United States, Liberia, Palau, Micronesia, Marshall Islands.

NEW ADOPTERS (in chronological order): Myanmar, Samoa.

VOLUNTARY ADOPTIONS: 2

approaches have neglected crucial phases of this process. Here I want to amend some of these deficiencies.

Let us start with a panoramic view of how the metric system spread throughout the planet. Figures 1.1 to 1.6 are six snapshots of the global pace of metrication from 1800 to 2024 over 50-year spans (except for the last 24 years). The maps display the geographical spread of the system highlighting the countries and colonies where it was officially adopted, either voluntarily or by coercion.

The criterion to define the beginning of metrication in a country is the year when the first legislation ordering exclusive and compulsory use of the metric system was passed. This excludes countries with legislations that made the metric system only optional (like the United States in 1866). In the maps, territories in black denote the countries and colonies that switched to the metric system. The captions list, in chronological order, the countries that adopted the metric system and indicate the number of voluntary and coerced adoptions in each period. **Names in bold** in the captions indicate countries where the metric system was introduced when they were under the control of a colonial power (coerced adoptions), and names in brackets [ ] are countries that eventually seceded from larger metric nations. The precise year-by-year list of metric adoptions can be seen in the appendix at the end of the book.

Broadly put, these maps show how during its first half-century of expansion (fig. 1.2), the metric system was limited to France, French colonies, and Frace's neighboring countries. By 1900 (fig. 1.3) the metric system had spread to almost all of Europe, and Latin America represented the first large group of non-European countries that adopted the system voluntarily (in Africa, at that point, all adoptions were the product of colonial imposition). In 1950 (fig. 1.4) the metrological state of the world was divided into two clearly defined factions; the United States and the Commonwealth of Nations were using imperial or US customary measures, while all the rest were metric. At the end of the twentieth century (fig. 1.5), after the United Kingdom and almost all former British colonies switched to metric, the US became a customary island in a metric world. So far in the twenty-first century (fig. 1.6), there have been two new voluntary adoptions: Myanmar and Samoa.

Among the present 197 independent countries existing in the world, 127 adopted the metric system voluntarily; in 38 the system was introduced when they were colonized territories; and another 27 seceded from countries that were already metric. Currently, there are five non-metric countries: the United States, Liberia, Palau, the Federated States of Micronesia, and the Marshall Islands.[26]

These maps give a broad idea of how and when the metric system extended around the world. Some general tendencies of diffusion are visible from this bird's eye perspective. But we need a closer look to figure out *why* and *when* some countries, and not others, went metric. In what follows I will trace geographical patterns by looking at countries as nodes in a diffusion network; next, I will look at the historical conditions that fashioned the dissemination of the metric system.

Vicinity and Diffusion Networks

The theory of "diffusion in networks" could give us some clues to analyze the global spread of the metric system.[27] This theory studies how behaviors, practices, opinions, and technologies spread through "social networks" as the members of a network influence other members to which they are tied.[28] The central proposition is that the choices made by the nodes in a network (individuals, social groups, organizations, states, etc.) are influenced by their network neighbors. People often do not care as much about the full population's decisions as about the decisions made by their closest ties (friends, clients, allies). For instance, people may choose a technology compatible with their direct collaborators, rather than the most popular gadget. Among the benefits of imitating others' behavior is that people have incentives to adopt an innovation when they must communicate with associates who have already adopted that technology (like fax, email, or WhatsApp). The benefits of adopting a new convention increase as more neighbors do the same.[29]

This theory can be used to analyze the metric diffusion around the world. The unit of analysis is the global network of countries, analyzed through the 216-year span between the end of the metric conference of 1799 in Paris and the year when the latest metric adoption was registered (2015). This network is contained in a closed space, planet Earth, but it is very dynamic. The number of network nodes has expanded and contracted continuously, and it has experienced numerous changes in its configuration through the decades. Empires have dissolved into dozens of independent countries; small countries have merged to form middle-sized nations, just to dismember again years later.

As more countries adopted the metric system it was less risky, more beneficial, and then more probable for other countries to adopt it as well. The chances that a country would switch to metric would grow if its associates had already made the change. We could advance the hypothesis that geographical vicinity played a role in shaping the patterns of diffusion of the metric system. If one of the purposes of switching to the metric system was to facilitate international coordination, it is probable that nation-states wanted to be on the same page as their neighbors. If this is the case, we should see that if one of your neighboring countries adopted the system your chances of adopting it should increase significantly.

A way to test this assumption is to see the role that geographical vicinity has played in the diffusion of the metric system. To this aim, I enquired how many years passed for every country to embrace the metric system after one of its neighbors had done it. Since this conjecture supposes that the spread of the innovation is the product of a conscious decision, I limited the analysis to cases of voluntary adoptions (which leaves out the countries where the metric system was imposed coercively).

As I said earlier, among the present 197 independent countries in the world, 126 adopted the metric system voluntarily; eleven of those switched to metric system without having a metric neighbor.[30] The remaining 115 countries that adopted the metric system voluntarily had at least one metric neighbor when they made the switch. On average these nations adopted the metric system sixteen years after one of their neighbors went metric. This number seems too high to show any relevant connection between metric acceptance and geographical vicinity. However, things look somewhat different if we consider that the median number of years is four, and that the value that occurs most frequently is zero years.

Nineteen countries went metric the same year as one of their neighbors. These simultaneous adoptions happened at different points in history and in diverse geographical contexts, from Brazil and Peru in 1862, to Burkina Faso and Mali in 1960, and Australia and New Zealand in 1969. Additionally, 62 countries adopted the metric system within five years after one of their neighbors did. Overall, two-thirds of the total—75 countries—made their metric transition within ten years after one of their neighbors. These patterns are relatively constant in all continents and suggest that vicinity was somewhat influential in international metric diffusion.

What remains difficult to understand, from this theoretical viewpoint, are the outliers. Why did Mexico (1857) wait 53 years before one of its neighbors (Guatemala) followed suit? Why did Guinea (1901) wait 75 years before Sierra Leone went metric? Why did Andorra (1934) remain non-metric for 85 years after both of its neighbors (France and Spain) made the transition? Why did Guyana (1971) accept the meter 100 years after the last of its neighbors did (Suriname in 1871)? Questions of this kind require another kind of explanation (as I show in the next section of this chapter).

There is another idea from network diffusion theory that is worth exploring. Since nodes in a network make choices based on the choices of their associates, diffusions could spread across the links of the networks like cascades. Some cascades run for a while but stop; in others every node in the network switches (complete cascades). In a network where nodes care about what their closest associates are doing, it is possible for a small set of initial adopters to trigger a domino effect that may spread the innovation through the whole network. However, tightly-knit communities in a network can hinder the spread of an innovation. Homophily (the tendency of nodes to bond with similar others) can serve as a barrier to diffusion by making it hard for innovations to arrive from outside of densely connected communities (clusters). Cascades stop when they run into dense clusters, they are barriers to cascades.[31]

The dynamics between cascades and clusters shed light on the state of global metrication by 1950 (fig. 1.4), when the large majority of nations outside the metric sphere were members of the Commonwealth of Nations that gravitated around the United Kingdom. It could also be useful to understand the present state of affairs (fig. 1.6) where, besides the United States, the other four non-metric countries in the world have been, at some point in their history, under American administrative control. The British and American clusters have halted the metric global cascade.

Despite its valuable insight, diffusion networks theory still leaves multiple loose ends. For one thing, all coerced adoptions are kept out of the picture. By looking at nation-states just as nodes in a network we learn very little about domination and conquest (important issues in the spread of the metric system). Network theory is useful for recognizing the value of interconnectivity, but not so much for understanding power; it becomes too easy to assume that all participants in an abstract network are equal in their resources and influence, and that their singularities come just from their position within a network. But some countries have had the ability to force others to adopt the metric system; besides, what happens inside every country could be more important than what happens outside. This is crucial to explain why countries with no metric neighbors, especially in the nineteenth century, decided to take the risk of making a costly metrological reform in favor of the metric system. Finally, network theory does not offer much help to elucidate why the United States has refused to go metric.

The Social Bases of Metrication

Since Russia still operated by the Julian calendar, which was thirteen days behind the Gregorian calendar adopted everywhere else in the Christian or Westernized world, the February revolution actually occurred in March, the October revolution on 7 November. It was the October revolution which reformed the Russian calendar, as it reformed Russian orthography, thus demonstrating the profundity of its impact. For it is well known that such small changes usually require sociopolitical earthquakes to bring them about. The most lasting and universal consequence of the French revolution is the metric system.
—ERIC HOBSBAWM, *The Age of Extremes* (1995)

As Hobsbawm suggests, the metric system has been adopted in many countries after they suffered sociopolitical earthquakes. These earthquakes have been revolutions, national unifications, massive sociopolitical change, international commercial agreements, colonization or imposition by a foreign power, and independence from colonial rule.

What it is true in all cases is that the adoption of the metric system was an imposition from the top—with different degrees of intensity, but always from the top. Some professions have adopted the metric system voluntarily, but no large segment of the general population has voluntarily abandoned its customary measures before their government decided to make the meter, liter, and kilogram the exclusive legal units. Metrication "from below" has been rather anemic and it can only be found in certain occupational groups (especially in scientific and technical areas), and that is one of the main reasons why non-mandatory transitions to the metric system (like the policy intended in the United States since 1866) have never achieved much.

REVOLUTIONS

Since the metric system was created by a social revolution (a revolution that not only replaced the governing political elite, but also brought a rapid transformation in the class structure), researchers have paid considerable attention to the links between revolution and measurement.[32] Not surprisingly a great deal has been written about the French revolution and why the metric system was created and implemented in such a tumultuous context.[33]

One of the problems of the political life of measures was multiple metrological sovereignty. This means that there was a lack of a single unified political hierarchy to regulate weights and measures and there were competing claims by two or more opposing parties over metrological authority on a defined geographical region or professional activity.[34] There usually was a proclaimed sovereign metrological authority, but in practice that authority was ignored or challenged. In Europe this situation ended, with the abolition of feudal privileges that deprived lords from the right to have the final say in metrological matters within their fiefs. Modern nation-states established a unified authority that could set a single system. This is why the birth of the metric system and the establishment of sovereign and autonomous metrological powers occurred simultaneously. When newer nation-states looked for the instruments to secure their metrological monopoly, the metric system was there as a ready-made system and with a proven record of success (more on this in Chapter 2).

Following a parallel path as revolutionary France, China adopted the metric system in 1912, the same year a revolution abolished the Qing Dynasty and established the Republic of China.[35] And Russia did so too in 1918, a year after the October Revolution ended tsarism and the soviets seized power (a few years later, in 1922, they metrified the rest of the republics of the USSR).[36]

Revolutions deeply alter numerous social and political institutions, and the metric system commonly becomes part of those efforts to reconfigure, or "regenerate," society at large.

NATIONAL UNIFICATIONS

In some cases, the metric system has been part of national unifications. With the purpose of linking together (and subduing) previously autonomous cities, regions, and provinces that had incompatible administrative and metrological structures, the introduction of the metric system as a shared, national system of measurement was an effective solution to achieve a greater degree of homogeneity.

This was the case in Italy, where the metric system was established as the standard for the whole peninsula in 1861, the same year when Rome was proclaimed the capital of the country and the Parliament declared Victor Emmanuel II as the first King of united Italy. Another instance of metrication as a means to stimulate national unification was Bulgaria, where the metric system became the official standard of measurement in 1888, three years after the Principality of Bulgaria and the province of Eastern Rumelia were united.

In Germany, similarly, the adoption of the metric system was parallel to the political and administrative reforms that culminated in the 1871 unification under Bismarck. When Friederich Engels observed Germany's unification, he underscored the political and economic functions of implementing homogeneous standards and destroying local barriers, and how that aided the creation of a modern (capitalist) country: "The government . . . removes the impediments to industry emanating from the multiplicity of small states; it creates unity of coinage, of measures and weights; it gives freedom of trade; it grants the freedom of movement; it puts the working power of Germany at the unlimited disposal of capital; it creates favourable conditions for trade and speculation."[37] The political centralization that accompanies national unification transformed loosely connected towns and provinces, with separate laws and systems of taxation, into "one nation, with one government, one code of laws, one national class-interest, one frontier, one customs-tariff," and one system of weights and measures.[38]

MASSIVE SOCIO-POLITICAL CHANGE

In many instances the metric system was adopted after drastic changes in political regimes, like civil war or the adoption of a new constitution. In Colombia and Mexico, for example, the adoption of the metric system happened the same year as the promulgation of new liberal constitutions (1853 and 1857, respectively). These reforms were concomitant with the separation of church and state, freedom of religion, universal male suffrage, reaffirmation of the abolition of slavery, and other liberal reforms.

What can be called, for lack of a better word, "modernizing" policies have also been an engine of metrication. It is not rare for governments to adopt or start a strict implementation of the metric system at the same time as other wide-ranging reforms are put into operation: tax and customs reforms, geographical-administrative rearrangements, modifications in land tenure, large standardization policies, and so on. Russia changed to the metric system at the same time as it abandoned the Julian calendar and followed the rest of Europe with the Gregorian calendar. Turkey decided to make effective its 1869 metric legislation as part of Atatürk's efforts to found a secular republic in the 1920s, which included, besides metrication, the employment of Latin characters and Arabic numerals, the introduction of surnames following the European style, standardization of the size and color of the flag, and so forth.[39] Myanmar switched to the metric system in 2013, in the midst of numerous structural changes in its political and economic institutions: the governing military junta was dissolved, some democratic liberties were granted, and a push was made to open the economy to more consistent international exchanges.[40]

By and large the metric system never comes into a country as a discrete, isolated reform; it is always part of an interrelated series of large social and political transformations.

INTERNATIONAL TRADE AGREEMENTS

International agreements have been a significant element in the spread of the metric system. Several countries switched to the metric system by signing multinational treaties. There are at least three of these cases.

In 1910 Guatemala, El Salvador, Nicaragua, Honduras, and Costa Rica signed a Central American convention relative to the regional unification of weights and measures and chose the metric system as their common standard.[41] The convention served two purposes, it pushed Guatemala to pass its first metric law and hard-pressed the other nations to start the actual implementation of their legislations from the 1880s and 1890s.

In 1967 Kenya, Uganda, Tanzania—three bordering nations—signed a joint government declaration to adopt the metric system in place of the British imperial system.[42] This declaration has been the only case in which a multinational treaty was signed with the exclusive purpose of implementing a simultaneous metric transition.

In 1969 the Caribbean Community (CARICOM, whose members are Antigua and Barbuda, Bahamas, Barbados, Belize, Dominica, Grenada, Guyana, Haiti, Jamaica, Montserrat, St. Kitts and Nevis, Saint Lucia, St. Vincent and the Grenadines, Suriname, and Trinidad and Tobago) took the decision to move as a group to the metric system. And a year later it was agreed by the Heads of Government Conference of the Caribbean Free Trade Association that all member countries would go metric.[43]

Other international agreements designed to facilitate economic integration have helped the advancement of the metric system. The First International Conference of American States (held in Washington DC in 1889) and subsequent Pan-American Conferences drove several Latin American countries to revive and enforce their dormant metric legislations and put pressure on the United States to harmonize its weights and measures with the rest of the continent.

British metrication in the 1960s came about in the midst of the United Kingdom's desire to have a more fluid participation in the European Common Market. Further on, the regulations by the European Union forced England and Ireland to go even further with their metric transition policies, as happened in Ireland in 2005, when road signs stopped showing miles in favor of kilometers, and car speedometers marked exclusively kilometers per hour.

Of course, not all international agreements looking for economic integration of a given region have been effective in advancing the metric cause. The signing in 1994 of the North American Free Trade Agreement (NAFTA; replaced in 2018 with the Agreement between the United States of America, Mexico, and Canada, USMCA) between two metric countries and a non-metric one did not push America to adopt the meter as its exclusive standard of measurement. The futility of NAFTA regarding metrication shows the limitations of any theory that may try to explain global dissemination of the metric system exclusively on the basis of geographical vicinity or international economic coordination. In fact, the NAFTA region presents several oddities. Mexico was the country that needed to wait the longest for one of its neighbors to go metric after it adopted the system. Fifty-three years elapsed between Mexico's first compulsory metric law (1857) and the beginning of metrication of Guatemala in 1910.[44] And Canada has the peculiarity of being the only country in the world that does not share borders with a metric country.[45] North America is the most anomalous region regarding international standardization of weights and measures.[46]

MILITARY IMPOSITION

Writing against Napoleon in his 1814 pamphlet *The Spirit of Conquest and Usurpation*, Benjamin Constant lamented:

> The Conqueror of our days, whether peoples or princes, wish their empire to present an appearance of uniformity, upon which the proud eye of power may travel without meeting any unevenness that could offend or limit its view. The same code of law, the same measures, the same regulations, and if they could contrive it gradually, the same language, this is what is proclaimed to be the perfect form of social organization. . . . The key world of today is uniformity. It is a pity that one cannot destroy all the towns to rebuild them according to the same plan, and level all the mountains to make the ground even everywhere. . . . The spirit of system was first entranced by symmetry. The love of power soon discovered what immense advantages symmetry could procure for it.[47]

Constant's words were meaningful in the context of post-revolutionary France and its methods of exporting metrication. Even before the meter's final magnitude was determined in 1799, France had been already active in implanting it in other European territories. While spreading the revolution through the invasion of neighboring territories, French authorities promoted the creation of new republics where the metric system was established as the official system of measurement—freedom, in the new social order, was to be measured in meters.

Starting in 1798, under pressure from the French ambassador in Turin, Piedmont began the introduction of the metric measures, with scientists and public administrators leading the march. In the Cisalpine Republic (Lombardy) it was determined that the future measures and currency would be arranged in a decimal scale, and in 1801 the metric system was officially introduced. These two experiments served as the basis for the metrological unification of what would be unified Italy decades later.[48]

In Geneva, which was annexed to France 1798, the compulsory utilization of the metric system began in 1801. In the departments located in what is today Belgium, administrative centralization came hand in hand with the meter and the kilogram, commencing with Brussels in 1799.[49] This endeavor served as the foundation for the future metrication, in 1816, of the United Kingdom of the Netherlands (later Belgium and the Netherlands).[50]

The introduction of metric units in these places was longwinded and thorny. Some progress was achieved in major cities, but very little elsewhere; and when the French armies marched back home after Napoleon's defeats, the new authorities abolished the metric system and returned (for a few years) to their old measures. These rejections of the metric system have not occurred again. Metrication has been a one-way street, once a country enters it, there is no return. The metric system is difficult to adopt and implement, but it is almost impossible to abandon.

COLONIALISM

The institutionalization of the metric system involved special difficulties because of the aspiration to universalism that helped to give it form. This universalism was consistent with the ideology of the revolution, and more particularly with the ideology of empire.
—**THEODORE PORTER**, *Trust in Numbers* (1995)

In 1927 Arthur Kennelly, an enthusiastic pro-metric engineer from Harvard University, noted that since 1800 a "wonderful sociological phenomenon" presented itself in Continental Europe. More than thirty countries officially adopted the metric system and abolished their old systems. The change, he said, was voluntary; in no case "did the change from national to metric measures come about in any one country, at the dictation of any other country."[51] In his naïve assessment of metric dissemination, Kennelly failed to remember the role of Napoleon's army in the first attempt of metric implementation in Belgium, Switzerland, and Italy. More importantly, he carefully overlooked what happened outside Europe, where the change to the metric system came frequently at the dictation of European countries.

Even though the voluntary adoption of the metric system by sovereign states is indeed one of the most notable themes in the global metrication process, it is astonishing that colonialism has not been a relevant topic in the historiography of the metric system. Colonialism was a substantial element in the globalization of the meter. It would be a stretch to say that it was the most crucial aspect, but its importance should not be underestimated. French colonialism in particular had a decisive role in advancing the use of the metric system outside Europe, predominantly in Africa.

Metrication by colonization was a process that lasted more than a century and helped to increase the number of metric territories all over the world, from Algeria and Senegal in 1840 to Macau and Timor-Leste in 1957. Overall, there are 38 countries (that is, one of every five countries existing today) where the metric system was introduced when they were colonies or administered territories of overseas powers.[52] And if we focus our attention outside Europe, we find that one out of every four existing non-European countries received the metric system as a colonial imposition.[53]

Breaking down these numbers by regions, we can see great variations in this colonial expansion. Africa was the main recipient of the metric "blessings" brought by European armies, with half of the African countries getting the meter in that way. This is by far the highest proportion of cases of colonial metrication in any continent. In Asia it was "only" 7 out of 46 countries, and in the Americas it was 3 out of 36. Oceania is a peculiar case; all eleven metric adoptions there have been voluntary, and the remaining three countries are non-metric (Palau, Micronesia, and Marshall Islands).

France played the leading role among the colonial powers that participated in the military export of the metric system. Out of the thirty-eight countries that received the metric system through colonial rule, eighteen were once ruled by France. Other contributors were Portugal (with seven), Spain (four), Belgium (three), and the Netherlands (two).

We are still waiting for detailed studies that can give us a better understanding of the nitty-gritty of the introduction of the meter in these colonies, but it is not difficult to imagine some of the main instruments of this process.[54] Based on other experiences, it looks reasonable to assume that rights for the possession of land and cadastral records were crucial for the introduction of the new measures of area, and that the payment of taxes in kind was the battering ram in the imposition of new grain measures.

The ideological justification for the imposition of the metric system on other populations sometimes took the form of a charitable donation to "less advanced" people. Take for instance the address by the French Minister of Agriculture at the final ceremony of the Exposition of the Second Republic, in 1849, where he declared that the products exhibited by Algeria though modest showed promise for future development, and in return for their raw products and native handiwork, France had aided Algeria to build wells, irrigation systems, and dams to increase agricultural production. Stressing the idea of reciprocity, the minister claimed that "the Arabs of the Middle Ages gave us their simple system of numbering and their admirable decimal system. We return the gift, so productive for the common good, by giving them our decimal metric system."[55]

Of course, the French did not limit themselves to giving their decimal system only to the Algerians. They shipped it to the rest of their colonies in Africa and their overseas territories in all corners of the world: the Indian Ocean (Réunion), the southwest Pacific (New Caledonia and Polynesia), Southeastern Asia (Vietnam and Cambodia), and the Near East (Syria). They did not have time to introduce the metric measures in Louisiana before it was sold to the United States in 1803, which prevented an interesting situation, with America adding a metrified territory to its lot. Neither were they able to give the meter to the rebellious Haitians (who ended up adopting the metric system voluntarily in 1920). But this did not prevent France from giving the Americas a taste of metrication, which was introduced in the island of Saint Pierre and Miquelon in 1824, and in Guyana in 1840.[56]

As of today, no former colony has rejected the metric system after their independence. This "voluntary retention" of the metric system is especially surprising considering that many independence movements have been notoriously nationalistic and a return to the "original" or "national" pre-metric measures seem to fit to their ideological outlooks. Newly independent countries have made conscious efforts to "return" to their precolonial dressing styles, languages, calendars, flags, and political icons; but metric weights and measures stand still. As illustrations, we can mention the former Soviet republic of Azerbaijan, where in 1991 (the very same year of its secession from the USSR), the government decided to stop employing the Cyrillic alphabet and ordered the exclusive use of Latin characters.[57] In other countries there have been attempts to inject nationalistic motives in the calendar, like in Turkmenistan where former president Saparmurat Niyazov promoted a law to change the names of the months; May became Magtymguly (after an eighteenth-century Turkmen poet) and June was renamed Oguz (honoring Oghuz Khan, the founder of the Turkmen nation).[58] In Libya—an Italian colony from 1911 to 1951— Muammar Qaddafi also changed the name of the months; February and August became Lights and Hannibal, the latter was an interesting choice, exalting Hannibal's figure, who in the third century BC occupied much of the Italian peninsula for fifteen years.[59] Curiously, it was the Italians who introduced the metric system in Libya in 1923, but no anti-metric retaliation was orchestrated by Qaddafi. Around the world the metric system has been impermeable to this kind of nationalistic expressions. The imagined communities of postcoloniality have not incorporated weights and measures into their imagery.[60]

Apropos of that, voluntary metric retention is a phenomenon not exclusive to former colonies. A recent event confirms that once the metric system is established eludes nationalist backlash. In 2022, in the wake of Brexit, England's prime minister Boris Johnson announced a possible return to imperial measurements allowing British shops to sell products exclusively in imperial units. Johnson had declared that measuring in pounds and ounces was an "ancient liberty" and signaled a "new era of generosity and tolerance" toward traditional measures.[61] However, in December 2023, the Department for Business and Trade announced that a consultation with more than 100,000 businesses, consumers, and academics "showed that 98.7 percent of respondents were in favour of using metric units when buying or selling products" and that the government therefore "decided not to make any legislative changes."[62]

Latin America: The Transcontinental Metric Bridge

An important period for the creation of new nation-states was between 1821 and 1845, on account of mainly the former Spanish colonies in the Americas (fig. 1.7), which played a decisive but neglected role in the international diffusion of the metric system.

In a way, Latin American countries, which switched to the metric system in the second half of the nineteenth century (fig. 1.2 and 1.3), can be considered the first truly deliberate adopters of the metric system. Metrication in these nations cannot be explained as a result of French imposition. Latin American countries did not exist yet when the Paris Congress on Definite Metric Standards of 1799 took place (so they were not part of the symbolic creation of the system); they were not France's neighbors (which eliminates the vicinity factor); and they were not French colonies. The big majority of the Latin American countries were voluntary adopters; the only exceptions were Suriname (a French colony), and Cuba and Puerto Rico (Spanish colonies). In other words, Latin American countries were the first voluntary adopters outside France's military and geographical area of influence, and they made the metric system a truly extra-European reality.

By 1850 (fig. 1.2) all adopters of the metric system were either France's neighbors or French colonies in Africa. Among the twenty countries that first adopted the metric system after France, eleven were from Latin America (nine of which were voluntary adopters: Colombia, Mexico, Venezuela, Brazil, Peru, Uruguay, Chile, Ecuador, and Dominican Republic). The other nine were countries that either participated with representatives in the finishing of the metric system in the 1799 conference (like Spain and the Netherlands), were France's neighbors (like Belgium, Luxembourg, and Monaco), or received the metric system by French military imposition (like Italy, Algeria, and Senegal).

Latin American countries were not only early adopters, but also the first that went metric free of any form of military coercion. They also greatly helped the cause of international metrication by creating, alongside

An Irresistible Force Meets an Immovable Object   57

Figure 1.7. Years of first adoption of the metric system in Latin American countries.

Western Europe, a critical mass of countries necessary for the system to be a convincingly multinational and multiregional metrological language. Overall, 16 out of the 35 voluntary adoptions made during the nineteenth century came from Latin America—with Bolivia, Argentina, Costa Rica, El Salvador, Nicaragua, Honduras, and Paraguay doing it during the later decades of the century.

For more than a century the metric system was, in terms of voluntary adoptions, a reality contained in Europe and Latin America. Europe's neighbor Turkey (1869) was the only country outside these two regions that adopted the system by its own will and did it during a period of increasing commercial exchange with Europe.[63] After Turkey, we have to wait until the 1910s when Siam, China, and Mongolia went metric.[64] In Africa the first voluntary adoption came with Egypt in 1951.[65] And the first adoption in Oceania came in 1969, when Australia and New Zealand followed England into the metric path.[66]

After Latin America and Western Europe embraced metrication, a new dimension was added to the situation in other countries. From that moment on, the importance of going metric was not just because it was considered a better system, but because it was assumed as "inevitable." Arguments like this newspaper editorial from 1902 became commonplace: "Conformity to one common system of weights and measures is not only extremely desirable, but certain to come. The only doubtful question is how long it will take."[67] For many people the implementation of the metric system was not a matter of if, but when it would happen. The common pattern of a self-fulfilling prophecy started to develop.[68] Actuated by the conviction that metrication was unavoidable, governments and industries around the world started to prepare for the change, thus facilitating the conditions for the initial diagnosis to actually become true.

Researchers have wrongly suggested that Latin American countries adopted the metric system simply following Spain's transition in 1849.[69] At first, this may look like a plausible assumption since all former Spanish colonies in the Americas began the implementation of their metric plans shortly after that year, starting with Colombia in 1853. But a closer look shows a different picture. Chile, for example, had passed its first metric legislation in 1848 (which made metric units optional); that is a year before Queen Isabel II signed the law to metrify Spain.[70] Mexico also commenced its plans for an eventual changeover to the metric system in the late 1840s; the Mexican Geography and Statistics Society presented to Congress in 1849 a recommendation to make the meter and kilogram the national standards, and later that year the first draft for metric legislation was discussed. Thus, the metrication plans in Latin America were simultaneous to those in Spain.[71]

For many Latin Americans of the time, embracing the metric system was seen as a way to get closer to the "civilized nations" (meaning mainly Europe), but they did it, in many instances, more promptly than their role models. This historical irony created paradoxical situations. For instance, when Archduke Ferdinand Maximilian Joseph of Austria arrived to the Americas in 1864 as "Maximilian I of Mexico" to head the Second Mexican Empire (with the backing of Napoleon III), he found a country with a complete metric legislation already in place, while his native Austria was just starting to experiment in its custom houses with a mixed system of customary measures rounded into metric equivalents (like the *pfund* of half kilogram and the *centner* of 50 kilograms) and a full-scale plan for metrication was still years into the future.[72] Maximilian—captured and executed in 1867 by the very same Mexican liberals who ten years before had passed the metric legislation in the country—died without seeing a metric Austria.

THE BRITISH EMPIRE

England opposed the metric system for more than one and a half centuries. Similarly, it took them 169 years to accept the Gregorian calendar after Pope Gregory XIII established it in 1582. Their repudiation of the calendrical reform made Voltaire scoff, "The English mob preferred their calendar to disagree with the Sun than to agree with the Pope."[73] When they finally adopted it, in 1752, riots exploded with people crying "Give us our eleven days!," as eleven was the number of days that the Julian calendar trailed the Gregorian calendar when the change was made.[74] England was also the country that stuck the longest with the Carolingian monetary system of 1 pound = 20 shillings = 240 pennies, a system dating from the times of Charlemagne's father. This came to an end on the "Decimal Day" (February 15, 1971), when decimal currencies were introduced in the United Kingdom and Ireland. England was also reluctant to enter the zone of the common currency of the European Union. And Britons will not start driving on the right side of the road any time soon—here again going against a French convention initiated in revolutionary France.[75]

This should not be interpreted as a simple "Anglo-Saxon" love for self-determination that is shared by England, America, and other "English speaking people." England is not in a defensive position trying to preserve its culture from foreign attack. Those are self-gratifying excuses that some Britons like to tell themselves, which a few observers have taken at face value.[76] The United Kingdom has not shied away from being an aggressive exporter of its own conventions—its fierce antagonism against the international aspirations of the metric system is a testament to that. Imperial ambitions are different from self-determination.

In response to France's meter, England undertook in 1824 a comprehensive reform of its own weights and measures that reduced the number of measurement units and greatly improved their exactness, creating a system able to compete internationally with the French invention. It was not by mere accident that they called these new measures the *imperial* system.[77] With the expansion of the British Empire during the nineteenth century, after the Napoleonic wars, the imperial system started a parallel run to that of the metric system. Since the meter emerged victorious from this clash, historians have focused their attention on the history of the meter, overlooking how the imperial system developed into a global system in its own right (thanks to British military and commercial power). The United Kingdom should not be considered a simple "late metric adopter," but a challenger of the metric system who developed a competing system of its own. England's adoption of the metric system in 1965 simply meant the defeat of the metric system's biggest competitor.

This does not mean that this conflict had been fought just outside the United Kingdom. In England and other parts of the empire there was political support in favor of metrication. A British metric movement existed at least since 1850, with associations like the Decimal Society pushing for adoption of the meter and for the decimalization of currency. Pressure also came from the outposts of the empire. During the Colonial Conference (an assembly of representatives of the British Empire) of 1902, New Zealand, Australia, Cape Colony, Transvaal, Orange River Colony, Sierra Leone, Nigeria, and Ceylon showed their support for metrication, and the Colonial Premiers passed a resolution favoring the adoption of the metric system for use within the empire.[78] London was not receptive to these demands and the imperial system stayed put until the 1960s. The British colonies had to wait until their independence to transition to the metric system.

It took a little more than a century, but by 1950 the globe was completely divided into two large metrological areas, with all the nations of the planet using either the metric or the English system (fig. 1.4). At that point the rest of the hundreds of measuring systems in the world had been displaced, partially tolerated in specific countries, or simply used surreptitiously in local communities. Seen from a larger perspective, the immense success of these two systems meant a great loss for humanity's stock of knowledge. The extinction of languages (when local dialects are replaced by one of the few dominant languages) represents losing entire social funds of knowledge; and so the death of local measures has meant the disappearance of collective experience accumulated by hundreds of generations. It is difficult to put in numbers how much has been lost.[79] The few catalogs of units of measurement in the world are vastly incomplete, but they may give us at least a vague idea. A survey made in Spain in the late 1950s listed approximately 793 non-metric, non-imperial units of measurement still in use then around the globe.[80] Another one made by the United Nations enumerated approximately 898 of those units.[81] It is foreseeable that most of those local measures that still exist will diminish in number during the coming decades.[82] This loss cannot be estimated only in terms of the number of units that have been swept out by the metric and imperial waves, but also by the antiquity and resiliency of some of the systems that are today forbidden from official transactions and may be doomed to disappear from everyday life in the not so distant future—like Japan's shaku-kan system, which was outlawed in 1966 after being used for a millennium, after its arrival from China in the tenth century.[83]

DECOLONIZATION

If colonization was an important factor in global metrication, decolonization was even more influential, particularly in the twentieth century following the decline of the British empire in the aftermath of World War II. The period between 1950 and 2000 (fig. 1.5) was by far the fifty-year span when the most voluntary adoptions of the metric system took place (71 new adoptions). Two interrelated developments account for this: the breakdown of the English control on its overseas dominions, colonies, and protectorates; and the subsequent birth of a large number of new nation-states.

The period between 1946 and 1975 was one of the most active in the creation of new nation-states, mainly due to the dissolution of empires.[84] Among these novel nation-states, those using a system other than the metric system quickly decided after independence to make the meter and kilogram their official standards. In that span alone 64 countries adopted the metric system voluntarily. Starting with Sudan (that went metric in 1954) and finishing with Myanmar (2013), all former British colonies that gained independence after 1945 switched to the metric system. Some did it the same year when they gained their independence (such as Somalia and

Kuwait), others one or two years later (like Cameroon, Nigeria, Swaziland, Qatar, and Brunei).

This process sped up metrication at a pace that had not been seen before. For example, it took European powers 84 years (from 1840 to 1924) to metrify half of the current countries in Africa; in contrast, the voluntary adoption of the metric system by the other half, after those nations became independent, took only 25 years, with 24 adoptions between 1951 and 1976 (18 of them former British colonies).[85]

Once countries gained independence, those interested in the introduction of the metric system saw new opportunities to push their plans forward. In India, where a movement in favor of decimal coinage and metric weights and measures existed since the beginning of the twentieth century, all attempts in favor of decimalization were blocked because colonial authorities preferred to wait for the metropolis to adopt it first. After independence, groups like the Decimal Indian Society found more receptive ears in the government to attain the reform, which finally came about in 1954.[86] Observers have noted several advantages for metrication in India and other Asian countries, like no emotional attachment to English units, an anticolonial stance that helped popularize the metric system, and the preexistence of the decimal number system—as someone put it, "Many Asian cultures prefer decimals to fractions. After all the Decimal system originated in Asia."[87]

An Immovable Object: The United States

The United Kingdom blocked any attempt at metrication in its overseas possessions and slowed down the global spread of the metric system for decades. By 1950 (fig. 1.4), the great majority of the countries outside the metric sphere were members of the Commonwealth of Nations; but the levee broke in the second half of the twentieth century. Today (fig. 1.6), with only five countries outside the metric ocean, an American imprint is the most visible characteristic of the metric holdouts. In addition to the United States, each of the other four non-metric countries (Liberia, Micronesia, Palau, and Marshall Islands) were at some point in their history under American rule. The hegemony of the United States over those countries was not as tight as the one maintained by England over its colonies, but has been significant enough to be considered one of the reasons why they are not metric.

Liberia was founded in 1821 by the Society for the Colonization of Free People of Color of America. It is estimated that between the foundation of the Society and the end of the Civil War, approximately 15 thousand African Americans were sent to Liberia. The Americo-Liberians took political control over the territory and the indigenous population.[88] In 1847 Liberia declared its independence (recognized in 1862), but the country bears numerous marks of its American past: it promulgated a constitution based on the principles of the US Constitution; the official language is English; the capital, Monrovia, was named after American president James Monroe; its currency is the Liberian dollar; its flag shows red and white stripes and in the upper left corner a blue square with a white star. During World War II, Liberia supported the Allied powers and received, in return, substantial investments in infrastructure from the US during the coming decades. The Americo-Liberian elite lost political control of the country in 1980, and two civil wars in the 1990s and 2000s wrecked the nation. Now that they are in a rebuilding phase and the American ascendancy over the country is not as heavy, Liberia has made initial moves to metricate. In 2018 the Liberian Ministry of Commerce announced that they were committed to adopting the metric system "to promote accountability and transparency in trade" and to help local manufacturers benefit from trade agreements with the Economic Community of West African States. To my knowledge, however, they have not passed legislation that makes the metric system exclusive and mandatory.[89]

Palau, Micronesia, and Marshall Islands became United Nations Trust Territories administered by the United States after the defeat of Japan in World War II. They gained their independence in the 1980s and 1990s but maintained a strong tie with the United States in the form of a Compact of Free Association under which the US federal government provided financial assistance in exchange for international defense authority. Curiously, the American occupation forces in Japan approved its transition to the metric system but did not do the same for the territories formerly controlled by Japan. Since the 2000s there is a policy of non-mandatory metric conversion in Palau, Micronesia, and Marshall Islands.[90] One may assume that if these four countries had not been under the United States influence, they would most likely have made their transition to the metric system at some point after World War II, as the rest of Africa and Oceania.

This leaves us with the great enigma, why is there no metric system in the United States? This is a country that many people, within and outside America, have waited to see embracing metrication for more than two centuries. A country that has created some of the most devoted metric activists. An ambivalent nation that has simultaneously assisted and deterred the international metric expansion.

The United States was one of the original signatories of the Metre Convention in 1875, a conference that cemented the leading position of the metric system in the world. Also, in the Pan-American Conferences of the 1890s the United States supported the plans for a full metric continent (plans that pushed Latin American nations to complete their metrication processes). The United States introduced the metric system in the Philippines when it took control of the islands after the 1898 Spanish-American war; and something analogous happened in Japan, where metric measures were finally introduced (after unsuccessful attempts in 1893 and 1921) during the Allied occupation with general Douglas MacArthur as supreme commander.

The pressure on the United States to switch to the metric system has been intense and unremitting. Living in a metric world has obliged the United States to make several adjustments and to open some doors to the metric system. During World War I, for instance, American forces faced problems in supplying French allies because the equipment of the Europeans was all metric, which forced them to start manufacturing metric provisions. The General Staff of the Expeditionary Forces in France had to implement the metric system in all their procedures: operation orders, map drawing, firing data for artillery, etc. All military personnel on the French front had to receive metric lessons. The National Bureau of Standards provided the Army with metric pamphlets, charts, comparison scales, and tables of equivalences; booklets with metric lessons were distributed among the soldiers.[91] In the newspapers of the time some observers claimed, exaggerating but pointing to the problem, that "had England and America been standardized on the metric system, making instant co-operation with France and the other allies possible, the war would have been shortened by two years."[92]

Peace among nations has also compelled the United States to go metric in certain areas. Influenced by the Olympic Games, in the 1930s the Amateur Athletic Union and the sports programs in several universities converted to the metric system for field events, as "few Americans were numbered among the world's record-holders on the Continent because the United States uses the meter only once in four years."[93] The change was soon effective in track competitions. But as happens with any metric transition, regardless of field or scope, the change was not made all at once; traditional races such as the Wanamaker mile, the Baxter mile, and the Millrose 600–yards were still run. In swimming the costs and inconveniences of replacing 50–yard pools with 50–meter ones were higher, and the transition was even slower.

Despite difficulties of this nature, the United States has clung tenaciously to its customary measures, which defy all theories and beliefs about the diffusion of metrication. From the point of view of its national history, under the assumption that metrication happens "more naturally" after periods of political turmoil, the predictable moments for the adoption of the metric system in the United States were the decades after independence, and after the Civil War. From the point of view of world history, under the assumption that metrication is often the result of incentives to achieve international coordination, the key moment for the United States to go metric was the 1970s, when England and former British colonies began their metric transitions and left America without metrological partners. In these junctures, the United States explored the possibility of becoming metric, but the change never came about.

For international standardization, America is more than just an empty spot on the world map. Some fields and practices (like aviation, electronics, and construction) rely at least partly on US measures even in countries that switched to metric long ago. This means that the lack of adoption of the metric system by the United States constitutes a problem for the rest of the world.[94]

The preponderance of American industries causes troubles, big and small. You may end up with funny-looking printed sheets if your computer is programmed with US paper sizes (letter, legal, ledger) but you have supplies with the ISO system (where each paper size is half of the area of the next larger size, and all are derived from A0, which has an area of exactly one square meter).[95] Or things may turn tragic. In 1999 a cargo flight from Korean Air crashed due to confusion when the flight crew was instructed to ascend to 1,500 meters but believed that the requested altitude was 1,500 feet. Given that the United States has dominated aviation since its beginnings, the foot is the most common unit to measure altitude, but not in all countries. This is the testimony of an airplane captain: "I often fly over Chinese, Russian, and North Korean airspace. The altimeters in our aircraft are calibrated in feet. When flying into metric airspace, we use a conversion [paper] card. When Shanghai Control clears us to descend to 3600 meters, we check the card and descend to the equivalent: 11,800 feet."[96] Think of that the next time you travel to China.

Between the extremes of barely noticing that letter and A4 are different paper sizes and risking the life of people in an airplane while doing conversions on the fly, there are several other industries and professions where customary measures invade metric territories and force people to switch between systems continually. The twenty-foot equivalent unit (TEU) is a general unit of cargo capacity used for container ships and is the main measure for the now omnipresent use of shipping containers in trains, ships, and harbors. The size of electronic equipment like cell phones, computers, and televisions is fabricated and commercialized according to the number of inches of the screen. In plumbing, pipes and faucets are also measured in inches. Feeding bottles and baby formula are sold in many countries with customary units, forcing first-time parents to learn promptly how to make calculations with fluid ounces. And if track and field in the US needed to adjust to the metric system, the internationalization of American sports like basketball forces other countries to build 6′ × 3½′ backboards with 18″ rings installed 10′ above the floor. As long as the US keeps its leading position in commerce and technology, these situations will continue.

One final point about the global consequences of a non-metric America and its international economic ascendancy. At the same time as US industries have flooded the world with non-metric commodities, the country has attracted millions of immigrants from metric countries, who have been pushed to learn, in practice, the intricacies of customary measures. This happened just partially in the nineteenth century. In that period a large portion of immigrants came from countries that had not yet started their metrication processes, like Ireland, Russia, and Canada; others, like Germany, were in the initial stages of the metric transition, which means that many of those migrants had not lived the metric system as their native metrological language.[97] Immigration trends in the twentieth century were different, though. Particularly after the 1965 Immigration Act that eliminated the system of quotas (known as the "national origins formula"), millions of immigrants from countries that have been metric for decades arrived in the United States. From 1965 to 2022 about half of all immigrants came from Latin America; other significant contributors were also firmly metrified countries, like China, India, and the Philippines.[98] This was a massive influx of proficient metric users, but despite their large numbers, they represent less than 14 percent of the US population. The vastness of the cultural and material landscape that was shaped with English measures has not been significantly altered by their presence. Interestingly, the customary environment imposes itself even in interactions between members of different metric communities; Filipinos buying in Mexican bodegas in Spanish Harlem still order cold meats by the pound—even if it would be easier for seller and buyer to interact using kilograms.

Figure 2.1. Public incineration of defective measurement instruments, ca. 1915. Photo National Bureau of Standards. Source: *National Geographic Magazine* 27 (1915).

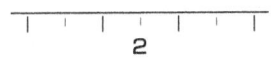

Leviathan's Foot

State Capacity and the Establishment of Measurement Systems in America

> The measurers include all of us. A full list would name every human art and craft. It is an army with world-wide front for mastery of nature and its countless possibilities. Measurement is the weapon, strategy, and tool of conquest; the password to the citadel of truth. To measure is to conquer.
> —HENRY D. HUBBARD, "Measurements of Tomorrow" (1927)

A burning pile of measures. Flaming scales, bushels, half pecks, gallons, quarts, and pints. Big and small, wide and narrow, cylindrical and square, all shattered in a fiery ring. It was a bonfire of short measures, a public spectacle to get rid of faulty measuring devices, the repetition of an old-age ritual to punish the instruments of fraud.[1]

In the first decades of the twentieth century, the National Bureau of Standards (NBS) circulated pictures of these burnings (fig. 2.1) in the press to make known the federal government's efforts to curb commercial wrongdoings involving the use of defective gauging tools in shops, stores, and markets. Local authorities also participated in these efforts. In New York, the Department of Consumer Affairs publicized a burning of 3,000 short bushel baskets in Wallabout Market Square (fig. 2.2), and distributed photographs of confiscated balances being thrown into the sea near Manhattan Island.[2] In Wilmington, Delaware, the sealer of weights told reporters that he had confiscated enough tricked measures and light weights to fill a float, and he planned to exhibit them in the city parade for the people to see how they were being cheated.[3]

These were histrionic actions in the middle of a crisis of commercial misconducts. Local and federal authorities sounded the alarm about massive accumulated losses for shoppers due to metrological scams. The NBS estimated that more than 100 million dollars were taken yearly from the pockets of consumers by retailers who used dishonest weights and measures.[4] Authorities were shorthanded and uncoordinated to face a problem of that magnitude. In 1909 only four states in the Union had efficient systems of inspecting weights and measures. Resources were insufficient. The Wilmington weights inspector pleaded to his legislature to pass a bill giving the sealer a salary (instead of only receiving a fee once a year for every instrument verified), so he could go over the territory three or four times a year; under the fee system he made only one visit to the stores a year, and as soon as he left the short measures were again put into service.

When inspectors needed to work across state lines, things got even gloomier. The supervisor of scales and weighing of the Pennsylvania Railroad System, for example, asked for a bill to concentrate more authority in federal hands. His employees, he explained, had "considerable inconvenience and expense" as a result of different regulations regarding lawful scales across the various states through which they operated. New Jersey, for example, had laws that made some scales incompatible with the laws in neighboring states like New York and Pennsylvania.[5] In other instances, false measures ran away from places with strict laws; the NSB identified the case of a hundred thousand short milk jars that were condemned in one state and then merchants sent them to another state with more lax laws.[6]

These all were scenes of a country with a state lacking infrastructural and administrative power. Authorities required qualified personnel, a larger budget, and homogeneous national regulations. Federal agencies like the NBS needed more muscle. Some of that could be fixed with money and training; but national coordination entailed, more than anything, a different conception of how the state should work. In other words, centralization and the establishment of a state monopoly on the legitimate means of measurement were required.

The centrifugal forces in American politics—the persistent obstacles against centralization—conspired against a proper solution. This chapter shows how the United States could not secure a centralized monopoly on the means of measurement, and how that spoiled its chances to

Figure 2.2. Personnel of the Mayor's Bureau of Weights and Measures prepares short bushel baskets for burning. Wallabout Market Square, Brooklyn, ca. 1911. Source: Department of Consumer Affairs, *What the Purchasing Public Should Know* (1911).

adopt the metric system as the exclusive and mandatory system of weights and measures.

Before that, something should be said about the reciprocal needs between state-making and metrological standardization, and about the different aspects that connect political power with measurement practices. These conceptual and historical considerations will be useful to understand why some countries have accomplished full metrication and others, like the United States, do not. These are models that show what the states *aspire* to do, but not necessarily what they actually achieve, regarding the administration of metrological matters. States may want to compel or force the people to use a uniform system of measurement, but those plans do not always translate into reality. Not all states are equally effective, patient, perseverant, and resourceful to mold the mores of millions of individuals.

Metrication Needs the States and States Need Metrological Standardization

The single most important factor to explain how it was possible that 95 percent of the world population uses meters, liters, and kilograms in their everyday activities is that nation-states officially adopted and enforced the metric system. This sections shows why modern nation-states have been vital for global metrication. It also explains why states took a special interest in controlling weights and measures—establishing a monopoly on the legitimate means of measurement—and what benefits states obtain from this control.[7]

Metrication needs the states and modern states need metrological standardization. States need metrological standardization because the uniform establishment of a homogeneous system of measurement gives them leverage to fulfill some essential functions: undermine the influence of local authorities, enhance the extraction of tax revenue, consolidate internal markets, make the population and the economic resources "legible," monopolize symbolic capital, reduce commercial frauds (heightening thus the administration of justice), and introduce homogenizing institutions that aid the creation of a common national experience.

Complementary, metrication needs the states. An impressive display of state power is required to enact a new measurement system.[8] Only modern states have been effective in helping, compelling, and, if necessary, forcing people to learn and employ metric units—i.e., to think and act metric. Science alone cannot do that, neither commerce and industrialization. And populations by themselves have never done it. Full metrication can only be achieved when at least two actions are combined: policing the employment of metric units and providing people with the intellectual and material means to learn the metric language. As the case of the United States shows, the effectiveness of science and industry to instill a metric way of thinking among the population at large is noticeably limited. Full metrication, in other words, requires compulsion—and compulsion is essentially a political issue, an issue of the state.

"Full metrication" does not refer to a situation in which a select group of people is fluent in using metric terms and decimal arithmetic—no matter how influential that group may be within a specific society. It does not mean either that a large part of the population of a country has some vague notions of what the metric system is. Full metrication—which is the most complex issue in the history of the metric system—means that most of the people use the metric system on the majority of the occasions in which a quantitative language is necessary to apprehend their environment in terms of length, volume, or weight.

Measurement and the State

The tomb of Meketre (ca. 1985 BC) contained, in a hidden chamber, several wooden replicas representing everyday activities in ancient Egypt.[9] One of these miniatures, exhibited at the Metropolitan Museum of Art, depicts a granary with scribes writing on boards and papyrus while workers measure the entry and exit of grain as it is accumulated in storage bins. The scribes recorded all transfers of grain and calculated available supplies for the making of bread and beer.[10] In this replica, it is possible to see, in a single glance, the operation of some of the necessary components for the functioning of large administrative and political organizations: writing, arithmetic, and standard weights and measures, all in the hands of a group of experts who work for a ruler to guarantee the efficient management of the collective means of survival.

As happens with other intellectual technologies, the development of systems of measurement has been intertwined with the flourishing of political and administrative institutions. Archeologist Gordon Childe located the origin of the standardization of weights and measures as part of a "revolution in human knowledge," when ancient Sumerians, Egyptians, and Indians formulated new methods for transmitting experience and organizing knowledge—a moment that represented the beginning of writing and of mathematics.[11] The administration of revenue by public servants, among other key state functions, required uniform measures, a system of numeral notation, and rules of counting.

Units of measurement had existed prior to this "knowledge revolution," but those units were usually based on bodily limbs and individual objects, which made them "personal" standards. Those measures were ineffective for a collective labor that demanded accuracy and cooperation among several workers. Standardization was needed. Fixed values were given to units like the finger and the cubit, and social standards were materialized on measuring rods and stone weights—in other words, they were made conventional. Like language and writing, standard measures rely on convention and social usage. Measuring by conventional standards, Childe observed, "is more abstract than the comparison of concrete individual objects. And all measurement involves abstract thinking. In measuring lengths of stuff, you ignore their materials, colors, patterns, textures, and so on, to concentrate on length. Ultimately this leads to concepts of 'pure quantity' and 'Euclidean space.'"[12]

Cohesive political units tend to have, to a greater or lesser degree, a unified metrological system. When centralized states began to conquer and absorb smaller political organizations, they forced their own units of measurement onto their newly acquired territories. Fittingly, Arnold Toynbee included weights and measures, money, and calendars in a list of "imperial institutions," like official languages and scripts, law, standing armies, and civil services. "Social currencies"—like standard measures of time, distance, length, volume, weight, and value—were necessities of social life and became a matter of concern to governments since their early manifestations. For the control of weights and measures, large states had to compete against local authorities; for the control of time regulation they had to compete against (or negotiate with) religious institutions (priests were usually the ones in charge of the calendars).[13] One of the *raison d'être* of governments is to ensure a modicum of social justice between their subjects and measures are involved in most business dealings. Standard measures concern governments of all kinds, but particularly "universal states," which, by their nature, are confronted with the problem of holding together a greater diversity of subjects than those found in a "parochial state," thus the former have a special interest in the social uniformity that standard measures promote when they are effectively enforced.[14] It is not surprising, then, that many of the measures used in Europe had Roman names, or that some of the most successful cases of political centralization in the history of Europe, such as Charlemagne's reign, were important periods for the standardization of measures.

The right to define measures is an attribute of authority.[15] Rulers make their preferred measures mandatory, keep the custody of standards, unify measures in a territory, and punish metrological transgressions. The struggles about metrological competence of the constituted power are manifestations of the rivalry between various organs of authority that aspire to control measures to bolster their standing.[16] In some social orders, competing institutions may use different systems of weights and measures as a means of asserting authority within a particular sphere; in medieval villages, for instance, one measure was used in the market, another for

Church tithes, and then another to pay dues to the manor. Fragmented political structures inhibit the establishment of uniform systems of measurement. As one of the earliest theorists of sovereignty, Jean Bodin, noted in 1576, "if coinage is one of the rights of sovereignty, so too is the regulation of weights and measures, even though there is no lord so petty that he cannot pretend to this prerogative by the authority of local custom, to the great detriment of the commonwealth."[17] This is why French kings like Philip the Fair, Philip the Tall, and Louis XI tried to have only one system of weights and measures, and equalize all the feudal measures throughout the kingdom—although implementation proved difficult due to the numerous disputes and lawsuits that resulted from it.[18]

Metrological autonomy is a symbol of sovereignty. In the Middle Ages, for example, when cities gained self-determination in metrological matters, they carefully protected that privilege to show their freedom and autonomy; but when a city was conquered, they were forced to use the conqueror's measures. This bond between measurement and sovereignty is suggested by the double meaning of the English word *ruler*, which means both "the person who rules and commands" and "measuring stick."[19] Symbolic representations of the link between sovereignty and measurement can be tracked down many centuries before our era in Mesopotamia, where archeologists found statues depicting governors with a graduated rule on their lap, a widely used representation of power that was still used in representations of European kings—like Henry, "the first King of England" pictured in an engraving holding a standard measure of length in his right hand and saying "weights and measures I corrected true."[20]

Absolute monarchs and modern states, in their quest for centralizing power, sought to gain "monopoly of metrological jurisdiction as one of the fundamental attributes of sovereign power."[21] Not accidentally, constitutions of modern states give the competence of weights and measures to the central authority. Despite its revolutionary upbringing and republican imprint, the metric system was quickly recognized by returning monarchs in Europe, after Napoleon's downfall, as a valuable administrative tool to increase their own power—Louis XVIII, in France, and William I, King of the Netherlands, for instance, decided to maintain it.

MONOPOLY OF THE LEGITIMATE MEANS OF MEASUREMENT

According to Max Weber's well-known definition, the state "is a human community that successfully claims the *monopoly of the legitimate use of physical force* within a given territory."[22] Subsequent theorists have extended these notions of state, monopoly, and power. Norbert Elias, for example, emphasized that equally essential for the states alongside their exclusive right to raise armies is their right to monopolize taxation.[23] He also highlighted how the setting of time became one of the monopolies of states—a crucial attribute that helps them to control the rhythms of social activity.[24] Along similar lines, Pierre Bourdieu argued that there is a "circular causality" with the development of armed forces and the collecting of tributes and taxes.[25] Raising armies asks for the exaction of economic resources to pay soldiers and finance campaigns; at the same time, the control of the means of violence makes the levying of taxes more feasible.

Along with these basic forms of monopoly, modern states have also tried to secure sole control of other key activities, like the monopoly to regulate the monetary system and the creation and coining of money.[26] Jurisdiction over these economic instruments and the establishment of efficient fiscal and metrological systems aided the operation of national markets.[27] Another of these state monopolies is the control of legitimate means of movement (mainly the movement across international boundaries), orchestrated by modern states and the international state system, which deprives people of the freedom to move across certain spaces and makes them dependent on state authorization to do so—an authority that was previously in private hands.[28]

States also try to establish a monopoly of the legitimate means of measurement. The authority to define units of measurement, store physical standards, sanction proper methods of measurement, appoint inspectors, and carry out punishments for metrological offences, which was theretofore held in the hands of various, uncoordinated authorities—cities, corporations, guilds, town markets, and so forth—was expropriated by central state authorities during the nineteenth and twentieth centuries. In the same way that states dispossess their domestic competitors of the instruments of physical violence and the right to use them, so they warrant their monopoly on the legitimate means of measurement by dispossessing social groups of their measuring rights and authority. This process can be described as the transition from multiple metrological sovereignty to a single sovereignty that holds the monopoly of the legitimate means of measurement.[29] Usually, a multiplicity of systems of measurement in a single territory reflects a multiplicity of sovereignties. That is why the adoption of the metric system is, many times, preceded by the destruction of old institutions and the start of a new central sovereignty.

This does not necessarily mean that states control measures effectively, only that they monopolize the authority to sanction measuring practices. Neither does this mean that all states have followed the same monopolization trajectory. The rights that are included in the monopoly over the legitimate means of measurement include the right to define the legal units of measurement—with the subsequent banning of all other units—making those units the only ones acceptable in civil contracts and commercial exchanges; the right to retain the physical standards in which the official units of measurement are embodied; the right to implement a traceability chain to secure the faithfulness of all standards and instruments approved by the state; the right to determine how objects should be measured, including the techniques and procedures of verification and the proper way to weigh and gauge commodities in the marketplace; the right to resolve how frequently standards should be verified and who should verify them; the right to designate inspectors to validate proper employment of weights and measures; the right to set fines and penalties for metrological wrongdoings; and the right to specify when and how the official measurement system should be taught in schools, which may include the right to make metrological education mandatory and even to select the teaching materials to be used in classrooms.

The continuity and authority of the state lessen some of the recurrent problems in measurement. While measures aim to function as immaterial magnitudes and ratios, they are embodied in physical, mutable instruments. Those objects break, bend, and get lost; they can also be stolen or intentionally destroyed. Many instruments (measuring rods, baskets, recipients, etc.) are made with perishable materials that do not resist friction, rust, mildew, or corrosion. Even modern instruments and materials suffer unwanted deformations and loss of mass. As Witold Kula put it, the efforts to warranty the minimal possible mutability in standards of measurement are part of "man's eternal struggle with the destructive power of time."[30] Among the few guarantees aimed at securing fixed measures are social control and supervision by the authorities. Some communities looked for the permanent accessibility of standards, constructing them with sturdy materials—stone, heavy metal—and locating them in town halls, markets, and public places where people could watch them. These standards were attached to building and monuments, so they were difficult to remove. To add reliability, authorities sealed the public instruments and, if they could afford it, they adorned them with artistic workmanship.[31] Time erodes all things, but political and communal care and vigilance mitigate and postpone the inevitable decay.

The state monopoly of the means of measurement involves three distinctive facets: the gathering and management of information; the imposition of a "logic conformism" in society at large; and the social production and distribution of knowledge. The implementation of the metric system—or any other exclusive system of measurement—in a country is a work of distribution of knowledge (as present and future users of the new system need to learn it) and production of knowledge (particularly state-sanctioned knowledge). This involves the writing and distribution of manuals, reports, catalogs of local measures, and tables of conversion. Weights and measures are intellectual tools, but to control them they must become an object of interrogation themselves; metrology is a science of the state.

Establishing a monopoly on the legitimate means of measurement serves many purposes for a state, such as undermining local powers (both political and economic); making the territory legible and homogeneous; consolidating an internal market; enhancing tax exaction and state revenue; reducing commercial frauds, which aids the administration of justice; establishing proper conditions for international trade; and strengthening nationalistic feelings and experiences.

One of the problems of the political life of measures in Europe and the Americas until the beginning of the nineteenth century (and, in some cases, decades later), was multiple metrological sovereignty: the lack of a single unified political hierarchy to regulate weights and measures in a given territory and the presence of competing claims by two or more opposing parties over metrological authority of a defined geographical region (town, city, or province) or activity (business or profession). There usually was a proclaimed—and nominally recognized—sovereign metrological authority, but in practice that authority was ignored or challenged. This situation ended in France on August 4, 1789, with the abolition of feudal rights, which deprived lords of the authority to have final say in metrological matters in their own estates. The revolution unified the French under a sole authority that set a single system of measurement and had the authority to settle all disputes. It is not a coincidence that the birth of the metric system and the establishment of this sovereign and autonomous metrological power occurred simultaneously. In the nineteenth and twentieth centuries, when other nation-states looked for the proper instruments to secure their own metrological monopoly, the metric system became a ready-made solution, legitimized by its scientific aura, and with a proven record of success.[32]

MEASUREMENT, KNOWLEDGE, LEGIBILITY

States are social frameworks of knowledge. The cognitive systems of the state accord primordial importance to "perceptual knowledge of the external world."[33] Modern states perceive the world from the point of view of assuring the regularity of its functioning and identifying economic resources in their territory. For this purpose, states quantify time and space, and establish standards of measurement, which became spatiotemporal references. States emphasize rational and technical types of knowledge to manipulate productive forces and control populations.[34] One of the key cognitive strategies employed by administrative apparatuses to know and manipulate the world is the creation and implementation of reliable and uniform systems of weights and measures.

The state is the culmination of the concentration of various forms of capital: capital of the physical instruments of coercion, economic capital, symbolic capital, and informational capital.[35] The state—as characterized by Pierre Bourdieu—concentrates and redistributes information and creates a "theoretical unification." It takes the vantage point of society in its totality and claims all operations of totalization and objectification through instruments like cartography and writing, which allow the accumulation of knowledge. The state also claims operations of codification—centralized in the hands of bureaucrats and men of letters—which allow cognitive unification.[36] States monopolize the legitimate use of both physical violence and symbolic violence. They exert the latter in the form of mental structures and categories of perception and thought. States shape mental structures and impose shared forms of thinking, cognitive structures, categories of perception, and principles of vision and division that social agents apply to all things of the world.[37] States have thus the ability to inculcate a *nomos,* and found a "logic conformism," a tacit agreement over the meaning of the world. This is achieved especially through schooling and the generalization of elementary education.

It is debatable how much the Bourdieusian account of the cognitive penetration of the state is on target for every empirical case—his description fits better in France than in US, for example. But this framework places the state's efforts to establish a homogenous system of measurement into a broader context. Systems of measurement are languages that shape the perceptual and cognitive structures of people; securing the existence of a sole system of measures reinforces a way of perceiving and classifying the world.

State efforts to homogenize weights and measures coincide with the rise of censuses and maps—as famously described by Benedict Anderson.[38] These instruments serve as a means for states to think of their domains in a "totalizing classificatory grid" that can be applied to people, regions, religions, languages, and products under their control. For the colonial state this was manifest in its ambition for "total surveyability," as it aimed to create a human landscape of perfect visibility. The condition of this visibility was "that everything, had (as it were) a serial number." This form of vision did not come out of the blue; it was the product of the technologies of navigation, astronomy, chronometry, surveying, photography, and print.[39]

Modern states use the standardization of weights and measures as a "tool of legibility."[40] The fragmented nature of customary weights and measures creates administrative incoherencies that work to the advantage of local power holders. Centralized states strive to introduce uniform measures that make territories, resources, and populations intelligible for administrators. These opposing interests promote a clash between local knowledge and state administrative routines.[41] Since traditional measures are local, contextual, and historically specific (as they are the product of indigenous understandings of labor and environment), centralized states cannot create coherent representations of their whole territories based on such standards. Local measures do not lend themselves to aggregation into a single statistical series and do not allow state officials to make meaningful comparisons.[42]

The illegibility of local practices of measurement not only creates administrative inefficiencies, it also compromises state security. For example, food supply could be in jeopardy without comparable units of measure due to complications in monitoring markets and contrasting regional prices for basic commodities; unstandardized measures could also thwart taxation because the state would not obtain equivalent information about harvests and prices. No effective monitoring or controlled comparisons are possible without standardized units of measurement.[43] For all this, the metric system—as the most widely recognized standardized system of measurement available—quickly became a perfect intellectual instrument for governments to crack down on local metrological dialects and launch an intelligible, homogenous, and universal language able to render territories, populations, and resources legible.

TAXATION AND MEASUREMENT

The need to improve fiscal capabilities and to create a fluid trade in increasingly integrated national markets prompted states to seek more stable systems of measurement. In the same way that unstandardized measures complicate the state's administrative apparatus for inspecting its territory, so they set hurdles for the collection of state revenue. A poorly standardized territory means that a uniform system of taxation must account for foreign and domestic customary variations in weights, measures, and containers.[44] To improve their revenue procedures, states have worked methodically to reduce the complexity and diversity of metrological standards and practices, with especial emphasis on the standards and containers used for excised goods.

That is why throughout the nineteenth and twentieth centuries governments usually coupled their plans for metrological reform with their strategies to perfect their revenue systems. Policies against metrological diversity were usually devised by the internal revenue offices. This, on the other hand, generated some interesting paradoxes. To seem fair, taxation should be universally implemented and governed by standards equitably applied; yet the imposition of such standards usually required illiberal methods that more times than not eroded diversity.[45] Crushing local units of measurement has had some extra economic benefits for central governments; unifying the territory under a single metrological system helps to connect more effectively remote areas with the regions more active economically. National unification of weights and measures thus brings areas together through metrology.[46]

This explains why some countries first showed interest in the metric system with their attempt to improve taxation on the import and export of commodities in commercial ports. In the United States, the first federal legislation regarding weights and measures was intended to aid the work of customs.

JUSTICE AND POPULAR PROTESTS

If states have to *impose* their measurement systems, it is because they encounter resistance along the way. This resistance may come from local chieftains, merchants, artisans, taxpayers, peasants, etc. This brings into play several state resources that may be scarce: present their actions as justified, convince the population that the change will bring fairer economic and political transactions, and if that fails, use force to break down dissent.

States should control weights and measures in a way that allows them to meet the social expectations of justice. It is not an accident that monarchs who corrected measures were named "fair" and "just." There is an enduring association between measurement and the imagery of impartiality and fairness—court buildings are still adorned with sculptures of Justice, a blindfolded woman bearing a sword and a balance.[47] But if measures are an attribute of political authority and a representation of justice, they also have been an ancient and perdurable symbol of despotism and dishonesty.[48]

A few examples can give us a clearer idea. The Old Testament repeatedly stresses that God loathes false measures, as in Proverbs 11:1, "The lord abhors dishonest scales, but accurate weights are his delight."[49] The New Testament (Luke 6:38) describes Jesus commanding "Give, and it will be given to you: good measure, pressed down, shaken together, and running over, will be given to you. For with the same measure you measure it will be measured back to you." And according to the Jewish tradition,

Cain—son of Adam and Eve, brother of Abel—"settled with his wife at a place called Nod, where he had children; indulging in every form of vice and violence, he grew rich and ended the simple life by inventing weights and measures."[50] The belief that weights and measures were invented by Cain or some diabolic entity was still discussed in literate circles in early modern times—and as late as the second half of the nineteenth century, an enlightened British critic of the metric system wrote that the meter came from "idolatrous and Cain-like Nations."[51]

These sorts of moral themes and images, which are part of conceptions of justice that cannot always be reconciled with the formal laws of the state, guide the action of protests and resistance of the people who are most affected by metrological changes.[52] Central governments make conscious efforts to eradicate local measures; but for many people, the forced substitution of a familiar system of measurement—especially when that substitution comes as a dictate from political authorities—is usually problematic and could spark explosive reactions. Popular reactions against state-mandated measurement reforms can take two distinctive (and sometimes complementary) forms: the idealization of the lost (old) measures and the deliberate obstruction and destruction of the new ones.

In 1768, in Tetbury, England, demonstrations and riots erupted against the imposition of the Winchester bushel, when the British Crown tried to unify the territory under the London measures. One day a traditional bushel of the town (which contained one gallon and a half more than the Winchester bushel) was adorned with trophies, ordered into the market with trumpets and bells, conducted to the church, and received "the acclamation of all orders and degrees of men, regular and secular."[53] In France, when metric units were reinstated as mandatory measures in 1840, a group of dockworkers in Clemency smashed metric measures, and authorities had to ask for the help of armed officials. The riot was initiated due to suspicions that the change to the metric system would create disadvantages for the workers and would open the town to harmful competition. Around that time this chorus was heard in the streets: "I'm not fan of our legislators' / Decimal / Systemical. / Long live the measures of yesteryear! / And damn the new weights and measures!"[54]

In the Americas, the introduction of the metric system also triggered violent rebellions. One of the most famous anti-metric revolts in world history occurred in Brazil, in 1874. In numerous towns and villages in the northern part of the country, rioters entered local markets and smashed kilogram weights, which had been recently introduced in economic transactions due to a new emphasis from the central government to enforce the years-long ignored metric legislation. Besides the destruction of *quilos* (kilograms), protestors prevented people from paying taxes, refused to answer census questions, and systematically destroyed records in tax offices, town halls, and notarial registers. Confrontations and violence were frequent. Due to the systematic destruction of weights and scales, the movement was called *Quebra-Quilo* (Smash the kilos).[55] In Mexico, at the very beginning of the effective introduction of the metric system, in 1896, in the southern state of Oaxaca, revolts exploded to oppose the use of meters and hectares (instead of traditional agrarian units) to calculate taxes on land property. Peasants entered district capitals shouting, "Death to all who wear pants," in reference to the mestizos and white elites who used European-style clothing. The unrest lasted three weeks and resulted in several government officials killed.[56] Authorities subdued these revolts with the army.

Violent rebellions against the introduction of metric measures—when they happened—appeared in the initial stages of the changeover. When the first shock is over, people resort to more discreet forms of resistance, like simulation, evasion, false compliance, feigned ignorance, and other "petty acts of insubordination" that constitute the "weapons of the weak."[57] After a while (years or decades) metric units became second nature to people.

States prefer to convince people of the usefulness of using standardized and centralized systems of measurement; but if the convincing fails, physical coercion is always a possibility. The resourcefulness and endurance of the states is, here again, crucial to implant and acclimate exotic measures in new lands.

State and Measurement in the United States

On January 6, 2011, for the first time in the history of the United States Congress, members of the House of Representatives read the Constitution on the floor of the House chamber. The event was an idea of republican representative Robert Goodlatte, who explained "Throughout the last year there has been a great debate about the expansion of the federal government, and lots of my constituents have said that Congress has gone beyond its powers granted in the Constitution."[58] This was a theatrical political performance, but it expressed a concern that returns time and again to the center of American politics. During the course of American history there has been a persistent ambivalence about the possible or proper role of concentrated authority.[59] This ambivalence is manifested in debates about the expansion of the federal government and in questions about what powers the Constitution grants to Congress and the executive branch.

Despite the prolonged political battle about the legitimacy of the federal government actions—and the complaints that legislators have assumed powers that are not explicitly mentioned in the Constitution, the US Congress has been reluctant to exercise one of the powers that the Constitution explicitly gives it. The charter indicates that Congress "shall have power… to coin money, regulate the value thereof, and of foreign coin, and fix the standard of weights and measures."[60] The unwillingness of the federal government to use this constitutional power and act decisively to set weights and measures is one of the keys to understand why the Unites States has not fully adopted the decimal metric system.

There have been multiple and ill-conceived attempts by the United States government to lead a national effort in favor of metrication.[61] Three periods were especially relevant: 1777 to 1836, from the *Articles of Confederation* to the joint resolution on weights and measures of 1836; 1866 to 1893, from the Reconstruction era to the final decade of the nineteenth century, punctuated by the Metric Act of 1866 and the "Mendenhall Order;" and 1968 to 1982, a period that revolved around the Metric Conversion Act of 1975 and culminated with the disbanding of the US Metric Board during the Reagan administration.

In recapitulating the history of the federal government's attempts to introduce the metric system I will focus primarily on these three periods. Then I will examine two crucial reasons that prevented the introduction of the metric system in America: the unwillingness to pass compulsory legislation and the absence of a centralized metrological apparatus.

COLONIAL AMERICA: A METROLOGICAL HODGEPODGE

The European immigrants who arrived in North America brought with them a wide metrological diversity, with groups from different parts of the old continent maintaining their measurement instruments and practices. This reproduced in America the European problems that resulted from lack of standardization. In eighteenth-century Europe there were at least 319 units called *pound* and 282 units named *foot*, all of different magnitudes.[62] In the British colonies lived units from Scotland, Ireland, England, Germany, and other countries, plus their local variations. It has been estimated that 100,000 measures coexisted in the thirteen colonies—a high but plausible number considering, for example, that in eighteenth-century France alone there were approximately 250,000 variants of measures.[63] This was a fertile soil for mercantile frauds and an obstacle for transactions beyond local markets.

Each colony acted independently in their metrological regulations, an arrangement that would cast a large shadow on the nation's future.[64] Numerous laws were introduced since early colonial days to fix weights and measures "according to the General Custom of England," but the problem persisted. 1776 arrived before any substantial solution could be found.[65]

MEASURES IN THE YOUNG REPUBLIC

The newly constituted United States of America set from the beginning some legal provisions to establish who would oversee the regulation of weights and measures in the country. The 1777 *Articles of Confederation* specified that "The United States in Congress assembled shall... have the sole and exclusive right and power of regulating the alloy and value of coin struck by their own authority, or by that of the respective States—fixing the standards of weights and measures throughout the United States—regulating the trade and managing all affairs with the Indians."[66] The Constitution, ratified in 1788, kept this stipulation, giving Congress the power to "fix the standard of weights and measures." Of course, the Constitution granted Congress that authority, but it did not say that Congress should exercise it—and it has not, despite being repeatedly urged to do it.

George Washington, who worked as a surveyor and mapmaker in his youth, was aware of the importance of fixing weights and measures and pleaded Congress for action. In his first annual address, in January of 1790, Washington told congressmen that "Uniformity in the currency, weights, and measures of the United States, is an object of great importance, and will, I am persuaded, be duly attended to."[67] In his second address, he returned to the issue, stating that "The establishment of the militia, of a mint, of standards of weights and measures, of the post office and post-roads, are subjects which I presume you will resume of course, and which are abundantly urged by their own importance."[68] And the following year, in his third annual address, he reiterated that "uniformity in the weights and measures of the country is among the important objects submitted to you by the constitution, and if it can be derived from a standard at once invariable and universal, must be no less honorable to the public councils than conducive to the public convenience."[69]

In 1790, in response to Washington's petition, the House of Representatives requested the Secretary of State, Thomas Jefferson, to study the subject and submit a recommendation. Jefferson had already showcased his inventiveness in tackling this sort of problem with the issue of currency—not to mention other metrological contributions embodied in his plan for the system of apportionment of Representatives in Congress, and the Public Land Survey System.[70] In 1784 Jefferson prepared his *Notes on the Establishment of a Money Unit, and of a Coinage for the United States*, which set the basis for the present American dollar.[71] A decade before the franc was established during the French revolution (1795), Jefferson set the blueprint for the first fully decimal national currency in modern history—pioneering what would become one the most prevalent monetary conventions in today's world. Jefferson justified this innovation arguing that "The most *easy ratio* of multiplication and division, is that by ten. Everyone knows the facility of Decimal Arithmetic."[72] He suggested that there should be a golden coin of ten dollars, a silver dollar, a silver "tenth of a dollar," and a copper "hundredth of a dollar." His proposal was a simple progression of 10, 1, 0.1, and 0.01. Contrary to the radical innovative spirit that characterized the French revolutionaries, Jefferson did not seek an entirely new form of money, as for him the new monetary unit should have a value "nearly of the value of some of the known coins" to facilitate its adoption by the people. To secure familiarity he proposed the Spanish dollar (minted in New Spain) to be the base of the American dollar. Overall, it was a creative mixture of innovation and continuity—characteristics that Jefferson later suggested in his weights and measures project.

In 1790 Jefferson presented to Congress his *Plan for Establishing Uniformity in the Coinage, Weights, and Measures of the United States*.[73] In truth, it consisted of two plans. The first one was a proposal to update the English customary system. The other was a novel decimal system that reorganized the progressions and subdivisions of measures and changed the magnitude of many of them, only retaining their old names. The latter was very similar to his plan for the dollar.[74]

The decimal architecture in Jefferson's proposal can be observed in the units of measurement and its subdivisions. For measures of length, he said, "Let the foot be divided into 10 inches; the inch into 10 lines; the line into 10 points; let 10 feet make a decad; 10 decads one rood; 10 roods a furlong; 10 furlongs a mile." For measures of capacity: "Let the bushel be divided into 10 pottles; each pottle into 10 demi-pints; each demi-pint into 10 metres, which will be of a cubic inch each. Let 10 bushels be a quarter, and 10 quarters a last, or double ton." And for weights: "Let the ounce be divided into 10 double scruples; the double scruple into 10 carats; the carat into 10 minims or demi-grains; the minim into 10 mites. Let 10 ounces make a pound; 10 pounds a stone; 16 [sic] stones a kental; 10 kentals a hogshead."[75]

With this plan, Jefferson—who carefully followed the development of metrological innovations in France (he had just returned from Paris, where he served as minister to France from 1785 to 1789)—beat the French to the punch, again.[76] The decimal (metric) system of weights and measures of the French revolutionaries was still two more years from being fully conceived, and another three years from being implemented. However, it was the creation of Condorcet, Lavoisier, and other French savants, and not Jefferson's plan, what became a universal measurement language. This was so, in good part, due to the form of political regimes that their respective revolutions brought about. In France, the revolution created a stronger central government; in the United States independence brought a confederation of semi-independent political units.

The recommendation made by Jefferson on currency and the one on weights and measures were based on the same principles: unification (the creation of a single system), simplicity, and decimal progressions and subdivisions. Despite its similarities, they had diametrically opposite fates. The currency project was accepted, set the stage for future monetary reforms around the world, and made the United States a pioneer in the use of decimal currencies.[77] The plan on weights and measures was ignored and started America's long-lasting attachment to English customary measures. Regarding metrological innovation, the United States represents a peculiar case not only because it did not accept the metric system, but because it has never accepted any significant change of system, metric or otherwise. Inertia has been the norm.

Jefferson's weights and measures project did not trigger any definitive action by Congress.[78] The *Plan for Establishing Uniformity* ultimately became the first chapter in a long—and ongoing—collection of studies, hearings, and proposals ordered by the United States government (at its different levels and branches) to clarify the condition of weights and measures in the country and to evaluate the convenience of reform. Altogether, it is a compendium to which some of the best minds of the nation have contributed, and that has become, after more than two centuries, a full library. It is one of the most futile collections of accumulated knowledge ever produced—so much brainpower invested, and so little to show for it.[79]

James Madison revisited the weights and measures problem during his presidency. In his 1816 annual message, he reminded legislators that "Congress will call to mind that no adequate provision has yet been made for the uniformity of weights and measures also contemplated by the constitution. The great utility of a standard fixed in its nature and founded on the easy rule of decimal proportions is sufficiently obvious. It led the government at an early stage to preparatory steps for introducing it, and a completion of the work will be a just title to the public gratitude."[80] He did not specify if the decimal system should be the metric system, the one designed by Jefferson, or another one; but he was asking for a drastic overhaul of America's metrological regime. In response, Congress asked John Quincy Adams, then secretary of state, to prepare a new report on the matter.

Adams's extensive and erudite report showed great praise for the metric system, saying, that "considered merely as a labor-saving machine, [the metric system] is a new power, offered to man, incomparably greater than that which he has acquired by the new agency which he has given to steam. It is in design the greatest invention of human ingenuity since that of printing."[81] Nevertheless, for him it was not a completed product yet, and ought to be perfected. Ultimately Adams thought that its introduction in the United States would create more problems than solutions. This position is understandable, considering the moment when Adams wrote his report. In those years the metric system was only being used in France in a version that was a mixture with the old regime measures (thanks to a reform made under Napoleon I in 1812, called *mesures usuelles*, that survived until 1837, when the original design of the system was restored), and very little had been done in terms of metrication in the rest of the continent. At that point in time it was not obvious that the metric system was on its way to global acceptance.

In the end, despite his interest in a reform of weights and measures based on some of the principles behind the metric system, Adams suggested to Congress a two-part plan: "1. To fix the standard, with the partial uniformity of which it is susceptible, for the present, excluding all innovation. 2. To consult with foreign nations, for the future and ultimate establishment of universal and permanent uniformity."[82] It was a compromise between the present and a hypothetical more perfect future, and between keeping whatever unity of measures existed in the country and the need to cooperate with other countries to establish a permanent international system. Ultimately, Adams did not think that the metric system, in spite of all its theoretical virtues, would be the answer for "universal and permanent uniformity"—in hindsight, he obviously underestimated it. History is full of irony. Adams wanted to collaborate in the future with other countries to create a global system, but the future brought something vastly different. The United States became the most tenacious obstacle for the world to have a truly universal system of measurement.

Toward the end of his report, Adams offered this moral, "If there be one conclusion more clear than another, deducible from all the history of mankind, it is the danger of hasty and inconsiderate legislation upon weights and measures."[83] This marked one of the biggest differences between the ways in which the American and French revolutions approached metrological reform. The French were in a rush to regenerate society and shape the future, and were not timid to enforce their will; Americans chose to wait and see, and never showed determination to solve the issue.

Waiting did not solve the pressing problems caused by poor standardization in the country. Facing inaction by the federal government, states took matters into their own hands, and one after another enacted local legislations to fix standards and set controls for measuring instruments. However, states simply legalized the standards that were then in use in their own lot, which only contributed to solidify their differences.[84]

An area in which the federal government showed resolution and effectiveness was in securing uniformity in coinage and in the administration of weights and measures for custom houses—something crucial for the efficient extraction of revenue. This institutional foundation opened the door for some national uniformity. Repeated reports in Congress indicated that there were considerable losses in revenue due to differences in the standards used in custom houses, and it was estimated that this loss in income in one week "would more than compensate the expense of establishing uniform standards."[85]

Some concrete actions were taken. In 1828, with John Quincy Adams now as president, the "Act to continue the mint at the City of Philadelphia" set a standard troy pound—acquired from London—for the Mint of the United States, which essentially became the fundamental standard of mass in the country. The Treasury labored to construct proper and uniform standards of weights and measures for the customs service. And finally, in 1836 a joint resolution was passed in Congress, which mandated the delivery of a complete set of weights and measures to the governor of each state in the union, "to the end that a uniform standard of weights and measures may be established throughout the United States."[86]

A key actor in the completion of these tasks was Ferdinand Rudolph Hassler, a Swiss geodesist with close connections with many of the scientists involved in the creation of the metric system. In 1793 Hassler studied in Paris with savants who were working in the design of the new measurement system, like the astronomers Lalande and Delambre, the mathematician Jean-Charles de Borda, and the great chemist Antoine Lavoisier. Before he emigrated to America in 1805, Hassler assisted Johann Georg Tralles in the execution of topological surveys in his homeland—Tralles had been Switzerland's representative at the 1799 Congress on Definite Metric Standards in Paris, where the calculations for the meter were completed.

Hassler was instrumental in two urgent technical projects of the United States government: the first attempts at standardization of weights and measures, and the establishment of the Survey of the Coast (which was the first scientific agency of the federal government). Hassler brought to the United States a standard meter made in Paris by the French Academy of Sciences in 1799 (which he had received from Tralles). His expertise in mathematics, surveying, and scientific instruments was greatly needed in an expanding country with a limited number of qualified technical personnel. In the 1830s Hassler examined the instruments used in custom houses, uncovered great discrepancies among them, and took the initiative to provide them with standards of length, mass, and capacity.[87] He was named the first superintendent of the Office of Weights and Measures in the Treasury Department.

The standardization of customary measures was the beginning of some sort of national uniformity. However, the joint resolution of 1836 did not contain a detailed plan, and many important issues about the administration and control of weights and measures were left unresolved. What legal requirements and specifications would regulate the day-to-day administration of weights and measures? What methods of testing and inspection should be established? Who had the authority to settle disputes between states? To whom should state inspectors of weights and measures report? There was no common framework or legal uniformity to solve these problems. If used properly, physical standards would only secure some homogenization, which was badly needed, of course; but not even that was fully achieved. The standards sent to the states by the federal government were in many cases neglected and some disappeared; others were stored carefully but were never used.[88] For metrological matters, *E pluribus unum* (Out of many, one) was not happening; there were rather, out of one county, many metrological sovereignties.

The expansion of the republic complicated things even further. The variety of weights and measures already present in the thirteen colonies was increased with the purchase of the Louisiana Territory from France (1804), the expulsion of Spain from Florida (1819), the annexation of the Republic of Texas (1845), and the annexation of the Mexican territories of Alta California and New Mexico (1848). This added to the American metrological pool a wealth of measures of French, Spanish, and Mexican origin.[89] This has had perdurable consequences. Even in today's *Texas Agriculture Code* the "Spanish *vara*" appears among the units for length and surface used in the state, and in Louisiana there are still parcels of land known as "*arpent* sections," following an old French unit of area.[90]

CIVIL WAR, RECONSTRUCTION, AND CENTRAL STATE AUTHORITY

If there was a propitious moment in the United States history for a full adoption of the metric system it was after the Civil War, during the Reconstruction era. It almost happened. Under the lead of northern Republican legislators, particularly congressman John A. Kasson from Iowa (chair of the House Committee on Coinage, Weights, and Measures), and senator Charles Sumner from Massachusetts, a plan was put in place to advance the metric cause. The newly created National Academy of Sciences was asked to evaluate the pertinence of adopting the system. In July 1866, sixteen months after the end of the war, a metric legislation was passed in Congress and signed by President Andrew Johnson, recognizing the meter, the liter, and the kilogram as legal (but *not exclusive*) units of measurement.

This was the Metric Act of 1866, also known as the Kasson Act. Kasson had worked in Lincoln's administration and had extensive experience in international agreements. His interest in metrication was very explicit, not only in his support for this legislation, but also in his participation as vice-president of the American Metrological Society (AMS) during the 1870s. The AMS was not an overtly pro-metric organization, but many of its members were prominent supporters of metrication in America.

Sumner was the other key player behind the Metric Act, a well-known abolitionist, leader of the radical Republicans in the Senate, and confidante to Lincoln. After the war, Sumner asked for the dismantlement of existing Southern governments, the reorganization of confederate states by Congress, and the allocation of homesteads for freedmen.[91] In 1866 he was the chair of a special committee to which all bills concerning the metric system should be referred. On July 27, Sumner asked the Senate to approve two bills and a joint resolution related to the metric system, which were taken up and passed. The joint resolution enabled the Secretary of the Treasury to furnish to each state one set of metric standards of weights and measures. It also authorized the use of metric weights in post offices.

The Metric Act stipulated that "It shall be lawful throughout the United States of America to employ the weights and measures of the metric system; and no contract or dealing, or pleading in any court, shall be deemed invalid or liable to objection because the weights or measures expressed or referred to therein are weights or measures of the metric system."[92] This legislation made the metric system *a* legal system of measurement in the country, but did not define it as the *exclusive* system. This was a notorious difference with most of the legislations introduced in other countries at the time, where *compulsion* and *exclusiveness* were always mandated.

In a speech in the Senate, Sumner delivered a spirited praise of the metric system, exalting its "cosmopolitan character," much in the vein of other metric devotees at the time all over the world:

> There is something captivating in the idea of weights and measures common to all the civilized world, so that, in this at least, the confusion of Babel may be overcome. Kindred is that other idea of one money; and both are forerunners, perhaps, of the grander idea of one language for all the civilized world. Philosophy does not despair of this triumph at some distant day; but a common system of weights and measures and a common system of money are already within the sphere of actual legislation. The work has already begun; and it cannot cease until the great object is accomplished.... The adoption of the metric system by the United States will go far to complete the circle by which this great improvement will be assured to mankind. Here is a new agent of civilization, to be felt in all the concerns of life, at home and abroad. It will be hardly less important than the Arabic numerals, by which the operations of arithmetic are rendered common to all nations. It will help undo the primeval confusion of which the Tower of Babel was the representative.[93]

Sumner's view of the metric system as an "agent of civilization" was in alignment with the rhetoric displayed by other politicians and scientists in Europe and Latin America. But his proclaimed determination to keep with the work in favor of metrication "until the great object is accomplished" barely amounted to passing a legislation that only made the metric system optional, not mandatory. He justified the softness of the law by saying that "the first step is taken there by making the metric system *permissive*, as is proposed in the bills before Congress. The example of Great Britain is of special importance to us, since the commercial relations between the two countries render it essential that these should have a common system of weights and measures. On this point we cannot afford to differ from each other."[94] At the time, however, England was in no hurry to abandon its yards and pounds; its economic position in the world was strong and it possessed an enormous empire where official weights and measures were defined from London—it took more than a century for the United Kingdom to adopt the metric system to participate in the European Economic Community in the 1970s.

The report of the Committee on Coinage, Weights, and Measures, headed by Kasson, provided some other arguments for the non-mandatory nature of the law:

> The metric system is already used in some arts and trades in this country.... Yet in some of the States, owing to the phraseology of their laws, it would be a direct violation of them to use it in the business transactions of the community. It is therefore very important to legalize its use, and give to the people, or that portion of them desiring it, the opportunity for its legal employment, while the knowledge of its characteristics will be thus diffused among men. Chambers of commerce, boards of trade, manufacturing associations, and other voluntary societies, and individuals, will be induced to consider and in their discretion to adopt its use. The interests of trade among a people so quick as ours to receive and adopt a useful novelty, will soon acquaint practical men with its convenience. When this is attained—a period, it is hoped, not distant—a further act of Congress can fix a date for its exclusive adoption as a legal system. At an earlier period it may be safely introduced into all public offices, and for government service.[95]

The promoters of the Metric Act justified the lack of compulsory legislation saying that there was a need to be in tune with England in terms of commercial standards. They also wanted to open the doors for the metric system to be used freely in all the states. They had the belief that Americans are particularly able "to receive and adopt a useful novelty" (a belief that was greatly overblown) and were confident that the metric system would be embraced voluntarily. Finally, they hoped that Congress would promptly set a date for the exclusive use of the metric system. None of that happened.

Little did they know that this was a wasted opportunity. No other moment in United States history was as favorable for a clean shift to the metric system. As it turned out, England was a century away from committing to the metric system (and even this eventual transition was not persuasive enough for the United States to make their own change, as we shall see); regular Americans did not voluntarily adopt the metric system; and Congress has not been able to set a deadline for a definitive metric changeover.

The timing for metrication, from the international vantage point, was very appropriate in the 1860s. Contrary to the circumstances during Jefferson's and John Quincy Adams's times, the metric system had now a tangible worldwide impact. It was clear for many observers that if one system of measurement was going to grow into a truly "universal" system it would be the metric and no other. More and more countries were defecting to the metric camp, with continental Europe and Latin America on their way to become complete metric territories.

The Reconstruction era was also an auspicious moment for metrication in America because it was inspired by principles and policies under which the metric system thrived elsewhere: national unification, expansion of the administrative capacity of the state, and economic modernization. For the first time in the history of the country, the federal government was in the position to consider and execute national-scale administrative undertakings, with the power of the states, especially in the South, being very limited. A revamped state had emerged from the ruins of the Civil War and it was penetrating into new areas of social life, like education, race relations, labor regulations, and economic development.[96] In this context, an aggressive metrological policy was highly plausible but never materialized. In the end, Reconstruction did not create an autonomous federal bureaucracy; even in the North, the forces of localism, division of powers, and distrust of government activism remained.[97] Even among Republicans doubts about a more proactive state persisted.[98] The kind of state needed to complete a national metrication program did not fully come to fruition.

Certainly, Congress legalized the use of the metric system, but it did not authorize programs to promote its use.[99] Few instrumental actions were taken. Similar to what had happened in 1836, federal authorities were content with sending a set of metric standards to all states. The results, not surprisingly, were very poor. And the federal government did little during the rest of the century, at least domestically.

If within its borders the United States was lethargic in introducing the metric system, internationally it was more active. In 1875 the Unites States signed in France, along with other sixteen nations, the Metre Convention. This treaty created the General Conference on Weights and Measures, the International Bureau of Weights and Measures (BIPM), and the International Committee for Weights and Measures.[100] In 1878, the Senate ratified the Metre Convention. Copies of the international kilogram and meter were assigned by lottery among the signers of the Convention. In 1890 the United States—which had paid for two sets of prototypes—was awarded kilograms number 4 and 20, and meters 21 and 27. These prototypes were state-of-the-art scientific instruments; few countries could afford them (and the US paid for two sets). Personnel of the Coast and Geodetic Survey brought from France the meter number 27 and kilogram number 20, carefully packaged in boxes sealed twice with red wax. These were received, unsealed, and exhibited in the White House in the presence of president Benjamin Harrison, in a ceremony with the secretaries of state and the treasury, and several guests from Congress, the federal government, the military, and the scientific community—among them, the director of the Mint, the chief of engineers of the US Army, the secretary of the Smithsonian Institution, the president of the American Society of Mechanical Engineers, and members of the American Metrological Society. These prototypes became the official national standards of length and mass and were deposited in the Office of Weights and Measures of the Coast and Geodetic Survey.[101]

With the metric standards from the BIPM, Thomas Corwin Mendenhall, superintendent of weights and measures, issued a directive, known as the "Mendenhall Order." With the approval of the Secretary of the Treasury, the order made the meter and kilogram the fundamental standards in the United States.[102] More than anything, this was just a technical issue. The Mendenhall Order indicated that the yard and the pound would be derived from the metric standards—this is why the yard and pound are officially defined relative to the metric system.[103] This helped greatly to bring conformity between the federal standards and those of other countries, but did not stipulate any change for regular folk, who could keep using customary units as usual.[104]

Overall, during the nineteenth century the federal government was reluctant to create an effective legal framework to secure uniform standards for the entire country and to secure an adequate provision of instruments and trained experts. The Coast and Geodetic Survey had the authority to make standards for the states, but did not have the power to issue certificates of verification. Besides, due to the lack of equipment and human resources, it could not verify all the instruments referred to the office of weights and measures (the entire staff of the Office consisted merely of a field officer, two scientific assistants, an instrument maker, and a messenger). Measuring tools and standards had to be sent abroad, mainly to Germany, for calibration.[105]

The most aggressive piece of legislation seriously considered in Congress came in 1896, with a law that proposed that the metric system would become the *only* legal system recognized in the United States, which received considerable support in the House of Representatives, but too many Congressmen were afraid of a possible adverse reaction by "angry farmers" in an election year.[106] Congress was again active during the first decades of the twentieth century, with around fifty bills related to the metric system proposed and debated, but nothing came out of them. That period signaled the beginning of a sturdy anti-metric movement, formed by a coalition of engineers and manufacturers that undermined the attempts for reform (as we will see in Chapter 4).

The alternative of using local governments to break down the opposition to the metric system by manufacturers and engineers was explored by people like Samuel Stratton, the first director of the National Bureau of Standards (NBS). In 1902 he wrote to one of the leaders in a pro-metric organization that "if any legislation is enacted concerning the use of the metric system by the public, it will probably be the states, and then only with reference to the common weights and measures as used in every-day business transactions."[107] But, yet again, neither the federal government nor the states passed any metric legislation.

In metrological matters, the United States has been, to use Rexmond Cochrane's expression, a country of "laissez-faire standards;" or not a country but a loose association of states, a commonwealth.[108] A compilation of state laws on weights and measures in the United States, dating from 1904, was almost 500-pages long, with dozens of chapters—one per each state and territory—of non-homogeneous legislations.[109] This arrangement did not provide adequate results at the national level.

To partially alleviate these difficulties, in 1901 the NBS (today the National Institute of Standards and Technology) was created, as the successor to the Office of Standard Weights and Measures of the Treasury Department.[110] The NBS was responsible for several scientific and technological tasks, many of them beyond weights and measures. It kept the national standards, tested the standards of the states, and supplied local officials with technical information and training; however, it had no enforcement authority or regulatory capabilities.

The lack of centralized enforcement served the interests of people hostile to metrication. That is why they systematically opposed reforms that would increase the power of the NBS. Anti-metric groups forcefully challenged legislation proposed by congressman Albert Vestal—chairman of the Committee on Coinage, Weights and Measures—that would allow the NBS to approve all types of weighing and measuring devices used in the country, and to enforce its exclusive use. Opponents of the bill banked on the states being "very jealous of their rights" to defeat the motion in Congress—as it ultimately happened.[111]

In 1905, the NBS hosted in Washington, DC, the first National Conference on Weights and Measures, a meeting of weights and measures officials at all levels of government and other people interested in metrology, which have met annually since then. The conference was convened with the hope of alleviating some of the problems caused by the lack of uniform standards and regulatory oversight. Typical of what may be called the "American model," the Conference is an unofficial organization that has no legal status or authority to enforce recommendations. To be effective, it relies on its reputation among all who participate in the distribution or control of commercial weighting and measuring instruments.[112] The resolutions of the Conference are merely recommendatory, "a code of specifications and tolerances ... or a model law that has been adopted by the Conference, can have no effect in any given jurisdiction until it is promulgated or enacted by competent authority within and for that jurisdiction."[113] This contrasts with most other countries, which have a more straightforward approach to control measures, setting mandatory regulations with a nationwide reach.

Considering the shaky legal ground on which the administration of weights and measures operates in the United States, this formula of nonmandatory coordination, led by the NBS, has worked reasonably well and has set general guidelines in measurement regulations. The National Conference on Weights and Measures became a long-living body and has developed important model laws that have been widely accepted by the states. The handbook *Specifications, Tolerances, and Other Technical Requirements for Weighing and Measuring Devices* has been issued since 1949 and regularly updated, and all fifty states in the union have adopted it as the legal basis for regulating measuring tools. As of 2002, 44 states had adopted a weights and measures law based on the uniform laws delineated in a handbook published by the Conference, *Uniform Laws and Regulations, in the Areas of Legal Metrology and Engine Fuel Quality*.[114] This is not bad, but it is not great either. Having six states that do not follow the same regulations as the rest of the country opens the door to multiple problems.

If we go back to 1901, when the NBS was created, the metrological order of the country looked rather bleak. Standardization across the national territory was weak, with every state having its own weights and measures laws; there was poor policing to prevent short measures and cheating (with abundant frauds in markets and small transactions); the only federal office dedicated to weights and measures had little power to verify the accuracy of standards (let alone keeping an eye on unfair commercial practices); and standards and instruments had to be sent abroad to determine their accuracy. The federal government was lethargic, offices were understaffed, and politicians were unmotivated to fix the situation. In short, at the beginning of the twentieth century, the United States did not have the proper legislation, bureaucratic apparatus, or expertise needed to provide the services and enforce the regulations necessary for a modern country.

This was unusual. Other countries at that point had already developed a more robust official metrological apparatus. Mexico, for example, had everything in place by 1901. The government had decided that metric would be the only sanctioned system; nationwide laws, codes, and procedures had been approved; the federal government founded a central weights and measures agency headed by a meticulous engineer with a team of bureaucrats, inspectors, and a laboratory of metrology; state verification offices; and, finally, a mandatory curriculum for teaching the metric system in all elementary schools. In 1896 the government had launched the first effective metrication campaign at a national level. Mexico had the system, the legislation, the experts, the inspectors, the instruments, and the teaching programs. The whole structure was working more or less effectively. By 1910 the penetration of the metric system in urban areas was near completion, and some advances had been made in rural areas. If we compare both countries at that point in time, it shows that a mandatory policy was more effective than the laissez-faire approach.

THE ASSOCIATIVE STATE

As the experiences of other countries have shown, the mandatory adoption of the metric system is a daunting and prolonged enterprise even when governments are unified, determined, and resourceful. Conditions in the United States, where the government itself was torn apart on the issue, only produced a long-lasting deadlock and failure to act.

Throughout the twentieth century, many people in the federal government supported plans for standardization but were unwilling to sponsor any mandatory legislation. Among them, the figure of Herbert Hoover, Secretary of Commerce, stands out. In the 1920s Hoover had the explicit intention of reducing the number of product sizes and simplifying the variety of manufacturing articles. He was interested in a larger plan to improve the nation's efficiency (a key word in many of his endeavors). The creation of the Division of Simplified Practice in the NBS was part of this project.[115] Simplify practices would help to protect customers and lower production costs. The grounds on which these principles were put in place were broad and affected numerous industries, from face-brick dimensions and asphalt grades to bread weight and food containers. As the newspapers reported,

> Secretary Hoover's Division of Simplified Practice has under way an endless campaign to eliminate needless sizes and styles. He has already brought about standardization of the sizes of bedsteads and bed springs,

and is now working on blankets, lumber, metal lath, and other things. At the same time industry itself, through the American Engineering Standards Committee, is carrying out more than 100 other standardization projects. The public is going to hear a great deal about standardization and simplification from now on.[116]

It would appear that Hoover, so fond of standardization, would be unambiguously in favor of metrication—and many members of the pro-metric groups thought so—as the metric system appear to be a perfect solution to avoid having around too many units of measurement. But Hoover, a trained engineer, did not consider that government mandates were an effective way to attain standardization. He considered that standardization cannot be achieved through the imposition of "arbitrary rules;" to be effective, he argued, laws and regulations ought to be based on agreements between all the parts and interests involved.[117] This applied to metrology as well. Hoover attended several annual conferences on weights and measures held at the NBS, where he stressed the same point underlining that the government would only help in the "voluntary adoption in commerce" of national specifications.[118]

This dictum was a central part of the idea of an "associative state" that characterized Hoover's political outlook, which consisted in businesses cooperating among each other and the government, giving priority to self-governing business associations.[119] This form of government and the work of the Division of Simplified Practice were successful in their standardization ambition for several industrial and commercial products; but it proved ineffective in securing a single system of measurement for the whole country, as the lack of agreements within industry made it impossible to find common ground. The lack of commitment of the federal government to push for a mandatory metrication policy created numerous problems for an effective national unification of weights and measures. Hoover's policies are a perfect example of this, but his position was part of a larger trend.[120]

A NON-METRIC ISLAND IN A METRIC WORLD

The United States failed to adopt the metric system in moments of massive socio-political change, which for many other countries marked the propitious juncture to push for comprehensive metric reform. However, in the second half of the twentieth century, a series of international developments brought a great deal of pressure to complete metrication. As shown in the previous chapter, by 1950 (fig. 1.5), apart from the United Kingdom and the Commonwealth of Nations, almost every country in the world had adopted the metric system. The globe was split into two parts, one metric, constantly expanding; and one using different versions of the English system (customary and imperial), contracting.

The non-metric block was confronted with the problems of increasing isolation and lack of uniformity in their own lot. Due to their relatively independent development, inches, feet, pounds, pints, and other units were not identical across different English-speaking countries. To fix this, in 1959 the United Kingdom, Australia, Canada, New Zealand, South Africa, and the United States agreed to a common yard and pound, called "international yard" and "international pound," which were defined in metric terms (the international yard as 0.9144 meters, and the international pound as 0.45359237 kilograms). They failed, however, to agree on a common pint—that is why a pint of beer in England is larger than one in the US.[121] International coordination made this agreement necessary. Among other things, the discrepancies between the inch in the United Kingdom and the Unites States became a problem during World War II, when mechanics found that some parts of aircraft engines, built following identical blueprints in both countries, were not noninterchangeable.[122] The United States also retained the "US survey foot" to avoid alterations in the accumulated survey data produced with the old definition—this survey foot was used until 2023, when it was phased out to facilitate the modernization of the National Spatial Reference System.

In 1965 the United Kingdom announced a ten-year plan to move to the metric system. Ireland and South Africa followed suit in 1967, New Zealand and Australia in 1969, and Canada in 1970. The cocoon that had protected the United States from metrological isolation was cracking. In response, a new push for adopting the metric system started in America.

In 1968 Congress passed the Metric Study Act, providing for a three-year investigation to analyze the impact of increasing the use of the metric system in the United States. The study, conducted by the NBS, was released in 1971. It was a twelve-volume report, presented as a report to Congress eloquently titled *A Metric America: A Decision Whose Time*

Has Come.[123] It was a comprehensive investigation that included dozens of public hearings and surveys on international trade, business, industry, consumers, national security, and education; there was even a volume dedicated to the history of metric controversies in the United States.[124]

The study presented by the Secretary of Commerce recommended, among other things, that the United States should change to the metric system "deliberately and carefully," through a "coordinated national program," and that "Congress, after deciding on a plan for the nation, establish a target date ten years ahead, by which time the U.S. will have become predominantly, though not exclusively, metric."[125] As in previous occasions, a non-mandatory policy was sought after. They kept doing the same thing but somehow expected a different result.

In a paper read before the American Philosophical Society, Lewis M. Branscomb, director of the NBS and the main figure behind the *U.S. Metric Study Report,* pondered the options of how to accomplish the transition to the metric system:

> We homed in on two alternatives. The first is *laissez faire*, a perfectly sound principle and indeed the one that should be recommended in the absence of contrary evidence. The Unites States follows no overall plan; everyone does their thing. We do not set a target date for being metric and government does nothing either to impede or to foster the change, but allows individuals and companies to make their own decision. That is of course what we are doing now. . . . The second alternative is a planned program of metric conversion based on an overall national program with a target for becoming predominantly but not exclusively metric. Within this framework, segments of the society would work out their own timetables and programs, dovetailing them with timetables of other segments. Such a plan would involve voluntary conversion; voluntary in the sense that it is not driven by any legislative or mandatory requirement, involuntary of course to the extent that if a consensus of the country decides to go that way then the commercial pressures to participate will be great for the minority of those who choose not; but it would become inconvenient for them.[126]

Branscomb showed here a noticeable disregard of reality and probability—

or to put it bluntly, he was painfully wrong. Suggesting that there was no evidence against laissez-faire metric policies was either misleading or plain ignorance. At that point in time, it was abundantly clear that voluntary transitions are not effective and that "commercial pressures" are not strong enough to defeat reluctance. Despite more than one hundred years of experience indicating that voluntary metrication is almost equal to no metrication at all, the option of a "legislative or mandatory requirement" was not even considered.

The NBS report was the framework in which discussions in Congress took place in the following years. Congress passed, in 1974, Public Law 93-380, the "Education Amendments," which included a section stating, "It is the policy of the United States to encourage educational agencies and institutions to prepare students to use the metric system of measurement with ease and facility as a part of the regular education program."[127] Again, the law was not a mandate, but a recommendation—and it came about more than a century after other countries in the Americas had initiated their programs of obligatory metric education.

The next year Congress passed the Metric Conversion Act of 1975, which declared that "the policy of the United States shall be to coordinate and plan the increasing use of the metric system in the United States and to establish a United States Metric Board to coordinate the voluntary conversion to the metric system." The Board was composed by representatives from scientific, technical, and educational institutions, and state and local governments. The Act required that each Federal agency "use the metric system of measurement in its procurements, grants, and other business-related activities, except to the extent that such use is impractical or is likely to cause significant inefficiencies or loss of markets to United States firms, such as when foreign competitors are producing competing products in non-metric units." Finally, the Act specified that it is the declared policy of the United States "to designate the metric system of measurement as the preferred system of weights and measures for United States trade and commerce," but also "to permit the continued use of traditional systems of weights and measures in non-business activities."[128] The federal government reduced its role to the coordination of a voluntary transition and to be an exemplar to the states by transitioning its agencies to the metric system.

As before, a coalition of scientists, educators, and government officials supported the idea of going metric. Against it were industrialists (like the Automobile Manufacturers Association and the American Iron and Steel Institute) and labor organizations (like the International Brotherhood of Electrical Workers and the United Brotherhood of Carpenters and Jointers of America).[129] The industrial sector feared carrying the burden of paying for retooling and retraining; labor believed that the switch could make some workers obsolete, particularly the older ones. The congress's plan did not assign federal funds to pay for those costs. As the *Metric Study* indicated in its recommendations, "in order to encourage efficiency and minimize the overall costs to society, the general rule should be that any changeover costs shall 'lie where they fall.'"[130] There was no mandatory dictate and no budget to help pay for the costs; they only provided a loose timetable. No carrot and no stick. Failure, as in 1866, was predictable.

Despite a great fuzz in media and schools about America going metric, the situation was not encouraging for the metric hopefuls. The groups who wanted or needed to use metric measures (like scientists and exporters) were already using it. Those who did not want to switch faced no obligation and received no incentives to change. And among the general public, indifference and skepticism toward the whole metric business prevailed. In 1971, according to Gallup surveys, 56 percent of the people said that they were not aware of what the metric system was, and 19 percent declared that they opposed its introduction in the United States.[131]

The toothless Metric Board could not do much and the arrival of the Reagan administration sealed its fate. In 1982 the president sent a letter to the chairman of the board assuring his support for the policy of voluntary metrication and asking for cooperation in the "orderly phaseout of the Board's activities as part of my program to reduce government spending and streamline its operations."[132] The Metric Board was dead. In the end, the government did not force the change, did not pay for it, and ultimately did not coordinate a voluntary transition. Fifteen years of research and legislative work were wasted, and the United States sealed its metrological isolation.

The only thing left for the federal government to try was the metrication of the federal government itself. This goal has not been pursued thoroughly but has not been abandoned either. Further legislation has advanced it, like the Executive order 12770, issued in 1991 by George H. W. Bush, which ratified the directive, already present in the Metric Conversion Act, for the Secretary of Commerce "to direct and coordinate efforts by Federal departments and agencies to implement Government metric usage." But results have been unimpressive. As Gerry Iannelli, director of the United States Metric Program (part of the National Institute of Standards and Technology) indicated in 1999, the government ran the "danger of building a metric island in the government, in a nation that is a nonmetric island, if you will, in a metric world."[133]

According to a study from 1994 by the US General Accounting Office, "since 1990, federal preparations for metric conversion have advanced dramatically, with more than 30 agencies having developed some combination of guidelines, transition plans, and progress reports that indicate a substantially greater commitment to metrication. However, they are still facing serious difficulties in putting their plans into practice. These difficulties include a procurement environment in which most products are nonmetric and in which federal agencies represent too small a share of the total market to stimulate private sector conversion."[134] The report indicated that government agencies cite private sector and public resistance to metric conversion as the reasons for their slow progress; and that the federal government "must have the support of the private sector and the public for successful conversion to the metric system." This is an interesting pronouncement, illustrative of the peculiar path that the United States has tried to achieve metrication. While in most countries the state has helped or forced the private sector and the citizenry to go metric, in the United States the federal government asks for the help of the private sector to metricate the government—the world turned upside down.

Mold the Citizens to the Law or the Law to the Citizens?

The controversies on compulsion, uniformity, and state intervention are prevalent in American history and can be found in other public debates similar to the metric controversy. In public health, for example, officials lack the legal instruments to ensure that parents will vaccinate their kids, something that raises concerns when a considerable number of children remain unvaccinated, which makes the whole population vulnerable to preventable diseases. However, studies show that the majority of people think that "parents, not the government, should make decisions about immunizing their children."[135] Equally, it has been impossible to secure uniform education standards across the country, with states and cities fiercely defending their autonomy to control school contents. As an editorial in *The New York Times* lamented, "The countries that have left the United States behind in math and science education have one thing in common: they offer the same high education standards—often the same curriculum—from one end of the nation to the other. The United States relies on a generally mediocre patchwork of standards that vary, not just from state to state, but often from district to district."[136] If we replace "education" with "measurement" in this quote, it may pass as a carbon copy of the metric debates—a battle that was ultimately won by the laissez-faire camp.

This issue appeared very early. In 1817 Thomas Jefferson wrote to John Quincy Adams, "On the subject of weights and measures, you will have, at its threshold, to encounter the question on which Solon and Lycurgus acted differently. Shall we mould our citizens to the law, or the law to our citizens?"[137] This question would have been unthinkable in France, Mexico, or many other countries—where state officials were convinced that they should mold the citizens to the law—but it has torn apart metric policymakers in America. So far, those who think that they should mold the law to the citizens have prevailed.

The question of compulsion has been one of the most poignant issues in the metric debate in America. Many opponents of the metric system were not strictly anti-metric, but rather adversaries of the compulsory imposition of any system. For many of them, the metric system is adequate and useful, but they did not want to be forced to use it. Some of them liked to use one system or the other depending on the circumstances. As Frederick R. Hutton, a member Society for the Promotion of Engineering Education, said to a metric proponent: "most of us very much prefer the present condition of affairs where anyone who wants to use the metric system is at liberty to do so because it is legal and standard. We do not want, however, to see the use of it made compulsory on anybody to whom such use will be inconvenient, costly, and unpopular."[138]

Loose Centralization and Aversion to Compulsion

Two of the main issues that impeded America's full adoption of the metric system—deficient centralized control on metrological matters and lack of compulsory use of a sole measurement system—were never an issue in other places. In many other countries there were controversies about what measurement system would be more appropriate, but everyone involved in high-level metrological discussions took for granted that it was the federal government's job to unify measures nationally and that the employment of official measures had to be obligatory, not optional.

Things were different in the United States. One of the trickiest questions in the American metric debate has been the definition of the locus of metrological authority. Should it be Washington, DC, the states, or cities and counties? It was agreed that the federal government should hold the national standards, but not that it should enforce, define, or police metrological dealings. On the other hand, local authorities did not took the baton; except for timid and short-lived attempts in Hawaii and Oregon (in 2013 and 2015, respectively), compulsory metric legislation has never been proposed in state legislatures.[139] With a steady antagonism toward centralizing metrological authority and with the states reluctant to change by themselves—due to the expense and considerable risk that going metric alone would bring—the chances to find effective leadership in the government to guide the transition to the metric system were very slim.

Other important reason for the "erratic course of adoption of decimal weights and measures" in the United States has been "an American aversion to compulsion."[140] The issue of forcing people to employ exclusively one system of measurement—the one chosen by the government—has fueled and galvanized metric opposition. State officials and metric promoters have not articulated convincing arguments to counter that and explain why the metric system (or any other one, for that matter) ought to be exclusive and compulsory.

The question of why the government should make the metric system mandatory if it was already legal and optional has been particularly delicate for the pro-metric camp. They never delivered a convincing answer to shape public opinion. Many of them have known that compulsion is essential to achieve full metrication, but emphasizing a policy of state-led imposition has been a thorny political strategy. The "repulsion to compulsion," sizable in American political culture, has weighed heavily against the metric hopefuls.

One of the main shortcomings for government agents and others interested in advancing the metric cause has been their deficiency in conveying the necessity for a mandatory policy within a democratic framework—which reflects, ultimately, a lack of creativity in their political philosophy. So, even if it was one of the crucial elements for metrication, metric enthusiasts usually avoid the topic of the obligatory use of the system. They always presented strong arguments about the international character of the metric system and about the benefits its adoption would bring to exports, science, and education; but whenever presented with the question of why to make the metric system obligatory (or as an anti-metric observer put it "If metric is so superior, why can't it win on its own, without compulsion?"[141]), pro-metric spokespersons usually tumbled. Their lack of clarity forced them to present odd arguments, like this by Edward Wigglesworth, member of the American Metric Bureau:

> We need a benevolent despot who would *compel* the use of the Metric System here after a fixed day. After a week no one would have any more trouble; after a month people would wonder how they could ever have used anything else, the labor of learning is so slight, the gain immense. All the poor peasants of Europe, the lowest classes of "effete despotisms," etc., etc., have been able to adopt it at once, and yet Americans, self-ruling, are really too lazy, while merely claiming to be so stupid so to do. Shame on a country which "to party gives up what was meant for mankind."[142]

This was a losing strategy in the battle to win public opinion (not to mention a gross misunderstanding of how long it takes for a population to transition from one measurement system to another). The anti-metric movement was more effective in building a vocabulary to emphasize the "anti-democratic" character of compulsory metrication. Samuel S. Dale, an active defender of customary measures, usually linked the metric system to its beginnings, underlining that the "bloody" French Revolution, regarding metrication, was "the inauguration of compulsion on a gigantic scale,"¹⁴³ it was the origin of "a decimal despotism imposed by the compulsory use of the metric system."¹⁴⁴ Or as he put elsewhere it in a more detailed way,

> [T]he experience of France has demonstrated beyond question that a people's established weights and measures cannot be changed by the power of law. . . . Drastic laws like those in France can compel government employees and to a certain extent, merchants in the marketplace to use certain standards, but they are powerless to compel the workman and the manufacturer in the privacy of his own workshop or factory to obey the law by thinking and using governmental standards. The force of habit is here superior to law.¹⁴⁵

Rhetoric like this was regular in the newspapers. As an anonymous contributor argued in a Wyoming daily, in response to some chambers of commerce, industrial organizations, and state legislatures that were "urging liberal metric legislation" to Congress: "Congress might be swayed into making the metric system the official system for the country, but the people of the United States are likely to continue to measure distance by the mile, instead of the kilometer, and to buy milk by the quart and not by the liter for many a day to come. Habits break more laws than laws break habits."¹⁴⁶

The durability of this line of thought has been remarkable. Take for example, the opinion early in the twentieth-first century of an engineer, who wanted to protect customary measures and created the internet site *Freedom to Measure*. He declared that the metric system was "creeping into the USA" and the heritage of Americans embodied in miles, feet, and gallons was in danger by the menace of "compulsory metrication." Those to blame were "unelected, unaccountable bureaucrats trying to take your way of measuring away from you. . . . Compulsory metrication is undemocratic. Who ever asked you if you wanted the metric system? . . . They are part of our heritage. Metrication will not only destroy part of our cultural inheritance, it will mean that a large percentage of Americans will be cut off from understanding measurement."¹⁴⁷

Since the metric legislation of 1866, opposition to the metric system in the public arena has centered its efforts on blocking any compulsory legislation. Pro-metric groups have not been able to articulate an adequate response to that position, and state officials have not shown determination to go for metric enforcement—determination that they had with the monetary policy, for example.

Besides the political aspect of the problem, the matter of why the introduction of the metric system must be mandatory to be effective is a real question. Why has voluntary adoption been so ineffective? Partly, because as any other language the metric system needs a whole community of speakers to coexist at once. Restructuring mental schemes and reconstructing the human-built world takes a prolonged and coordinated effort. For a carpenter who adopts the metric system it is not enough to measure wood with meters and centimeters, she needs to communicate with suppliers, assistants, and clients: if they do not understand metric units she will not be able to actually switch entirely to the metric system. She should also face the intrinsic problems of the material landscape that she inherited, which is fashioned with non-metric units, like instruments gauged in inches, nails of lengths like 1", 1.25", 1.5", 1.75", 2"; and hardwood lumber sold in thicknesses measured in quarters of an inch. The materiality of a world constructed with customary measures forces people to think and talk in non-metric terms.[148]

Something like that happened with the timid experiments carried out in the early 1970s by the US Metric Board, when a few road signs were switched to metric in some Arizona and Ohio highways.[149] This created confusion and displeasure among drivers, who were not skilled in using metric units and whose car's speedometers indicated miles per hour, not kilometers per hour, which made estimations of speed and distance complicated. In a related but unsynced effort, in 1979 some gas stations, mainly in California, adjusted their fuel pumps to start selling gasoline by the liter. It was another futile, short-lived initiative. Effective transitions are coordinated and sustained. Different results would have been obtained if road signs, gas pumps, and speedometers had been metric all at once in the entire country.

Voluntary metric adoption does not work because it punishes people willing to take the risk of making the change alone, while the rest of the material and social landscape is still working with the old measures. Merchants and industries that stepped forward and adopted the metric units voluntarily got isolated in the middle of a sea of users who did not understand or regularly use metric measures.

This helps to explain why the adoption of the metric system has always been an imposition from above; the metric system has never been the system of measurement consistently used by regular people prior to the official adoption mandated by the authorities. The meter replaces local forms of measurement (in processes that are slow and contested, but successful in the long run) and that creates social tensions and, in some cases, even violence. Thus, the coercive abilities of the state have been central in the effective introduction of the new system of measurement. And because the change to the metric system in a country requires a sustained effort during various decades, the consistency of state policies is crucial.

What, then, is necessary to make the transition from one system of measurement to another in a given territory? Multiple national experiences of transitioning from pre-metric to metric measures teach us that the following procedures are required: 1) pass laws that prohibit the use of old units and instruments of measurement; 2) possess the ability to enforce those new laws; 3) prevent, control, and repeal social and political reactions against the imposition of the new system; 4) make the new system understandable for the people; 5) provide the technical and scientific tools to make the metrological transition possible. All these are tasks that only states (not markets or scientists) have been able to achieve consistently.

Anti-metric writers were right when they said that "the force of habit is superior to law" and that "habits break more laws than laws break habits." But this is only true in the short run. The habit of thinking in terms of inches and pounds cannot be abandoned in a couple of months, not even in a couple of years. Metrication is a slow-moving train; it is powerful, but it needs a lot of time to reach the final station. Laws can break habits, but only in the long run. Like the construction of medieval cathedrals, the thorough adoption of the metric system in a country is never something a single generation can achieve.

A Troubled Lingua Franca

Scientists, Engineers, and Educators in
the Battle of the Standards

> A common obstacle to human progress is what may be called "intelligent prejudice," ... an obstinate conservatism which makes people cling to what is or has been, merely because it is or has been, not being willing to take the trouble to do better, because already doing well.
> —THOMAS C. MENDENHALL, *Popular Science Monthly (1896)*

> We cannot break completely with the past, because not only must we take account of public repugnance, but scientists themselves have a tradition to which they remain tied.
> —HENRI POINCARÉ, *Commission de décimalisation du temps (1897)*

If the American Metric Association organized in the 1920s "metric luncheons" in the Hotel Astor, in the 1970s and 1980s, also in New York, antimetric groups organized a weights and measures festival in Central Park, a fundraiser in the Dakota building, and a "Foot Ball" in Battery Park. The occasions were planned by the painter and self-described "radical traditionalist," Seaver Leslie, who was also founder of Americans for Customary Weight and Measure. The New York socialite partied in those gatherings, repeating catchphrases like "Stand up for the foot!" and "Don't give an inch!" To attend the Foot Ball partygoers paid fees based on customary 12-base measures: 36 dollars for a couple and 24 for a single person. One of the amusements was the Most Beautiful Foot Contest, where participants exhibited their painted or bejeweled feet in front of celebrity judges with the hope of winning the first prize: a paid travel to the pyramids of Egypt.[1]

A French lawyer who was present at the ball declared with amusement, "It's fantastic to see Americans giving so much energy to something that is not important. It is much better than selling atomic nuclear bombs."[2] But Americans did consider the metric system important. A representative of the International Brotherhood of Electrical Workers attended the dance; like other unions, they worried that metrication could outdate workers' skills. The main organizers were not so worried about their job security but believed the meter was a threat to American culture. Leslie went to great lengths to convince reporters that the old relation of customary measures with the body made them more "human." An inch, he said, was the average thumb breadth, a yard the distance from the tip of the longest finger to the nose, a cup the volume of blood in a person's heart.[3]

These events were a response to the last significant push of the federal government to adopt the metric system. The signing of the Metric Conversion Act of 1975 by President Gerald Ford marked the beginning of what many thought would be the definite transition to the metric system. Even though the campaign suffered from the same glaring weakness as previous iterations (lack of compulsion), this time the government at least launched a publicity campaign that drew many to believe that the metric transition was unavoidable. Schools and media conveyed the impression that metrication was only a matter of time. Books like *Prepare Now for a Metric Future* and *Metric Power: Why and How We Are Going Metric* were common in bookstores.[4]

Artists and cultural personalities who disliked the idea of a metric America—like Leslie, Stewart Brand, William Burroughs, and Tom Wolfe—provided the press with abundant news bites and were photographed wearing t-shirts with anti-metric slogans (fig. 3.1).[5] However, contrary to the arguments articulated by anti-metrics from previous decades, their postures were purely defensive and nostalgic, when not mystical and esoteric—a person who influenced some of these outspoken anti-metric American personalities was the writer John Michell, a student of the Stonehenge "sacred measures" who the *New York Times* described as "a self-styled Merlin of the 1960s English counterculture."[6] They did not outline a grand reform that would surpass the intellectual ambitions of pro-metric scientists (like Herbert Spencer in the 1890s); and they did not scheme an international expansion of customary measures (like Frederick

Figure 3.1. Novelist William Burroughs in his New York apartment, known as The Bunker, wearing a "No metric" t-shirt in 1979. Photo by Victor Bockris. Getty Images.

Halsey in the 1920s). More than advancing new ideas, they were reciting eulogies for inches and pounds.

The opposition to the metric system in the 1970s and 1980s differed in other important aspects from that of previous decades. The familiar narrative of "the English-speaking countries versus the metric system" was no longer available; England, Canada, Australia, New Zealand, and South Africa had already committed to go metric. Also, intellectuals in the 1970s and 1980s were indeed passionate about customary measures, but their opinions were not as consequential as those of experts at the beginning of the twentieth century, because they were not leaders in technical or scientific fields, and they made few alliances with strong economic or political interests.

Compared with their counterparts from the nineteenth and early twentieth centuries, these intellectuals were small potatoes. Those involved in public discussions were people whose main reason for prominence in those discussions was their flashy personalities and the eccentricity of their pronouncements—they created great material for easy journalism. Even the caliber of their opponents had diminished. The most visible metric advocates at the time were not great scientists, but rather science communicators like Isaac Asimov.[7] The momentous metric confrontations between top-notch figures in science and technology were a thing of the past.

Meters and yards had been the object of prolonged debates that embroiled some of the most recognizable faces of the scientific world during the eighteenth, nineteenth, and early twentieth centuries. The pertinence of using pounds or kilograms was fiercely discussed among figures like Nicolas de Condorcet, Antoine Lavoisier, Pierre-Simon Laplace, Gaspard Monge, James Watt, Gabriel Císcar, Andrés Manuel del Río, Francisco Díaz Covarrubias, John Herschel, Auguste Comte, Herbert Spencer, Charles Piazzi Smyth, Dmitri Mendeleev, Maria Montessori, James Joule, Lord Kelvin, James Maxwell, Marie Curie, and a myriad of other notable scientists, educators, and inventors around the world.[8]

In the United States, towering intellectual figures also participated vigorously in the debates over metric adoption. Polymath statesmen like Thomas Jefferson and John Quincy Adams, Ivy League presidents like Frederick Barnard, philosophers like Charles Sanders Peirce, scientist-functionaries like Ferdinand Rudolph Hassler and Thomas Mendenhall, science philanthropists like Andrew Carnegie, educators like Melvil Dewey, sociologists like William Graham Sumner, physicians like Charles H. Mayo, industry innovators like Henry Ford, and inventors like Elihu Thompson, Thomas Alba Edison, Alexander Graham Bell, and George Eastman raised their voices to support or oppose the metric system.

The battle of the standards was a heavyweight fight. Gradually, in the twentieth century, the metric system became the taken-for-granted system of measurement in most of the scientific world and was officially adopted by most countries. This reduced the sense of urgency among the scientific community to participate in public debates about metrication, and the topic fell into the hands of lesser-known figures. However, the intensity and acrimony of these quarrels have remained over the decades. Since the United States is one of the few places where the discussions on the convenience of using metric units are not purely academic, those disputes keep resurfacing periodically.

In this chapter we will follow scientists, engineers, and inventors in their struggles around two interrelated issues: how radical the innovations in the reforms of weights and measures should be, and how to deal with the knowledge and expectations of lay people in metrological dealings. These questions triggered bitter disagreements. Scientists and engineers care about their own past and traditions, so when the arrival of the metric system confronted them with the prospect of a sudden and thorough change, some of them embraced the cause of innovation; others sought refuge in their national, religious, and professional traditions. Some believed that the practices of regular people embody a wisdom and empirical practicality that experts ought to respect. Others considered that what was better for people was the methodical and unmitigated use of rational schemes; in their view, if people resisted those plans it was because they did not know better, and it was the responsibility of experts to enlighten them. Chronic systematizers, like Melvil Dewey, and scientists who valued arithmetic simplicity, like Lord Kelvin, were on one side; obstinate traditionalists, like Charles Latimer, and thinkers who saw craftiness in how things had been done for centuries, like Herbert Spencer, were on the other. Their disputes took place in the intersections of innovation versus tradition, and general knowledge versus specialization.

The Moving Distance between Special Knowledge and General Knowledge

Creating and perfecting methods and instruments of measurement is one of the oldest and most prolific ways in which scientific and technical creativity is displayed. Measurement is the job, and the joy, of many experts.

Way before Sudoku became a popular pastime, mathematically minded people used measurement as a distraction, a *divertimento*, and even a coping mechanism.[9] Conceiving measuring systems can be a hobby. In 1958, as part of a fraternity pledge, a group of MIT students measured the Harvard Bridge between Boston and Cambridge using their buddy Oliver R. Smoot as a rod.[10] The freshman repeatedly lay down on the curb of the bridge while his pals marked with chalk the length of his body. The task was completed after several hours, and the result was that the bridge was equal to 364.4 smoots ± 1 ɛar. They used the plus-minus sign to show uncertainty and wrote the *e* in *ear* with an epsilon to indicate a possible error in measurement.[11] The festive and ingenious invention of the smoot has been widely celebrated. A plaque on Harvard Bridge commemorates the fiftieth anniversary of that occasion, and *smoot* has been included in the *American Heritage Dictionary*.[12]

Executing, improving, and creating units and methods of measurement can be a charming pastime, but is not a trivial activity. Most of the time, experts embark on such tasks driven by practical necessities. In doing so they need to consider not only what is technically possible, but also the customs and traditions that could be altered in the process of updating or overhauling a metrological convention.

Metrology is a field where the points of contact between special knowledge and general knowledge are particularly important. Experts in astrophysics or biochemistry can create their own vocabulary and definitions without much regard for non-specialists. Other disciplines are more constrained. Computer scientists use a binary number system (one with only two digits, *1* and *0*) and do not worry about the ability of regular people to understand what number is expressed by "11011111101" in binary notation; however, if they work for commercial developments, they may be confined to the limits of the most common operative systems in the market, like Microsoft Windows and Mac OS. Similarly, scientific metrology is tied in some crucial aspects to the understandings of non-experts. Most people would be unable to explain why temperature in laboratories is measured in kelvins and not centigrade or Fahrenheit degrees, or what some of the base units of the International System of Units (SI) like the mole, are for. Even the objects that are measured with some of the SI units are incomprehensible for non-experts, like magnetic flux (measured with the weber), absorbed dose of ionizing radiation (measured with the gray), and catalytic activity (measured with the katal).[13] Of course, metrologists do not need the general population to comprehend what the Planck constant is or how a Kibble balance works, but when they try to update or redefine basic measurement units like the meter or kilogram, they have their hands tied by the public's expectations of what a meter and kilogram are.

This has not always been the case. Traditional units of measurement were usually created through piecemeal processes. Local customs, cultural intricacies, technical limitations, and necessities of particular trades combined to compose systems that further on were adjusted and systematized when economic and political conditions required it. Modern systems of measurement, on the other hand, had been expressly designed by scientists. Experts, instead of regular folk, are today the main social constructors of measures. This means that they do not just invent new units and instruments of measurement, but also that they participated actively in the implementation, justification, maintenance, and upgrading of the current metrological regime.

Usually, units of measurement do not change radically. It is common that metrological reforms are made by adjusting existing units, instruments, and methods of measurement—the metric system was an exception. Units of time reckoning (second, minute, hour, day, week) attest to that process of lasting continuity and repeated updating. Scientists play a significant role in the complicated task of revising traditional units of measurement while preserving enough of their old elements to keep them recognizable and operational in everyday life.

The story of the carat—the unit of mass used for gemstones—illustrates the process of redefining and fruitfully standardizing a unit of measurement within a relatively narrow professional community. The carat is an old unit, it has been registered in the English language since the fifteenth century, coming from the Italian *carato*, the Arabic *qīrāṭ*, and the Greek *kerátion*. Its modern history is an exemplary case of how old units can be modernized without making a radical cut from the past.

The activities related to precious metals and stones have always had an acute sense of the importance of precise measurement, small errors could mean big losses. Discrepancies among multiple carats were undesirable. Prior to the twentieth century, the carat had diverse magnitudes: 213.5 mg in Turin, 205.8 mg in Portugal, 205.1 mg in Russia, 196.5 mg in Florence, 192.2 mg in Brazil, etc.[14] In the 1870s some jewelers suggested the use of a carat of exactly 205 mg, but this solution had the risk of interfering with numerous national metrological laws that forbade the use of non-metric units in any commercial transaction.[15]

The answer to this predicament was proposed in the 1890s by mineralogist and vice-president of Tiffany and Co., George Kunz.[16] Kunz was interested in reducing the transaction costs in the commerce of gemstones, but he was also akin to the idea of international metrication (he was the first president of the American Metric Association). Kunz mixed his knowledge and interests in both metrology and precious stones, and invented the so-called "metric carat," which ultimately helped to standardize that whole field. The metric carat was defined as exactly 200 mg.[17] This gave the carat a perfect metric equivalency and a decimal notation. The scheme proved so effective that today the "metric carat" has simply become the "carat."

This solution solved various problems simultaneously. It preserved the old name of the unit, which was familiar to jewelers and their clients; it offered a magnitude of the unit that was close to the old carats—avoiding a radical change that could be confusing and prone to be rejected by users; and because the "metric carat" is an exact aliquot part of the gram, it did not interfere with metric legislations. Probably a more logical solution would have been to drop the name *carat* altogether and simply employ the metric units (milligrams) in the commerce of precious stones, but what prevailed was a compromise between continuity and reform which was quickly implemented among a reduced group of professionals.

The creation of the meter, liter, and kilogram is an example of a different path to successful metrological innovation. In its case, no compromise with the past was sought. It was invented at the end of the eighteenth century by members of France's learning societies. In the beginning, it was only understood and employed by fellow savants; today 95 percent of the world's population uses it every day. Since its creation, the metric system made the incredible transition from being a "special knowledge," possessed only by a reduced group of specialists, to becoming "general knowledge," knowledge that is routinely divulgated to all members of a society and is socially defined as being relevant to everyone.[18] It was not an easy or expeditious process, but the metric system became part of the social stock of knowledge of humanity.

There were plenty of difficulties to get there. Some of them had to do with how special knowledge (including both the concepts and instruments) has modified, displaced, or abolished different provinces of general knowledge. Others were related to how the prevailing metrological knowledge of populations helped or impeded the utilization of the new system.

Systems of measurement are linked to the mathematical abilities of the members of a given society. Different languages express plurality, numeration, and quantity with diverse levels of abstraction and precision.[19] The sociological analysis of measures must be related to the historical evolution of the human capacity of counting and quantifying, on the one hand; and with the creation, accumulation, and distribution of knowledge, on the other. Since measurement implies quantification, the history of measurement is closely related to the history of mathematical abilities. A system of measurement in a given society is always shaped by the arithmetical capabilities of its members.

The cognitive elements in measurement explain in part the difficulties of changing from one system of measures to another. The transition from traditional to metric measures are slow and painful because regular people usually do not feel it is either necessary or useful.[20] Part of the difficulties in accepting the metric system is its decimal character. Most pre-metric systems are based on a combination of twelve-, sixteen-, and sixty-base divisions, like the division of the day in two dozen hours, and the hour in sixty minutes. These number systems offer multiple advantages for people with little formal mathematical training, because it is simple to understand that half of a dozen is six, half of the half is three, and the third part of a dozen is four. In a decimal number system, it is simple to calculate the half of ten (five); however, without some training in the use of the decimal point it is complicated to express a quarter (2.5), or a third (3.333 . . .).

An additional difficulty in metrological transitions is the alteration in the expected relationship between numbers and previous experiences. It is hard for one not accustomed to it to say naturally that 24 degrees is a pleasant temperature or that a basketball player is 2.1 meters tall. It takes a generation or more for the native speakers in one set of measures to appropriate the new units and feel them as "natural."[21] That is why, in reforms like the 200 mg metric carat, experts tried to create the fewest possible alterations vis-à-vis previous experiences—the metric reform was the opposite, and that created multiple issues for its adoption.

Measurement standards are both physical tools and mental categories. A change of system alters both. When anthropologists Malinowski and De la Fuente studied a rural market in Mexico in 1940, they showed the great difficulty experienced by people adapting to the new system, and they explained why persons who grew up using old units stick with them as long as they could:

> The Mexican government has been, for some time past, exercising a more or less persistent, and at times energetic, pressure for the general introduction of the metric measures. So far, success has been limited; indeed, in some ways it complicates the picture and introduces one more element of confusion and elasticity in the standard of comparison [for maize prices]. From the point of view of the whole system of maize marketing, this usually means that the poorer and less educated or less keen market agent is exploited by one who can manipulate figures, as well as measures, more rapidly and skillfully.... The poor Indian and peasant prefer the *almud* [a traditional measure of grain], not because they are "conservative" or "dislike innovations," but because this measure enters into all their domestic calculations in a manner which has been standardized for centuries, and they are accustomed to calculate with it. Thus, they know how many tortillas can be produced from one *almud*, or how many cups or bowls of *atole* (maize drink); in short, how many *almudes* per week their budgets require.[22]

This is a snapshot of the practical and cognitive implications of any metric changeover. Metric units are not easy to use by people accustomed to defining quantities in non-metric measures, and the decimal progression that coordinates the grouping and division of units is different from the arithmetical logic of many customary systems. During the transition period, the metric system interferes with everyday thinking. This "cognitive noise" generated by the abrupt change in standards is not rare; but despite the complications, national metric transitions have eventually succeeded.

Schooling, the multiplication of metrically manufactured products, and technological change all have contributed to make meters, liters, and kilograms the "normal" units of measurement among larger populations— as it was said in the previous chapter, the state is crucial in making sure that schools teach the fundamentals of the metric system and in forcing economic actors to produce and sell metrically designed products. These channels for the social distribution of metric knowledge are critical to connect the worlds of experts and laypeople. But the costs of naturalizing the metric system can be considerable. Experts create measuring devices and machinery that make centuries-old skills based on bodily and personal experiences obsolete. Historian Manuel Moreno Fraginals illustrated one of the thousands of instances of that process:

> [The sugarmaster] was basically an artisan, guided by judgment which years of living with sugar had acutely sensitized. Calculating everything by sight, smell, and taste he knew if enough lime had been added, when clarification or defecation was completed, and when the conglomerate had reached the sugaring point. His eye measured the correct amount of beating and the moment to air the sugarloaves. The tragedy of the sugarmaster, his loss of supremacy, came when vacuum pans were introduced and his judgment was replaced by physical measuring apparatuses.[23]

Scientists know that metrological reform is inextricably linked to the acquired knowledge of the population. This is why their debates about new systems of measurement gravitated around two possible paths of action: create new systems that break completely with the existing systems or simply update the old measures. As we shall see, the traditions of experts (as the customs of people) have played an important role in the history of the metric system. The changing balance between innovation and continuity in technical fields has marked the position of many experts regarding metrication.

The Metric System and Its Modern Competitors

Despite the cheerful, heartrending, and practical uses of measurement, scientists' efforts to create instruments, techniques, and standards of measurement is a serious and competitive business.

One of the epic and tragic aspects in the history of the metric system is the marginalization and obliteration of hundreds of local measuring systems—and their corresponding thousands of units and instruments of measurement—around the world when the meter, liter, and kilogram were adopted in country after country since the nineteenth century. As cultural creations, systems of measurement had been incredibly variable. Not only did all civilizations develop broad conventions and instruments to measure nature, society, and the self; weights and measures were adapted to local settings to meet specific necessities and interests. Until the end of the eighteenth-century, a common feature of systems of measurement was their variability. For better or worse, the metric system changed all that.

A lesser-known aspect of the process of the global diffusion of the metric system is its confrontation against other systems created *ex professo* by scientists and engineers to replace it.[24] The overwhelming triumph of the metric system in scientific circles over these challengers helped in its ultimate global success. Experts in different countries (but not in America) were able to present, most of the time, plans for metrological reform centered around the metric system and avoided intestine disputes to define what system should be used to supplant traditional measures. Since the second half of the nineteenth century, lack of national metrological uniformity and absence of international coordination in weights and measures were illnesses that found everywhere one and only one prescription: metrication.

The inadequate understanding of this phenomenon and the scarcity of investigations on who, when, how, and why tried to challenge the metric with newly invented systems may create the impression that those plans were not a factor in the history of metrication.[25] Misinformed commentators have thus argued that "left without competition as the only scientifically based, decimalized measurement, the French meter has taken its present place as the simplest, most accurate means of measuring everything between the dimensions of a quark and a black hole."[26] This is a widespread misconception. There was plenty of competition for the metric system. For once, at the end of the eighteenth century, there were several plans of metrological reform that came about shortly before or simultaneously with what the French savants were doing in the early 1790s. Just to mention a few of them, there were plans formulated by Cesare Beccaria in Italy, by James Watt in England, and a couple more by Jefferson and William Waring in America.[27]

Once the metric system was established and started its global dissemination many other proposals were advanced to defy it. Between 1851 and 1889 numerous plans were designed in England, the United States, and other countries with the intention of displacing the metric system.[28] Some were new measuring methods based—like the metric system—on a decimal principle, but with units of different names and magnitudes. For example, C. E. Macqueen (secretary of the Liverpool Financial Reform Association) published in 1855 *The Advantages of a Complete Decimal System of Money, Weights and Measures*; and in 1857 a joint committee of the Chamber of Commerce and the American Geographical and Statistical Society produced another such proposal.[29]

Other plans proposed a different arithmetic for dividing and grouping units and multiples, like John William Nystrom's hexadecimal (base–16) *Project of a New System of Arithmetic, Weight, Measure and Coins* from 1862 (Nystrom was a Swedish nautical engineer who migrated to the United States, where he invented a sophisticated calculating device and penned a popular *Pocket Book of Mechanics and Engineering*, which had more than twenty editions); and Alfred B. Taylor's octonary (base–8) plan published in 1887 in the *Proceedings of the American Philosophical Society*.[30] Other plans were centered on creating natural units—but not based on meridian arc on the Earth's surface from the equator to the north pole, like the metric system. W. Wilberforce Mann, for example, proposed in 1872 a *Decimal Metric New System Founded on the Earth's Polar Diameter*, that was designed for the "adoption of all civilized nation as the one common system."[31]

None of these schemes represented a challenge for the metric system outside the realm of speculation. The ideations of solitary experts were no match for the political and economic might behind meters and kilograms. In 1890 C. S. Peirce, who at the time worked as a metrologist in the United States Coast and Geodetic Survey, reviewed one of these plans—Edward Noel's *Science of Metrology or Natural Weights and Measures: A Challenge to the Metric System*—and showed great skepticism toward the proposed plan and even compared the undertaking with a foolish intention of combating a force of nature.[32] Peirce made a neat explanation and prediction of why challenging the metric system was not a wise idea:

> We can . . . make a reasonable prognosis of our metrological destinies. The metric system must make considerable advances, but it cannot entirely supplant the old units. These things being so, to "challenge" the metric system is like challenging the rising tide. Nothing more futile

can well be proposed, unless it be a change in the length of the inch....
Mr. Noel's system is nearly as complicated and hard to learn as our present one, with which it would be fearfully confused, owning to its retaining the old names of measures while altering their ratios.... The scheme is not without merit, and might have been useful to Edward I.[33]

The only competitor that represented a veritable threat to metric hegemony was not one created by isolated experts, but one made by scientists working for an empire.[34] It was the "Imperial system," the British response to the audacious French invention. The international strength of the imperial system was anchored in the infrastructure of the actual British empire. The English government not only reshaped and perfected the old units of measure, it also manufactured the necessary instruments and physical standards, and—crucially—increased the power of inspectors and the metrological officer corps to secure a total regulation of weights and measures.[35]

A clever quip among linguists says, "A language is a dialect with an army and navy."[36] This is true in linguistics as in metrology. The metric and imperial systems of measurement were the only languages of weights and measures with an army and navy. The metric system had Napoleon's army, one of the most feared in European history; the imperial system had the Royal Navy, one of the most imposing in naval annals. The scientists' elucidations to confront metrication were mere metrological dialects, paper airplanes flying in library halls. The historical relevance of a measurement system does not spring from the beauty of its ideas or its mathematical elegance; it originates from the administrative, military, and economic resources of its social carriers.

The metric system succeeded in replacing hundreds of customary methods and units of measurement around the world that had existed for centuries. No modern, scientific measurement system designed in the nineteenth century was able to challenge it seriously. The metric system achieved, thus, a double victory.

Metrology does not have a history of its own, its existence is determined by social structures, economic interests, and collective beliefs. We should keep this in mind when we reconstruct and analyze the polemics and discourses of the experts who got involved in the metric fights in America.

Systematizers and Traditionalists

The gap separating expert and common knowledge and the tension between innovation and tradition were two pillars that articulated the metric debate in the scientific community in America.

In the final decades of the nineteenth century, experts in different fields coalesced in two camps that formed organizations and publications to articulate their ideas and influence the public debate. On one side there were engineers and self-described "practical men." They saw themselves as individuals who "made things," not mere "theoreticians." They defended the preservation of customary measures, self-determination in technical matters, and the right to define their own strategies to standardize manufacturing processes. Many of them mixed these technical ideas with nationalistic and religious ideals. As they would put it, the metric system may be useful for astronomers who observe eclipses, but not so much for men who build machines and railways.

On the opposite side, there were educators, university professors, and exact scientists who embraced the metric system enthusiastically. For them, internationalism, global communication, and intellectual elegance were paramount. They wanted to leave behind the cumbersome conventions that obstructed the "advancement" and "progress" of school children, universities, laboratories, and industries. The only explanations they could find for why people cling to yards, gallons, and pounds were ignorance and backwardness.

The first camp was supported by engineering bodies, like the American Society of Mechanical Engineers; the second by scientific organizations, like the National Academy of Sciences and the American Association for the Advancement of Sciences. Of course, this division in two factions with clearly defined ideologies had exceptions. Multiple anti-metric engineers were themselves university professors and never invoked nation, tradition, or religion in their defense of customary measures. Also, many pro-metric experts worked in factories and workshops and were motivated purely by pragmatic reasons. Two organizations, led by idiosyncratic personalities, embodied these opposite worldviews; Melvil Dewey's American Metric Bureau (founded in 1876) and Charles Latimer's International Institute for Preserving and Perfecting Weights and Measures (founded in 1879).

MELVIL DEWEY AND THE AMERICAN METRIC BUREAU

Melvil Dewey (1851–1931) is mostly recognized for his work as an educator and for the design of the decimal classification system for libraries—widely used around the world to catalog and shelve books according to a thematic arrangement. Lesser known is Dewey's five-decade-long effort to introduce the metric system in the United States and his role as founder of the first organization exclusively dedicated to that goal, the American Metric Bureau (AMB).

Most scholars have overlooked Dewey's significance for the American metric movement. His status as the patron saint of modern librarianship has eclipsed his vigorous role in favor of metrication and his participation in several other campaigns aimed at improving education and promoting effectiveness in the use of material and human resources. Besides advancing plans for library rationalization and wider use of metric measures, Dewey was a passionate adherent of simplified spelling, abbreviations, shorthand, home economics, reform of the Gregorian calendar, and "efficiency in all aspects of human life."

According to his autobiographical notes, Dewey's interest in metrological reform first emerged at the age of fifteen, when he became frustrated by the lack of simplicity in customary weights and measures and the way they were taught in math classes, particularly in relation to compound numbers. This is Dewey's own description of his initial interest in the metric system; it was written with his modified spelling—reading it in its original form could help to grasp the spirit of Dewey (or "Dui," as he liked to spell it) and his reforms:

> In skool in Adams Center I rebeld agenst compound numbers. I told the teacher that jeometri taut us a strait lyn was the shortest distance between 2 points & that it was absurd to hav long mezur, surveyor's mezur & cloth mezur; also absurd to hav quarts & bushels of diferent syzes & to hav avoirdupois, troy & apothecari weits with a pound of feathers hevier than a pound of gold. I spred out on my attik room table sheets of foolscap & desyded that the world needed just 1 mezur for length, 1 for capasiti & 1 for weit & that they should all be in simpl decimals lyk our muni. I was puzling over the names to giv the new mezures when I red that Senator John A Kasson of Iowa had past in

Congress a bil legalizing the metrik sistem.³⁷ I lookt it up at once, found that it met my plan ideali & the next week went to our vilaj lyceum & gave a talk on the great merit of international weits & mezures. From that day I became a metrik apostl.³⁸

Dewey's inclination for simplicity and effectiveness was reinforced after an accident that almost took his life. In 1868—when he was seventeen—his school caught fire and he helped to save as many volumes as he could from the library, inhaling a great deal of smoke in the process. Dewey developed a deep cough that caused his doctor to predict that he would not live for more than two years. The prospect of a premature death made Dewey obsessed with time-saving methods.³⁹ Efficiency became a credo that shaped all his professional interests.

In 1873, while a student at Amherst College, Dewey wrote in a reading notebook, "My heart is open to anything that's either decimal or about libraries."⁴⁰ At the time he was studying library economy, and his research culminated in the creation of his classificatory system, an invention that permanently linked his name with libraries and the decimal principle.⁴¹ Dewey's decimal classification for libraries follows a logic similar to that of metric weights and measures. In the metric system every unit is divided into ten subunits that can be subsequently divided into ten more parts as well (as the meter, which is fractioned in ten decimeters, one hundred centimeters, one thousand millimeters, and so forth). It is also similar to the design of the American dollar—the decimal currency, invented by Thomas Jefferson—in which an eagle coin had ten dollars, the dollar ten dimes, the dime ten cents, and the cent ten mills.

In devising his system of library classification, Dewey conceived a scheme to arrange books in a specific and repeatable order within ten main classes, each class with ten divisions, and each division with ten sections. The whole system was then constituted by ten classes, 100 divisions, and 1000 sections, all of them with an assigned number. With this format, each field of knowledge, discipline, or topic has a traceable place, prearranged under an easily readable general framework. In 1876, Dewey patented the system and published the first edition of his decimal classification and relative index for libraries. He explained there that by applying this system the usefulness of libraries "might be greatly increased without additional expenditure;... with its aid, the catalogues, shelf lists, indexes, and cross-references essential to this increased usefulness, can be made more economically than by any other method."[42] This was vintage Dewey, doing more with less to improve the means of education.

Also in 1876, when he was twenty-five years old, Dewey founded three associations, each focused on one of his main reform plans: the American Library Association (ALA), the Spelling Reform Association (SRA), and the American Metric Bureau. He was secretary of all three and oversaw almost all the associations' tasks. That the three organizations had considerable impact in their respective fields—and the ALA is still in operation—speaks of Dewey's determination and sense of opportunity.

The three programs of reforms sought by Dewey—library efficiency, spelling simplification, and the thorough introduction of metric units in the United States—were part of an overarching vision that was the driving force behind his work: efficient education of the masses. Libraries, Dewey thought, should be well-organized to provide readers, especially those of the lower classes, with readings to shape their taste and intellect. Metrological and spelling modernization would save pupils in elementary and secondary schools considerable time that could then be used to teach other topics—or as reformers put it, the metric adoption would be a "labour-saving" innovation in education that could prevent "brain fatigue."

The AMB was the first explicitly pro-metric association in the United States. The American Metrological Society (AMS) had started its operations three years earlier, but it was not openly pro-metric. One of the objectives of the AMS was to achieve international standardization of weights and measures, and while most of its members were keen on the idea of securing the exclusive use of metric units, that was not part of the stated objectives of the association.

As secretary of the AMB, Dewey was in charge of basic operations and received no salary. He edited its official publication, the *Metric Bulletin*—later called *Metric Advocate*—first issued in July 1876. F. A. P. Barnard accepted the position of president of the AMB. Barnard was one of the most influential scientists in the country and president of Columbia College (now Columbia University).

The stated mission of the AMB was to disseminate information concerning the metric system to urge its adoption, and to bring about actual introductions wherever practicable. To this end it published articles, books, pamphlets, and charts; distributed scales and measures; and promoted the teaching of the system in schools. It aimed, overall, to urge the matter "upon the attention of the American people, till they shall join the rest of the world in the exclusive use of the International Decimal Weights and Measures."[43]

The AMB concentrated its efforts on building public awareness of the necessity for metrological reform and trying to ensure that the metric system would "be taught in the schools, introduced into factories, shops, and homes, used in markets and stores."[44] As part of his interest in education, Dewey negotiated with the powerful Fairbanks Scales and other companies to provide the AMB with low-cost metric materials, with the expectation that this would be a good investment for companies once the metric system had been adopted nationally. Dewey then distributed metric supplies to schools and educational institutions interested in teaching and popularizing the new measures.

Dewey combined his duties as secretary of both the American Library Association and the AMB. As chair of the standards committee of the ALA, Dewey was instrumental in setting the standard for library catalog cards in metric units: the Harvard College size of 5 × 12.25 cm and the postal size of 7.5 × 12.25 cm. He promoted the idea that library catalogs should specify the size of books and other printed matter in centimeters (instead of using the old book size terms such as folio, quarto, and octavo) and was ready to provide libraries with these metric supplies through the Metric Bureau.

In 1876 Dewey appointed two committees of twenty educators to study the impact of metrication and spelling simplification in school teaching. They reported that using only metric units, abolishing compound numbers, and implementing "scientific spelling" would save the average child three to four years between first grade and graduation from college. The metric reform alone would save a full year of academic education for every student. According to Dewey, "the importance of this is simply incalculable at a time when we find impossible to get our limited courses the subjects it seems necessary to teach. The school life of the average child is now so very short that a system which causes such a waste is a national crime."[45] The results of this study became one of Dewey's favorite talking points in the weights and measures debate.[46]

Dewey's reputation made him a point of reference in numerous public debates about metrication. In 1904, when metric legislation was discussed in Congress, James H. Southard, chairman of the Committee on Coinage, Weights, and Measures in the House of Representatives, solicited Dewey to appear before the committee and present his views. Dewey was unable to travel to Washington, but he sent a letter to be read in the hearing. The missive summarized Dewey's view on the metric system as a civilizatory achievement and his harsh opinion about the "obstructionists":

> [People who oppose the metric system have] protested against railways, steamships, free schools, free libraries, good roads, laborsaving machinery of all kinds, in short against the things which have made modern civilization what it is. . . . There were similar protests [against] the adoption of standard time throughout the United States. Learned disquisitions on the trouble, expense and impossibility of such a change were not wanting and yet it was effected with the greatest ease and has been of untold benefit to the country. A still closer illustration is the adoption of the decimal currency in place of the £, s, and d. . . . If the arguments of obstructionists had been listened in these cases, it would have seriously set back the course of civilization and economic development. . . . A really important reason [to favor metrication] is the economic waste. Computation have repeatedly been made and verified to show the countless millions wasted each year by English speaking people because we have so cumbrous a language of weights and measures.[47]

Dewey followed with attention the public debates on the metric system and was more than willing to defend the cause. In 1925 he sent a letter to the *New York Herald Tribune* to refute an anti-metric article by a member of the American Society of Mechanical Engineers who argued that the metric system was only of interest to a "little group of theorists." Dewey rebutted with one of his favorite points when talking about measurement reform: first underlining the qualities of the metric system and then placing the adoption of the metric system in line with other intellectual innovations in history, like the calendar and the adoption of Hindu-Arabic numerals:

[the columnist] says "in practice one seldom works with 10s." He writes exactly as did the Englishman who refused to adopt Arabic numerals and protested that I, V, X, L, C and M were so simple and plain that it was absurd to urge an Englishman to learn Arabic. I found in an old book this same argument saying that "everyone knows instantly that V means five, but these iconoclasts would destroy the beautiful simplicity of English measures and tell us that 5 is to be the new character. Away with so foolish an innovation!" And this stupid talk resulted in England being long behind the Continent in adopting Arabic numerals. The mind can hardly grasp what it would mean if we were dependent on Roman numerals today for our computations. It would bring the business world to a standstill, and yet exactly the same arguments were used for them that are now urged against completing the work and having the measures of length, capacity and weight, like measures of numbers, simple decimals.[48]

Until the end of his life, Dewey devoted time and energy to helping organizations that promoted social and practical improvements. In 1918 Dewey was named president of the National Efficiency Society, described by him as "a nation-wide movement, not for promoting efficiency in the field where engineers have already done so much, but rather for extending the principles of efficient engineering to all other relations of life."[49] And a month before his death, he joined the World Calendar Association, an organization that sought to modernize the Gregorian calendar, an effort similar to the metric reform but for the calculation of time.[50] Dewey's role in these other enterprises, though secondary, was a natural extension of what we may call the "Dewey doctrine," a form of social engineering focused on the proper and proficient utilization of resources to bring permanent benefits for the majority of people.

Through the decades Dewey's entangled ideas on measurement, efficiency, and language were—knowingly or not—repeated several times by other metric proponents, some of which displayed marked similarities with Dewey's rhetoric (such was the case with science fiction novelist, popular science author, and hardcore metric advocate Isaac Asimov).[51]

Dewey is a relevant figure in the history of the metric system in America for two reasons. First, his endless vigor to promote metrication; besides being the creator of the AMB, Dewey was the only person with active membership in the four major pro-metric associations in the United States from the 1870s to the 1930s.[52] Second, he can be seen as the embodiment, in its purest form, of the worldview and mindset that infused the actions of many other "metric apostles." This worldview combined the desire for social change and collective improvement with the geometric spirit of simplicity, order, and rationality.

CHARLES LATIMER AND THE INTERNATIONAL INSTITUTE FOR PRESERVING AND PERFECTING WEIGHTS AND MEASURES

Prior to the 1870s there had not been organized movements in the United States to uphold the retention of customary measures.[53] This changed thanks, in part, to the initiative of Charles Latimer (1827–1888), a chief engineer of the Atlantic and Great Western Railway with experience in railroad companies in New York, Pennsylvania, and Ohio. He invented a system of naval signals by lights and a method to return to the track derailed trains.[54] Along with his engineering work, Latimer had a lasting interest in subjects like numerology (the study of the occult significance of numbers), mesmerism (a form of hypnotic induction involving "animal magnetism"), and water-witching (the practice of using a forked stick to locate underground water, minerals, and other hidden substances, a topic on which he wrote an entire book).[55]

The first overtly anti-metric organization in the United States was the International Institute for Preserving and Perfecting Weights and Measures, founded in 1879 by Latimer, with the Scottish astronomer Charles Piazzi-Smyth as "counselor" of the organization. The manifested aim of the organization was to deter the progress of the metric system through the publication of books, charts, and other materials. In its publications, the Institute emphasized the monetary cost of switching measurement systems, and portrayed metric promoters as atheists, "closet philosophers," and importing merchants whose interests were distant from those of common people.[56]

The main interest of the Institute, however, was something bigger than the opposition to the metric system. They were believers in a movement called "pyramidology." Latimer and his associates put most of their energy into spreading information about issues like the measures of the pyramids, the origin of the "sacred cubit," and the "god-given" English customary weights and measures. Following Piazzi Smyth, the idea that connected their rejection of metrication and their enthusiasm for the "pyramid lore" was that English measures had a divine origin. These measures, they said, were used and incorporated into the great pyramid of Giza by its "Hebrew builders," and later transmitted to the English people.

To make sense of these ideas, we need to trace the trajectory of Piazzi-Smyth and his interest in "pyramid metrology" and "British Israelism," two doctrines that had a considerable number of followers at the time. Piazzi Smyth was a professor of astronomy at the University of Edinburgh, son of the founder of the Bedford Observatory, and godson of the Italian astronomer Giuseppe Piazzi. Piazzi Smyth was a respected scientist, astronomer royal for Scotland (one of the few publicly funded positions in astronomy), and a pioneer of high-altitude observations with telescopes.[57]

In the 1860s, midway through his career, Piazzi Smyth showed interest in the pyramids of Egypt. Egyptian culture had for a long time been an object that attracted the attention of specialists and laypeople in Europe. Archeological sites and recovered objects, like the Rosetta Stone, fostered intense debate in the popular and scientific presses. The pyramids, in particular, aroused great curiosity. Above all, the pyramid of Giza had long been envisioned as a source of profound secrets and untold wisdom. Its structure, some people thought, embodied geometrical principles superior to those of any other ancient culture and were so advanced that could not be attributed to human intelligence alone.

Piazzi Smyth was intrigued by the measurement system used to build the pyramid of Giza and by the repeated suggestions that the pyramid, and the objects in its interior, were standards of measurement. He was following ideas that had been floating in the air for quite some time. Napoleon's cartographer and archaeologist Edme-François Jomard, for example, speculated that the sarcophagus in the King's chamber was actually a measure of capacity. Piazzi Smyth was absorbed by this and other archeological propositions, like the belief that the builders of the pyramid had been familiar with the number *pi* and designed the pyramid to show future generations the solution to problems like the "squaring of the circle."[58] All these ideas were agglutinated in what was called "pyramid metrology."

Pyramid metrology was based, first, on the notion that the pyramid of Giza was not built as a tomb for the pharaohs, but as a storehouse for divinely inspired weights and measures. Pyramidologists claimed that God trusted Noah (and not idolatrous Egyptians) with the construction of the pyramid, and that the units of measurement to build it were the "sacred cubit" and the "pyramid inch." A second pillar of this doctrine was the idea that the modern British people (and the "Anglo-Saxon race" in general), inherited as a racial patrimony these measurement standards virtually unaltered. They assured that the pyramid inch and the English inch were essentially identical. Assuming its divine origin, pyramid metrologists tried to demonstrate, based on science and religion, the inherent superiority of the English system of weights and measures over its metric rival.[59]

The writings on the pyramid by British essayist and publisher John Taylor (who had published the astronomical work of the decorated astronomer John Herschel, another critic of the metric system and one of Piazzi Smyth's scientific idols) exerted great influence on Piazzi Smyth.[60] Taylor followed the lead of seventeenth-century scientists like Isaac Newton and the mathematician John Greaves.[61] Newton had made some investigations on the "sacred cubit of the Jews" and the dimensions of the Temple of Solomon.[62] In the work of these scientists it was suggested that the British inch was commensurate with $1/_{25}$ of the "ancient cubit." This led Taylor and others to propose that the inch of the imperial system had an ancient and divine origin. Going even further, Taylor claimed that while in Egypt, the Israelites built the Giza pyramid with the purpose of preserving for humanity the sacred standards of length and capacity described in the Old Testament.[63] Even at the time, for many people these were far-fetched ideas; but with the names of prestigious figures like Herschel and Newton backing parts of these speculations, pyramid metrology gained some acceptance in intellectual circles.

In 1865 Piazzi Smyth interrupted his work in the Calton Hill Observatory in Edinburgh to travel with his wife to Egypt and make detailed measurements of the orientation, angles, and size of the Giza pyramid. This part of his work, strictly confined to the measurement of the physical dimensions of the pyramid, earned him the Keith Prize of the Royal Society of Edinburgh. But he went beyond that. He wrote several books, pamphlets, and articles based on a combination of his onsite research and Biblical speculations, with the purpose of confirming that the pyramid was indeed a metrological monument "devoted to memorializing a system of weights and measures."[64] These works were widely read. The better-known were *Our Inheritance in the Great Pyramid* (first published in 1864 and expanded over the years) and the three-volume *Life and Work at The Great Pyramid*.[65]

Mixing his studies of the pyramid with religious conjectures, Piazzi Smyth articulated a vision that interconnected metrology, the pyramid, and "British Israelism" (or "Anglo Israelism," as it was also called). The latter was a vigorous religious movement that spread through both sides of the North Atlantic. The supporters of British Israelism believed that British people (the "Anglo-Saxon race") descended directly from the lost ten tribes of Israel. As their racial heirs, the British were God's chosen people and the inheritors of great blessings. This ideology, also known as "The Identity," had at the beginning of the twentieth century around two million adherents in the British Empire and the United States.[66] Some prominent leaders of British Israelism in England (like Edward Hine) and the United States (like C. A. L. Totten), got close with Piazzi Smyth. He wrote in their publications and, conversely, Hine and Totten produced some works on pyramid metrology.

For British Israelites, metrology was of great importance. Totten called metrology "the veritable science of sciences," and Piazzi Smyth considered that "next in importance after language and religion comes metrology ... as a necessary practical method of justice between man and man, and a foundation for all social systems, organized civilizations, and lasting human government on earth."[67] The affinities between British Israelism and pyramid metrology were abundant and meaningful. They were mutually reinforcing theories.[68] They shared important concepts and worked from similar presuppositions: the authenticity of Biblical prophecies, the Hebrew origin of the British, and the importance of preserving English culture.

According to Piazzi Smyth, the pyramid and its metrological system "can only be attributed to a certain amount of Divine interference."[69] So, for him, there were only three alternatives among the existing systems of measurement: Hebrew, French, and Egyptian. The first (embodied in the pyramid), was the product of "revealed religion." The second (the metric system) was the result of "modern science without any religion." And the third (the ancient non-Hebrew measures), was based on "old idolatry."[70] All measures that were not grounded on revealed religion were "metrologies of the idolatrous and Cain-like Nations."[71]

Piazzi Smyth believed that "Anglo-Saxon measures" came so exceedingly close to the "Hebrew standards" that the coincidence could not be accidental.[72] The alleged connection between English people and the lost tribes of Israel became an argument to explain historical metrology. Piazzi Smyth maintained that the English pound, pint, inch, ell, and acre were the traditional representatives, through divine providence, of the "Israelite measures of old." The English quarter, according to him, was the fourth part of "the Ark of the Covenant," where the manna that had formed the bread of the Israelites during their forty years' sojourn was kept.[73]

Though controversial in scientific circles (his opinions on the pyramid triggered a bitter debate that ended with his resignation from the Royal Society of London), Piazzi Smyth's writings were popular reads in England and the United States. It was in America where Charles Latimer embraced the cause of pyramid metrology.

After reading various works by Piazzi Smyth in 1878, Latimer was convinced of the divine origin of English measures and the need for a concerted effort to halt metrication in the United States. Latimer became the driving force behind the International Institute for Preserving and Perfecting Weights and Measures. Along with the organization of the Institute, Latimer penned a large number of anti-metric works. One was the most famous book of the movement at the time, *The French Metric System or the Battle of the Standards: A Discussion of the Comparative Merits of the Metric System and the Standards of the Great Pyramid*, published in 1880.[74] He was also the editor of the *International Standard*, a bimonthly magazine devoted to "the preservation of Anglo-Saxon weights and measures," which lasted from 1883 to 1887. Latimer traveled several times to Washington, DC, to lobby against proposed metric legislations. The Institute, for example, wrote to the US president during the Meridian Conference of 1884, where Greenwich time was being defined, to avoid any possible negotiation regarding metric adoption in England and the United States. On that occasion, Totten, who was a member of the Institute, advocated that a prime meridian should pass through the great pyramid.[75]

Totten, who was the first national-level figure of British Israelism in the United States, got quickly involved with Latimer and the International Institute.[76] He even wrote the theme song of the organization, *A Pint's a Pound the World Around* (which included lines like these: "For the Anglo-Saxon race shall rule / The earth from shore to shore. / Then down with every 'metric' scheme / Taught by the foreign school.")[77] He was a West Point graduate who taught, as non-affiliate faculty, some courses on military tactics at Yale. His academic credentials were handy for the Institute to provide the movement with a patina of academic excellence. In 1884 Totten published *An Important Question in Metrology*, which he dedicated to "English-Speaking peoples of the Earth, who already possess the world's gates of commerce."[78]

Like numerous other anti-metric figures in the United States, Latimer and Totten were a blend of cultural preservation with recalcitrant Americanism. Rejection of the metric system was associated with the defense of Christianity and the American way of life. The preservation of customary measures was in line with the protection of American political and economic interests. For example, during the initial attempts to build the Panama Canal, Latimer begrudged Colombia commissioning a French company to build it. He demanded its construction by and for American interests, and—of course—using English customary measures to prevent the invasion of French measurement units. He also resented that the Statue of Liberty—a gift from France to the United States in 1886—was measured "in French milli-meters" and not "in good earth-commensurable Anglo-Saxon inches."[79] Another member of the Institute predicted that America will "supplant the mongrel civilization of Spain and Portugal, and give uniformity of laws, land, manners, customs and measures to the western world. The inch is mightier than the metre."[80] In the metric debates in the United States, arguments to combat metrication have frequently married practicality with narratives of patriotism, Americanism, anti-foreignism, and traditionalism.

It would be unfair to deem Latimer an adequate representative—or even an approximate type—of other anti-metric intellectuals. Contrary to Dewey, who seems to embody some characteristic traits of many other metric enthusiasts, Latimer and the pyramidologists are outliers in the anti-metric landscape. Life for advocates of metrication would have been easier if all their opponents were like Latimer, because it would have been simpler to label them as a bunch of fanatics—Barnard had a field day mocking what he labeled "the imaginary metrological system of the pyramid."[81] However, other anti-metric scientists and engineers, like Coleman Sellers (chief engineer of William Sellers & Co., professor of mechanics at the Franklin Institute, and president of the American Society of Mechanical Engineers), were leaders in the development of the machine-tool industry and the design of industrial standards.[82] Even in Latimer's time, many metric opponents were to a great degree preoccupied with the technical aspects of measurement conversion, voiced strictly practical concerns, and pushed their nationalistic and religious opinions to the background.

The battles around pyramid metrology were colorful and eye-catching. Today the premises of the pyramidologists look so farfetched that the whole episode seems like a headline of yellow journalism sneaked into the annals of the history of science. However, at the time these were ideas sponsored by prominent engineers and scientists, accepted by an educated public, and published in recognized outlets, like *Van Nostrand's Engineering Magazine*.

Pyramid metrology epitomizes an interesting problem: how scientists and engineers imagined and reimagined the past of their professions, practices, and instruments. Historian Simon Schaffer considers that Latimer, Piazzi Smyth, and other nineteenth-century metrologists were active in "the invention of traditions," cultural forms designed to portrait changing present situations under the guise of immemorial custom.[83] But maybe it is more accurate to say that they were performing acts of genealogical imagination.[84] Pyramid metrologists were mentally linking past and present generations by constructing schemes of descent, drawing lines of succession, and assembling an uninterrupted lineage from the materials of a remote and disunited history. Like other forms of genealogy, this was a "system of organizing legitimacy," an attempt to enhance one's social status by establishing decent from prestigious ancestors (in the next chapter we will see other attempts to reengineer the history of weights and measures in the United States).[85]

Pyramid metrology made some noise in the public conversations about metrication in America, but it never got sustained social support. When Latimer died, in 1888, the organization and its magazine vanished. It was an entirely different set of intellectual arguments that really shaped the metric debate in America.

Herbert Spencer, Lord Kelvin, and the Transatlantic Battle of the Standards

The decisive episode in the scientific clashes around the metric was the debate between Herbert Spencer and Lord Kelvin, which had a lasting impact on both the United Kingdom and the United States. In 1896 there were discussions in the British Parliament on a proposed law that would enforce the compulsory use of the metric system.[86] In March and April of that year, sociologist Herbert Spencer published in London's *The Times* four anonymous letters (signed by "A Correspondent") opposing metrication.[87] The letters were soon after printed as a pamphlet and sent to all members of the House of Commons, some members of the House of Lords, and all representatives of the United States Congress, where a similar bill was being considered.

The interest aroused by Spencer's text in the United States warranted its republication, a couple of months after it appeared in London, in the *Popular Science Monthly*.[88] For this reprint Spencer consented to allow his authorship to be known.[89] In 1899, when the subject resurfaced in public and legislative debates, Spencer came back to the topic publishing another four installments.[90]

In his editorials, Spencer developed two sets of arguments, some intellectual, others political. The intellectual arguments revolved around the flaws of the metric system and its decimal principle, the virtues of duodecimal systems, and the integration of metrological policies in the context of the historical development of human knowledge. The political arguments focused on a critique of the metric system based on the affinities between metrication, compulsion, and state power.

Spencer was proud to say "I yield to none in the love of method and system." Contrary to other critics (like Latimer and Totten), he underscored several of the positive aspects of the metric system. Spencer's defense of customary systems of measurement was not a vitriolic, nationalist rejection of "foreign" metric measures—something that separates him from several anti-metric intellectuals who came before and after him.

According to his *Autobiography*, Spencer had thought about the benefits of a "12-notation" number system since the 1840s when he sketched his "Ideas about a Universal Language," which included a "Memoranda Concerning Advantages to be Derived from the Use of 12 as a Fundamental Number."[91] It is interesting to notice that he started to develop his preference for a duodecimal system while reflecting on universal languages, one of the favorite topics of metric proponents—a coincidence that shows how close Spencer was in several of his intellectual interests with nineteenth-century metric advocates.[92]

Spencer's analysis of the metric system was distinctive from other pioneering sociologists in France and the United States. Auguste Comte, in his *Course of Positive Philosophy* (written during the years when the metric system was being reintroduced in France, after the failed experiment with the *mesures usuelles*) praised the metric system and considered its diffusion an example of the beneficial influence of science on society, and prognosticated its adoption in all "civilized nations."[93] For him, the metric system was an "admirable system of universal measures, which was begun by revolutionary France, and thence slowly spread among other nations. This introduction of the true speculative spirit among the most familiar transactions of daily life is a fine example and suggestion of the improvements that must ensue whenever a generalized scientific influence shall have penetrated the elementary economy of society."[94] In America, William Graham Sumner—who held the first professorship in sociology in the country, at Yale University—did not relate the metric system to the progress of science and society, as Comte. Rather, he saw traditional weights and measures as part of the most perdurable mores in society and as "primary folkways in their earliest forms." Writing at a time when intense debates over metrication were taking place in Congress, Sumner downgraded the effectiveness of the new measures: "The metric system was invented to be a rational system, but the populace has insisted on dividing kilograms and liters into halves and quarters. Language, money, and weights and measures are things which show the power of popular custom more than any others."[95]

Spencer believed in the evolution of human thought (like Comte), but also praised the achievement and perdurability of non-specialized forms of thinking (like Sumner). He shared with the "metricists" the desire to improve the methods of measurement and to terminate "the chaotic character of our modes of specifying quantities." For him, scientists were entrusted in the French revolution to form a rational system of measurement for universal use; it was a great idea and, despite its defects, it had been carried out admirably. Spencer saw the new system as a notable innovation for scientific purposes. Their great advantage was that calculations registered by a scientist in one place will be intelligible to scientists in other countries. Spencer did not call into question the advantages of decimal calculations and the use of a "rationally-originated" system. He saw merit in the promise to end "the present chaos," but objected that the metric system was inconvenient for purposes of daily life and that the conveniences it achieved in others activities could be attained without entailing the inconveniences.[96]

Spencer's main purpose was to show that there were strong grounds for a "rational opposition" to the metric system on the basis that it is "ill-adapted for industrial and trading purposes." For him, one of the principal defects of the metric system was its decimal character, which brings a lack of appropriate divisibility that is crucial in everyday calculations. Spencer underlined that lay people have a need for easy divisions into aliquot parts, like using thirds and quarters, but the number ten cannot be divided by four or three without using decimals. This deficiency was "great and incurable."[97]

Spencer was convinced that duodecimal systems (base–12) were superior to decimal ones. The great divisibility of twelve made it perfect for solving practical problems. The 12–group, he said, has an enormous advantage over base–10 systems. Ten is divisible only by 5 and 2; twelve by 2, 3, 4, and 6. This is why different cultures developed the habit of dividing into twelfths:

> we have 12 ounces to the pound troy, 12 inches to a foot, 12 lines to the inch, 12 sacks to the last; and of multiples of 12 we have 24 grains to the pennyweight, 24 sheets to the quire. Moreover, large sales of

small articles are habitually made by the gross (12 times 12) and great gross (12×12×12). Again, we have made our multiplication table go up to 12 times 12, and we habitually talk of dozens . . . a general system of twelfths is called for by trading needs and industrial needs; and such a system might claim something like universality, since it would fall into harmony with these natural divisions of twelfths and fourths which the metric system necessarily leaves outside as incongruities.[98]

According to Spencer, when people count by twelves instead of tens, they encounter fewer troubles with fragmentary numbers, and this produces an "economy of time and mental effort," a practical advantage of greater importance than the advantages of the "theoretical completeness" characteristic of the metric system.

Spencer was thus in favor of a base–12 measurement system. He was aware, however, that one of the virtues of the metric system was that it put in agreement our decimal numbers system with a decimal system of weights and measures. To do that the creators of the metric system eliminated duodecimal units and subunits of measurement; in other words, they put metrology in tune with the dominant numbers system. Spencer, on the other hand, proposed reordering things the other way around: the numeration system should agree with the customary measurement system. What was needed, he said, was "a small alteration in our method of numbering to make calculation by groups of 12 exactly similar to calculation by groups of 10."[99]

Contrary to many critics of metrication who argued that a transition to the metric system was too radical and complicated, Spencer thought that the metric system was not radical enough. Like the metric inventors, he wanted to synchronize the numbers system with the measurement system; but while the French opted for transforming measures, Spencer sought to change the whole numbers system and make it duodecimal. In that way it was possible to have the facilities of a method of notation like that of decimals and with the facilities of duodecimal division; this solution, he said, "needs only to introduce two additional digits for 10 and 11 to unite the advantages of both systems."[100]

Only to change the numbers system! This was a reform of the highest magnitude. If people find it complicated to change from yards to meters, changing number systems would be unmanageable. A dozenal number system, like the one proposed by Spencer, "only" requires two additional digits for ten and eleven. Let us take a look. Suppose, for instance, that X is ten, E is eleven, and *10* is a dozen. Numbers would be like this:[101]

one	1
two	2
three	3
four	4
five	5
six	6
seven	7
eight	8
nine	9
ten	X
eleven	E
twelve	10
thirteen	11
twenty-four	20
one hundred twenty	X0
one hundred forty-four	100

Spencer knew that this would be a radical and profound change, but for him that a reform would be too difficult to be accepted by people at a certain point in time does not mean that it cannot be successfully carried out by future generations. One reform can be more rational and perfect than others (like using a duodecimal rather than a decimal *number* system), even if it is unpopular or conceived as impractical. Spencer's plan was bold and sanguine, and presupposed a historical-developmental theory of human knowledge. Spencer was confident about the intelligence of future generations and their ability to fulfill a plan like this. As he put it, "Does not experience teach us that the impossibilities of one century become the facilities of the next?"[102]

Spencer hoped for a greater human agreement—across generations and geographical areas—on the importance of having divisibility in numbers to perform calculations. His plan for numerical reform was not only a leap into the future, but also a link with the past. In Spencer's view metrication represented a rupture with the shared wisdom of humanity—a break for the worst: "the ancient wise men of the East and the modern working men of the West, have agreed upon the importance of great divisibility in numerical groups.... And yet, though the early man of science and the modern men of practice are at one in recognizing the importance of great divisibility, it is proposed to establish a form of measure characterized by relative indivisibility!"[103]

Spencer thus asked for patience and to wait for more favorable conditions to settle the problem of the numbers and measurement systems in the most perfect way. The adoption of the metric system, he said, would entail an extended period of trouble and would increase the obstacles to the future adoption of a more "perfect system." So, for him, it was preferable to stand the evils of the current system for the time being.

The idea behind this vision was that the metric system may be an improvement compared to current weights and measures, but it would be a hindrance to future perfection—do not let the good be the enemy of the perfect, he could have said. Spencer synthesized his position by saying that "rather than establish a fundamentally imperfect system based upon 10 as a radix, it will be better to wait until we can change our system of numeration into one with 12 as a radix; and then on that to base our system of weights, measures, and values: tolerating present inconveniences as well as we may." What he wanted was to leave things as flexible as possible, so that the greater knowledge and higher intellectual capacity of the future should meet with fewer obstacles.[104]

Foreseeing the consequences of global metrication and the difficulties of dismantling a future metric world, Spencer hoped that England and the United States could avoid metrication: "if in the United States as well as in England and its colonies, governments prompted by bureaucracies, but not consulting the people and clearly against their wishes, should make universal this gravely defective system, very possibly it will remain thereafter unalterable."[105] He considered that if trade within each nation and international commerce was unified under the metric system, the obstacles for a his radical plan—despite its superiority—would be insuperable. Thinking in terms of path dependence, he wanted to avoid the metric system to be permanently locked in.

The second set of arguments presented by Spencer against the metric system was political—similar in tone to his famous book *The Man versus the State*.[106] He saw the spread of the metric system as a result of authoritarianism: "the imposing list of the countries that have followed the lead of France . . . headed 'Progress of the Metric System' [should] have been headed 'Progress of Bureaucratic Coercion.'"[107] Spencer stressed that the manifest weakness of the metric system was that it could only be introduced by way of compulsion, its adoption always came from official will, instead of popular will.[108] And even using coercive methods, the results of metrication had been mixed. Despite French citizens being forced by the State to use francs and centimes, "people can still talk in sous and ask for fourths." Even a century after the revolution, metrication was not complete in the metric system's birthplace. "Doubtless 'ignorant prejudice' will be assigned as the cause for this," Spencer said, "but one might have thought that, after three generations, daily use of the new system would have entailed entire disappearance of the old, had it been in all respects better."[109]

Besides compulsion, Spencer saw other undemocratic aspects in metrication. For example, its planning was always the works of minorities (mainly scientists and exporters), with the inconveniences being shared by all. The opinion of the majority was never considered—"name any country where the metric system has been put to the popular vote?," he asked. Stressing this point, he declared that:

> Ten thousand persons intend to make 20 million persons change their habits. The ten thousand are the men of science (by no means all), the chambers of commerce, and the leaders of some trade unions—leaders only, for the question has never been put to the vote of the mass. The 20 millions are the men and women of England with those children who are old enough to be sent shopping. Ten thousand is an over-estimate of the combined bodies who are forcing on the metric system, and 20 millions is an under-estimate of the numbers to be coerced.[110]

The people most negatively affected by compulsory metrication were people of narrow incomes, wholesale dealers, shopkeepers, hucksters, buyers and sellers of small quantities who would need to replace their measuring appliances, and then relearn new methods of exchange and calculation. They were, Spencer said, people with "experimental knowledge" in matters of buying small quantities for small sums, involving fractions of measures and money. This majority, however, was excluded from the decision-making process of adopting the official measuring system. Forcing the metric system on the population at large, instead of just allowing its voluntary use, was a form of "secular popery," a universalistic imposition that would change the daily habits of men and women.[111]

Spencer hoped that the democratic character of the British society, its colonies, and America would function as an antidote to compulsory metrication. He considered that "among nations less disciplined in freedom than ourselves there is scarcely a thought of resisting *l'administration*;" but the will of the "English people"—less submissive than "continentals"—could not be so easily overruled by state officials. He predicted that should the attempt be made to force the metric system on them, the opposition would be so strong that the attempt will have to be abandoned.[112]

At the end of his life, Spencer was so determined to stop the introduction of the metric system in his homeland, that he left a clause in his will directing the republishing his anti-metric pamphlet in case the question of metrication would come to Parliament again.[113]

Spencer's crusade was quite unique and had various ramifications. At the heart of many of his arguments, there were ideas identical to those of metric activists like Dewey: systematicity, universal languages, intellectual improvement, and economy of time and effort. However, they disagreed completely on what system of measurement was more appropriate to achieve those goals. The promoters of the metric system aspired to innovation and reforming collective habits. Spencer found rationality in people's traditional thinking; he defended, for lack of a better concept, "rational populism"—as condensed in his formula of the intellectual agreement between "the ancient wise men of the East and the modern working men of the West." In the end, the final score for him was mixed. The United States rejected metrication, as he wished, but his fears of a metric United Kingdom and a metric world came true—the metric system did lock in.

SPENCER'S ANTI-METRIC LEGACY

The legacy of Spencer's anti-metric stance was large but mixed. On the one hand, the core of his argument—changing the present decimal number system for a duodecimal one—had no chance of succeeding. Despite the attention his plan received in some newspapers, few people took it seriously.[114]

On the other hand, even if Spencer did not defend customary weights and measures as such and he described the existing state of affairs in British metrology as highly imperfect, his arguments against the metric system were widely used by people interested in preserving the old units of measurement. Assuming that the enemy of my enemy is my friend, anti-metric groups used Spencer's name in their campaigns to stop pro-metric legislation. This meant that they presented heavily edited versions of Spencer's writings, cherry-picking excerpts where he questioned metric measures and ignoring his plan for a duodecimal number system and his praise for the metric system's systematicity.[115] The influence of Spencer—an engineer himself—was mostly visible in the arguments displayed by mechanical engineers in the United States, who picked up the idea of a "rational opposition" to metrication (as we will see in the next chapter).

A third and unexpected impact of Spencer on the anti-metric movement came from an interpretation of the history of metrology through the eyes of Spencerian evolutionary theory. Or more precisely, through a broad understanding of which system of measurement ought to perish or prevail based on the "survival of the fittest"—the famous phrase coined by Spencer in the 1860s in his *Principles of Biology*.[116] Spencer himself did not frame the metric issue in those terms, but anti-metric propagandists lumped two ideas that Spencer presented in separate contexts—the opposition to the metric system and the "survival of the fittest"—to create a single line of argument. See for example this description of Spencer's position in the *Baltimore Sun* in 1920: "[It is] a plea to the English-speaking world not to abandon their fundamental standards being, in Spencer's view, a 'survival of the fittest,' developed through centuries of evolution, and he warned against abandoning them for the artificial metric standards."[117] Or this reasoning by a textile industrialist: "We now have a system that has resulted from a process of natural selection since man first began to measure and to weigh. It is a part of our language, gauges our ideas of length, breadth and thickness of every object, and is bound intricately and inextricably to all our complicated industries." And then, following Spencer's libertarian leanings, he concluded that "the only safety is in leaving the people alone to use what system they please without interference from the government. Let the government protect our standards by keeping models of the yard, pound, etc., but let the government control end there."[118] The translation of ideas from evolutionary biology into principles of political economy had been widely popular in America throughout the second half of the nineteenth century, the novelty is that now they were used in metrology.[119]

Spencer's anti-metric articles generated multiple and loud reactions. The first response against Spencer in the United States was an article by Thomas Mendenhall that appeared in *Popular Science Monthly* (the same magazine where Spencer's texts were published in America).[120] Mendenhall, president of the Worcester Polytechnic Institute, was not only a well-known scientist—he served in 1889 as President of the American Association for the Advancement of Science—but one of the most experienced metrologists in the country. From 1889 to 1894 Mendenhall was superintendent of the US Coast and Geodetic Survey, at the time the office in charge of weights and measures in the country and one the most important scientific appointments in the federal government.[121] Under his direction, in 1893, the "Mendenhall Order" was published, which made the meter and kilogram the fundamental standards of length and mass in the United States.

Mendenhall pointed out that when Spencer's articles were first published anonymously they aroused little interest among metrology experts; but he admitted that when it was revealed that the great Herbert Spencer was the author, there was astonishment in the scientific community— astonishment produced by the great reputation of Spenser and the oddness of his arguments. Mendenhall did not find much merit in Spencer's critique, but felt obliged to respond due to "the great influence which everywhere and always goes with the name of the distinguished author of these letters."[122] He delivered a harsh rebuttal. He first showed that Spencer was following, in an incomplete and almost dishonest way, arguments presented by John Herschel in the 1860s.[123] Then he showed that divisibility of the decimal system was preferable in many operations and that in several fields (such as land surveying and engineering) there was a tendency to decimalize customary units, like the avoirdupois pound, the foot, and the inch. Mendenhall was methodical and addressed Spencer's main ideas. He even acknowledged that the transition to the metric system would be slow and would come with "considerable hardship." He recognized the perdurability of habits, and described how a century after the introduction of the decimal currency in the United States some old monetary units and denominations continued to be used; nevertheless, "no one would now think of giving up" the decimal currency. Similarly, Mendenhall concluded, in the long run it will be better to adopt the metric system than to continue with "our illogical, brain-destroying, time-consuming system of weights and measures."[124] The only loose end in his article was that he did not address Spencer's anti-compulsory argument—the chink in the armor of all metric campaigns in America.

Mendenhall's reply was doubled by university professors such as E. E. Slosson in *Science*, and Oscar Oldberg in *The Bulletin of Pharmacy*.[125] Considering Spencer's reputation in American intellectual circles, it was important for pro-metric scientists to show their support for the metric system.

The rebuttals to Spencer had merits, but their effect was limited. Spencer had added new elements to the public discussion on metrication in the United States. He did not frame the problem as a confrontation of nationalistic, religious, and conservative ideas against universal, scientific, and rational principles—as had happened before in the metric debates. Spencer had no interest in trying to show that imperial measures were God-given units of measurement; he did not even pretend to show that customary measures were better than the metric ones. He provided a thoughtful and technical critique of metrication—a striking contrast with the groups that had just been harshly discredited for their persistence in linking English measures with Egypt's pyramids.

Spencer's position was impactful due to his immense influence on American intellectual life since the 1860s. In the three decades after the Civil War it was difficult to be active in any field of intellectual work without some knowledge of Spencer's work.[126] Prominent metric advocates, like Frederick Barnard, on the occasion of Spencer's visit to New York in 1882, said that the English sociologist was not only "the profoundest thinker" of his time, "but the most capacious and most powerful intellect of all time," a man who's name could only be compared in the history of science with Newton's—fortunately for Barnard, he did not live to see how his idol publicly bashed the metric system and campaigned to stop metrication in America. With Spencer throwing his weight against the metric system, pro-metric thinkers found a formidable adversary that could not be dismissed as a quack. Fortunately for them, some unsolicited help was on its way to America.[127]

LORD KELVIN AND THE METRIC SYSTEM

An interesting twist in the debates around Spencer's anti-metric stance came directly from England, where a figure equal in size to Spencer entered the scene. Sir William Thompson, better known as Lord Kelvin, refuted Spencer and used his own reputation—at the time he was arguably the most famous physicist in the world—to support metrication in the United Kingdom and America. As Spencer, Kelvin delivered his message directly to lawmakers in both countries. It is extremely rare for the most prominent sociologist and the most celebrated natural scientist of an era to get entangled in a public debate, but that is what happened in the 1890s when Spencer and Kelvin clashed on both sides of the Atlantic over the metric system.

Kelvin was famous for articulating a view of knowledge in which measurement and quantification played a central role. As he famously declared in 1883, "I often say that when you can measure what you are speaking about, and express it in numbers, you know something about it; but when you cannot measure it, when you cannot express it in numbers, your knowledge is of a meager and unsatisfactory kind."[128] His scientific contributions to measurement were so profound that there is a unit of measurement named after him (the *kelvin*, which measures temperature and is one of the seven base units in the International System).

Kelvin was a harsh critic of English customary measures and, according to witnesses, it was one of his favorite topics while lecturing. Referring to what he called "the British no-system of measurement" he lamented: "It is a remarkable phenomenon, belonging rather to moral and social than to physical science, that a people tending naturally to be regulated by common sense should voluntarily condemn themselves, as the British have so long done, to unnecessary hard labour in every action of common business or scientific work related to measurement, from which all the other nations of Europe have emancipated themselves."[129]

Shortly after Spencer published his anti-metric articles in 1896, Kelvin wrote a letter of his own to *The Times* defending the idea of England going metric. Kelvin contrasted the "uniform simplicity" of the metric system with the "monstrous complexity" of British measurements.[130] This was not the first time that these two colossal figures quarreled publicly on this topic. In 1895 Spencer wrote a lengthy letter to Kelvin, who had made pronouncements favoring metrication.[131] Kelvin replied to Spencer:

> Dear Mr. Spencer—It is the uniform simplicity of the French metric system which gives it its great advantage. . . . Miles, furlongs, perches, roods, yards, feet, inches, acres, square yards, square feet, square inches, cubic yards, cubic feet, cubic inches, cause enormous loss of efficiency to English engineers, and really involve a great national loss in useless labour, month after month, and year after year. . . . I wish you could be convinced, and give your powerful influence to a reform which is much needed, and from the want of which we in England, all of us, suffer every day of our lives.[132]

In April of 1902, Kelvin appeared before the House Committee on Coinage, Weights, and Measures in the United States to endorse the adoption of the metric system in America—the *New York Times* described the event as "the remarkable spectacle of a member of the British nobility appearing before a Congressional committee and urging the passage of a house bill."[133] Kelvin voiced his frustration with England's lack of action, but expressed his hope that if the United States would take the lead in adopting the metric system, his own country would follow suit.[134] With that Kelvin was pointing to a persistent problem, Americans waiting for England to go metric and Britons hoping for the United States to take the first step—an stalemate that lasted until the 1970s, when the United Kingdom finally went metric.

A couple of years after his visit to the United States Congress, Kelvin participated in the debates in the House of Lords, pushing again in favor of strong metric legislation. There, unlike so many other metric proponents like Mendenhall, Kelvin explained clearly that a compulsory law was absolutely necessary: "experience has proved that the change from the system that has been in use in this country to a new system cannot be made over the whole country voluntarily. It is a case for compulsion, and I think the Legislation will be thanked by the country for having applied compulsion."[135]

With his call for a mandatory metric policy Kelvin was hitting the nail on the head. The essential problem with countries like the United States was the lack of compulsory regulations, as all voluntary transitions to the metric system that have been intended have ultimately failed—a fact well-known by Spencer and the antagonists of the metric system who made the opposition to any compulsory legislation the centerpiece of their public campaigns, as we will see in the next chapter.

Metrology, a Ventriloquist Knowledge

In trying to make a balance of the metric debates in America, it would be tempting to ridicule metric opponents like Charles Latimer for their exotic beliefs. Pyramid metrology, British Israelism, and the biblical origins of the "Anglo-Saxon race" have, today, a rancid smell. However, it is important to stress that they were not a bunch of yahoos throwing stones at the windows of university buildings, as some have tried to portray all people who oppose metrication. As historian Stephen Mihm argues, many of the members of anti-metric organizations like the International Institute for Preserving and Perfecting Weights and Measures were renowned figures in their fields of expertise (like railroads and metalworking trades) and had worked in the technical subtleties of measurement and standardization.[136] Their opposition to metrication was not an unintelligible act, an irrational outburst of people blinded by ideology. Many of them were defending technical instruments and innovations of their own making, as well as their economic interests and the right to choose whatever standard they considered preferable.

However, Mihm's argument begs the question of why rational people espoused wacky ideas that were discredited in their own time. Even if they had a legitimate preoccupation with technical standards, their interest in the "sacred cubit" and the "wisdom of the pyramid" was not superficial or accidental. We may find an explanation to these apparent antithetical positions if we consider that science (metrology included) is a "ventriloquist knowledge."[137] Culture speaks through science. Scientific activities are shaped by political, economic, religious, and ideological views. Multiple extra-scientific interests are expressed in science under sublimated forms. It is plausible to say that Latimore and company were experts defending technical matters with legitimate scientific grounds and their passion for the arcane was incidental; but it is equally conceivable that they were nationalists, nativists, or racists who employed scientific topics—like weights and measures—to rationalize their ideological inclinations.

One wonders if their arguments about the technical advantages of customary measures were but a façade to express their "true" motivation: the protection of American traditions and identity. Different generations of accomplished engineers who have opposed metrication have articulated their demands with a vocabulary that allows such interpretation. In the 1880s Charles Latimer asked for the "preservation of Anglo-Saxon weights and measures." In the 1920s textile industrialist Samuel S. Dale spoke about the "foreign attack on our weights and measures." In the 2010s Jon Bosak—an expert in the standardization of computer data languages who led the creation of the Extensible Markup Language (XML)—published a book on the origins of customary measures with the hope of inspiring "some respect for these elements of our cultural tradition, and such consideration would lead naturally to a defense of the future of the Customary System in an 'age of metrication.'"[138] The rhetoric of a cultural inheritance that ought to be defended from external threats runs steadily through decades of anti-metric literature produced by technical experts.

It seems that in some cases, like Latimer, metrology was more a sublimation of religious values. In the case of other anti-metric figures, like Spencer, nationalism and chauvinism played a more subordinate role. In the coming chapter, we will see how the most important men in the fight against metrication followed Spencer's vision.

4

Searching for a Perfect Language for Commerce

Measurement and Economy

> We are absolutely opposed to the metric system. It is the most vicious thing that was ever suggested.... The only people who are for it are internationalists, college professors, and high-brow failures.
> —AMERICAN PROTECTIVE TARIFF LEAGUE,
> *San Francisco Chronicle* (1921)

> We are now a great world trade nation. We are pressed by necessity to speak the language of world trade. Prices vary, but standards of measure must be identical and uniform. The language of all trade is price per quantity and item. We do not speak the language of world trade in Latin America. We are needlessly tongue-tied in one of the richest markets of the world.
> —*SOUTH ATLANTIC PORTS* (1926)

Breaking the Yardstick

Days after congressman Fred A. Britten from Illinois introduced H.R. 10—also known as "the Britten Metric Bill"—into the House of Representatives in December 1925, he took a yardstick and broke it in half with his knee in front of photographers (fig. 4.1). "I have no further use of it," he said, confident that his bill will pass.[1] Britten's bill was not a conventional metric legislation. It did not ask for a complete banning of customary measures in favor of metes and kilograms. It called for the use of metric units in the buying and selling of commodities, starting January 1, 1935; but it allowed an exception to manufacturing and land surveying.

Figure 4.1. Congressman Fred A. Britten breaking a yardstick after the introduction in Congress of the "Britten Metric Bill." January 30, 1926. Photo: *Measurement* 1, no. 3 (1926).

For more than two decades manufacturers had been one of the strongest groups fighting metrication in America and that provision was a concession to soothe their position.

Despite Britten's theatrics and the manufacturing compromise, what caught the eye of the press was something else. The truly eccentric part of H.R. 10 was that it wanted to rename metric units with more familiar—and grandiose—names: the meter was dubbed "world yard;" the liter, "world quart;" and the half kilogram, "world pound."[2] Instead of simply saying that the metric system would be mandatory in commerce, Britten and the pro-metric groups that supported the bill presented the idea as a simple change of approximately 10 percent in the traditional yard, quart, and pound rounding them into common metric units. They even started to call the metric system the "dollar-meter-liter-gram system"—desperate times call for desperate metric measures.[3]

The bombastic nomenclature proposed by Britten was more than a publicity stunt. It was an effort to make metric units more palatable to the public. And more profoundly, it was part of a conception that saw the metric system not just as a tool to gauge physical magnitudes, but as a language—an idiom of quantity, an instrument of communication, a transnational code, a vocabulary of trade, a lingua franca of science, an Esperanto of commerce.

In the end, the Britten bill was not approved—the yardstick proved more resilient than anticipated.[4] Nonetheless, the whole episode encapsulated key aspects of how divergent economic interests clashed for the definition and regulation of measurement standards.

In America, economic debates about metrication revolved around what standards are better for selling dear, reducing expenses, and making profit. But these interests found affinities with larger visions of politics, science, and culture. As we will see, the metric system was married with an ideology of internationalism and laissez-faire markets, and with a type of expert knowledge inclined to the creation of abstract and universal languages. Anti-metric forces, on the other hand, defended customary measures as part of a discourse of nationalism and protectionism, and a vindication of practical knowledge.

This chapter shows how economic actors in the United States struggled to define what standards of weights and measures benefited them the most. Three aspects are analyzed in depth to illuminate the economic dimensions of the metric question in America. First, the path taken by the United States to establish a nationally homogeneous currency with a policy that did not have a parallel plan to standardize weights and measures. Second, the diplomatic proposals that tried to use the metric system as the basis for a "universal" currency. And, finally, the international treaties that sought to create a free market in the Americas. Before that, the chapter presents some historical and conceptual considerations to clarify the pivotal but changing role of weights and measures in economic life.

Economy and Measurement

Systems of measurement are economic institutions. Weighing and measuring are among the fundamental cognitive processes in economic activities; as economic instruments, measures are a form of "social knowledge" crucial in production and exchange.[5]

Economic exchange requires measurement. When different products and commodities are sold or bartered it is necessary to know how much of a product is being bought or exchanged. Commodities are sold using measures of weight, length, and volume: bushels of grain, yards of linen, gallons of petrol. Multiple other economic activities require the use of accepted conventions about measurement as well. Taxation (particularly in societies where taxes are paid in kind), and virtually all spheres of production need effective forms of measurement. For economic actors, it is crucial to know the extension of land required to sustain a community, the amount of meat that can be sold in a week, the weight of wood that can be transported by a mule, etc. As economist Douglass North has shown, throughout history measurement has occupied the attention of people in their efforts to improve exchange and to take advantage of each other in that process; "the very terms price and quantity imply the ability to measure those two dimensions."[6]

Economic systems depend on an "economy of knowledge." They need a certain degree of accumulation of knowledge and intellectual abilities, like literacy and numeracy, or "quantitative literacy"—there is a cognitive embeddedness in economic phenomena. This embeddedness refers to "the ways in which the structured regularities of mental processes limit the exercise of economic reasoning."[7] Socially shaped cognitive processes make economic reasoning possible. The generalized use of money, for example, requires individuals equipped with intellectual skills to grasp the abstraction involved in transforming qualitative differences into quantitative ones.[8] Thus, every economic system presupposes a particular form of social distribution of knowledge. The historical development of intellectual technologies has made possible the existence of complex economies. The invention of writing, for example, extended the possibilities of management, commerce, and production; it transformed the methods of capital accumulation and changed the nature of individual economic transactions.⁹ The development of fixed, clear, and stable systems of weights and measures has had analogous consequences.

Conversely, the genesis and development of the systems of measurement are tied to particular forms of economic relations. Measurement of quantity is "an operational use of number," and its functions are defined partially in economic terms; the institution of measurement "must have 'utility' before one may expect to find it in any given culture."[10] Systems of measurement are developed and adopted within specific sets of socio-economic relations. A system of measurement used by peasants in a self-sustained village varies greatly from one used by merchants in a big commercial Renaissance city, like Venice. Their exactness, complexity, and level of standardization diverge considerably because they respond to different economic needs.

To a greater or lesser extent, systems of measurement adjust to the requirements of the economic and social systems in which they are used. Local, autarkic economies preoccupied with self-maintenance develop local systems. A global economy favors an internationally accepted system of measurement. Of course, the correlation between measurement and economic needs is not perfect (there are other determinant factors to explain the global spread of modern systems of measurement); but there are numerous commonalities and mutual influences. Large-scale capitalism contributed to the global expansion of the metric system and the metric system facilitated the operations of the world market.

MEASUREMENT, TRANSACTION COSTS, AND ASYMMETRIC INFORMATION

Measurement is intimately linked to transaction costs (the costs involved in specifying and enforcing the contracts in exchange) and to the problem of asymmetric information (which refers to buyers and sellers exchanging goods on the basis of different amounts and quality of information about cost-to-measure attributes of services and goods).[11] To reduce transaction costs it is important to have the accurate specifications of what is being exchanged. A crucial element is the cost of measuring the attributes of goods and commodities that are traded; when weights and measures are intricate and imperfectly standardized those costs are higher.

The diversity of weights and measures makes the search for information laborious, uncertain, complex, and irregular. Conversely, uniform and trustworthy weights and measures reduce the costs of measurement.[12] Improvements in measurement reduce transaction costs between suppliers and customers. Establishing standardized weights and measures helps the negotiation and enforcement of contracts in commerce. Thus, metrological normalization aids economic processes in a similar way as the development of units of account, mediums of exchange, merchant law courts, and notaries. All these institutions lower information costs and provide incentives for contract fulfillment.[13]

A common source of market failure is asymmetric information between buyers and sellers, where the buyer cannot differentiate good products from bad ones. This frequently arises because measurement is complicated and costly. As measurement improves, buyers can measure the characteristics of products more easily, which reduces asymmetric information.[14]

If we link these ideas with what was said in Chapter 2 about states creating the conditions for homogeneous weights and measures in large territories, one can see how the intervention of political authorities provides favorable conditions for good information in the market—even if that intervention may benefit some actors more than others.[15]

ECONOMIC RATIONALIZATION AND MEASUREMENT

Another important area for understanding measures in the economy is the Weberian problem of rationalization. For Max Weber, rationalization means that social actions are disciplined, systematic, rigorous, and methodical; it implies that certain areas of social life are directed by regularity, calculability, and coherence.[16] One of the main aspects of rationalization in the economy is calculability. A system of economic activity is formally rational to the degree to which the provision for needs can be expressed in numerical, calculable terms.[17] Money, technology, free labor, capital accounting, and double-entry bookkeeping are institutions that helped the development of rational capitalism. Money, in particular, was decisive for economic calculability. It rendered possible the assignment of numerical values to goods and services involved in economic exchanges.

As Weber put it, money has been "the propagator of calculation."[18]

Calculation is based on quantitative and impersonal systems. Here lies the importance, for economic activities, of using numerical terms that are unambiguous and without "a wholly subjective valuation."[19] Capitalist markets are based on this numerical and impersonal character. A market situation is confined to the exchange of money because money allows "uniform numerical statements" about social relations. Moreover, instead of evaluating goods exclusively in terms of their importance for the present moment, monetary calculation makes possible the systematic comparison of future opportunities for utilization of those goods.[20]

In precapitalist economies, the calculation of economic activities went beyond monetary terms. There the role of weights and measures was crucial. Weber saw how weights and measures were important for the development of the routine of quantified calculation—in particular calculation in kind.[21] He observed that bars of bullion were weighed instead of coined; they were treated as money and used for payment and exchange. The fact that these bars were weighed (and thus quantified) was enormously important for the development of the *habit of calculation*. The development of more precise methods of measurement runs parallel to greater sophistication in economic calculability.

WEIGHTS, MEASURES, AND MONEY IN PRE-CAPITALIST ECONOMIES

The functions of measures and money in pre-capitalist economies varied from the functions they have today.[22] This could be seen, for example, in prices and land measurement.

The historical relation between money and measures is useful for understanding the evolution of the functions played by money.[23] In a modern economy, as contemporary consumers know, the relation between the quantity of money and the quantity of a commodity is that the quantity of money is variable while the quantity of the commodity is fixed. For example, when the cost of milk changes, what varies is the price of a fixed quantity of milk; so, a *gallon* of milk that today costs 3.26 dollars, a year later may cost 3.53. Variable money for a fixed quantity of a commodity; we are familiar with this formula.[24] In pre-capitalist economies, however, this relation was inversed: the quantity of the commodity was variable while the quantity of money was fixed. Monetary prices of essential products like bread, butter, and cheese were fixed and could not be altered. When the cost of a commodity raised or dropped what varied was the quantity of the commodity. The monetary prices of bread in Europe and European colonies in the Americas, up to the eighteenth century, did not change. The oscillations in the price of bread were regulated by modifying the weight of the loaves. In New Spain, for example, a loaf always cost half a *real*, but the weight of the loaf could vary; so, the higher the price of wheat and flour, the smaller the size of the loaf. Sometimes half a *real* could buy an 18-ounce loaf, in other occasion only a 13-ounce loaf.[25] Prices were expressed in the quantity of the commodity, not in the quantity of money. The price as a mechanism that reduces to a common denominator all factors in commercial operations is a relatively recent phenomenon.[26] Weights and measures were something like the money of the past—they fulfilled some of the functions that money has in today's economy.

Money became the universal commodity equivalent, as we know it today, only after the advent of capitalism; and even then, the practice of adjusting prices through measure still exists.[27] Sometimes, when the cost of goods increases, manufacturers decide to maintain the price of a product, but they reduce its size, as a discreet way to pass the rising costs on to consumers. This strategy is known as "downsizing." Studies on consumer behavior show that buyers are more sensitive to changes in price than to changes in quantity.[28] So it is more profitable for companies to reduce the amount of product than to increase the monetary price. This practice is widely employed in times of economic crisis and high inflation. Organizations that monitor the prices and sizes of products, like Consumer Reports, noted that after the "Great Recession" of 2008 companies downsized their products while charging the same price. Yogurt cups went from 6 to 4 ounces (with the product full of air bubbles to fill up the container); coffee cans went from 11.3 to 10.3 ounces; paper towel rolls went from 90 to 80 sheets per roll, etc.[29]

Modern systems of measurement are based on a mostly quantitative conception of measure. However, in older systems qualities were also considered in determining measures. The qualitative element in measurement was present, for example, in the measurement of land. Today, all acres are geometrically equal (4,840 square yards), no matter if they are used to measure land that is in a wasteland or in a wheatfield. Pre-metric measures of land, on the other hand, were defined partly by geometrical standards, but also by their quality and productivity. A common unit to measure land was determined by the "amount of seed." In some provinces of France, for example, the *setier* was a measure for dry products like corn and wheat (similar to the English bushel), but *setier* referred also to the necessary amount of land required to sow a *setier* of seed. A *setier* of fertile soil was geometrically smaller than a *setier* of less fertile soil. With this, "two plots of unequal area might thereby be 'equated', that is, shown to have virtually the same productive potential."[30]

Another category of land measure in pre-modern Europe was derived from the relation between time and labor, the "labor-time for plowing." In Spain and its colonies *huebra* was defined as the land that a single person could plow in one day, and *yugada* as the land that could be plowed in one day using a pair of mules or oxen. The geometrical extension was not the prevailing factor of measurement; what mattered most was the relation of labor to land.[31]

These practices of measurement survived until the origins of capitalism. In England, for example, the change occurred after Henry VIII expropriated and sold a huge portion of the land of the Catholic Church. It was also the period when the enclosures of land started. These processes changed the practices that gave sense to the old systems of measurement. Once land was exchanged for cash, its quality—its relation with "labor-time for plowing" and its ability to support people's life—became less important than how much rent it could produce.[32]

CAPITALISM AND MODERN SYSTEMS OF MEASUREMENT

Many of the economic settings in which pre-metric measures were used resemble what is known as the "bazaar economy"—a market where information about the price and quality of goods is poor, scarce, maldistributed, inefficiently communicated, and thus very valued because there is neither product standardization nor fixed standards of money and measures. It is a system where almost nothing is packaged or certified, everything is approximative, and the possibilities for bargaining along non-monetary dimensions are vast.³³ The primary problem of the participants in the bazaar economy is not balancing options but finding out what those options are.

In economic contexts of this nature, measures acquire meanings and functions that are usually part of communal, tacit understandings. Studying a local market in Haiti, anthropologist Sidney Mintz described a unit of measurement for oils and underlined the subtleties that give robustness to measures in economic life:

> The principal measure for oils is the "little bottle" (*ti-poba*) or, more specifically, the "little polish" (*ti-glòs*). Filled to the neck, the *ti-glòs* holds 2.84 fluid ounces. It is a short, stubby square bottle of heavy white (or rarely, blue) glass with a wide round neck and thick lip. These bottles originally held a liquid shoe polish, whence the name; the polish is no longer sold in Haiti in such bottles, but the measure remains. It is convenient because it fits quite neatly with three important larger units, its contents are a popular selling quantity, its design and massiveness make it durable and readily filled and emptied, and its form is so distinctive that no other bottle would ever be mistaken for it. A customer buying any oil can tell immediately that the seller is using genuine "little polish," and thereby knows exactly how much he is getting.³⁴

Despite the different degrees of precision of the multiple measures used in this market, buyers and sellers know what they are getting and have an informed comprehension of the relationship between quantity and price.³⁵

This homegrown way of setting units of measurement, however, has limitations when commercial circuits get larger, which makes folk understandings too opaque for buyers and sellers from other communities. In such cases, the useful aspects of local knowledge became hindrances for long-distance trade (and state formation). One of the problems with local measures is the incommensurability between their units, which renders economic calculations complicated or impossible. Converting units from one region to another requires the intervention of experts in reckoning and it is an obstacle for large-scale commercial networks. Not surprisingly, the development of capitalism (and nation-states) were determining factors in the unification and simplification of measures.³⁶

The utility of standardized weights and measures for economic life can be better appreciated when standardization does not exist. Take as an example this description by a chief trader in Madras, India who suffered mightily for the lack of transparent information:

> I never can tell what I am buying nor how much I am selling. My agents inform me that rice is at so much the *seer*, while in another quarter it is double that price. I take advantage of the opportunity, invest largely, and expect great profit. When the transaction is closed, I find I have lost greatly. The *seer* in the first place was perhaps less than half the size of that in the other. No two villages have the same measures, and to ensure success, I should need an agent in every place, each with infinite opportunity for deception.[37]

When systems of measurement are defective and not shared by all the parts comprised in an economic exchange, it is complicated to grasp accurately what and how much is being produced and traded. Problems like these tend to become even more prevalent when commerce involves people from more distant places. When trade was increasingly conducted across several countries or continents, merchants needed more and more time to figure out the names, magnitudes, and idiosyncrasies of measures used by both local producers and intermediaries. They needed to worry about the quality of the products they acquired (there was little product standardization and the qualities of commodities varied from year to year and from place to place) but also about how much they were buying. In eighteenth-century New York, for example, merchants complained frequently of the differences in measure and the imprecision of the terms used in contracts. One of them, who was ordering fifty barrels of rice from South Carolina to be sent to London, made this remark: "By barrels I suppose is meant your half Tierces of about 4 bushels each. We have a variety of Wooden Vessels called in general barrel to that with us the Term has no determinate meaning without prefixing the name of the Commodity they are to contain and generally used for, as Flour, Pork, Bread, etc. which are all different."[38]

Reacting to what already was a globally interconnected economy and to the volume of international trade, experts prepared manuals that provided information on weights, measures, and exchange rates between monies of different countries. The publication of these manuals represented a major development in history to reduce information costs.[39] One of these was Tomás Antonio de Marien y Arróspide's *Tratado general de monedas, pesas, medidas y cambios de todas las naciones* (General treaty on coins, weights, measures and exchanges of all nations).[40] A manual like this was conceived as a practical and time-saving device. However, its fact-filled 600 pages with an infinity of charts, equivalencies, and logarithms look, to today's readers, cumbersome and perplexing. Its complexity, however, is just a pale reflection of the intricacies of thousands of local metrological conditions that traversed the world economy of the time. Despite its limitations, this treaty was a colossal accomplishment that eased the toils of long-distance traders. But probably the most significant figure in this book packed with thousands of numbers is the one stamped on its cover indicating the year of publication, "MDCCLXXXIX" (1789), exactly the year when the history of metrology changed forever, the year that started the revolution that made Marien y Arróspide's treaty obsolete due to the invention of the decimal metric system.

Since the creation of the metric system in the French Revolution, the scientific and political elites conceived it to be a universal language and a tool of economic exchange. A single and rational system of measurement was an effective means to go beyond local markets and facilitate economic interconnectivity among the different parts of a country and among countries.[41]

Contrary to the variability, inexactness, and lack of standardization that characterized many pre-modern measures, modern systems of measurement are fixed, more exact, and globally standardized. The expansion of these systems and the development of the monetary economy made the practices of the bazaar economy clumsy and burdensome. Capitalism and industrialization (along with state centralization) paved the way for the spread of more homogenous and transparent systems of measurement. The level of exactness achieved by the metric system and its highly quantitative character meshed well with other developments in modern societies where objectivity and quantification are amply used.

With the development of capitalism a series of institutions started a process of standardization at the same time as they increased their presence in social life. Parallel to this, the conditions of large-scale capitalism involved the destruction of the obstacles to the free economic transfer of goods.[42] The standardization of currencies and measures, through conventions like the "gold standard" and the metric system, entailed the demolition of local and regional instruments in favor of national and international standards of value and measure.

THE UNIVERSAL INTERDEPENDENCE OF NATIONS

Today it is difficult to conceive a world without standardization in weights and measures, but the world prior to the nineteenth century differed greatly from ours. Merchants and travelers found different measures in every region and town. In a famous passage from the sixteenth century, the Swiss trader Andreas Ryff complained that while traveling in Germany he found that each community has its own money, laws, regulations, and weights and measures; in Baden alone, there were 112 different measures of length, 92 square measures, 65 dry measures, 163 measures for cereals, 125 for liquids, 63 for liquor, and 80 different pounds.[43] Today, one person can travel from one continent to another and always use the decimal metric system. The metric system bolstered the universal interdependence of nations and has become itself a symbol of how the intellectual creations of individual nations become common property.

In their vivid description of the effects of capitalism on history, portrayed in the *Communist Manifesto*, Karl Marx and Friedrich Engels underlined the global character of capital and its relentless consequences for social life. Capitalism undermined local relations and economies; it transformed not just Europe, but China, India, the Americas, and the rest of the world. Local traditions were replaced by cosmopolitan customs. Instead of regional and national seclusion and self-sufficiency, capitalism brought the interdependence of nations in material and spiritual production. "The intellectual creations of individual nations become common property. National one-sidedness and narrow-mindedness become more and more impossible."[44]

The material side of the "universal interdependence of nations" is well known: global circulation of commodities, international division of labor, and expansion of capital throughout the planet. This material intercourse required specific intellectual means, certain languages that could allow the existence of this global interaction. A universal world cannot rely on the narrow-mindedness of localities and nations. Capital demanded universal codes to facilitate the communication and coordination among groups, classes, nations, and regions that in the span of a few decades in the nineteenth century got closer to each other. The metric system was one of these universal languages. Its universality was founded on exactness, abstraction, and standardization. It was no accident that the first century fully ruled by capital was the period when the metric system started and consolidated its planetary expansion.

The third quarter of the nineteenth century was the first time when the world was actually "unified;" not with the capacity to work as a planetary unit in real-time, as today's economy, but intercontinental structures started to operate more fluidly than ever before.[45] The telegraph, the steamship, and the railway linked the most remote parts of the world. Goods and people moved massively from one corner of the globe to another. There was an unprecedented increase in trade, capital flows, and migratory movements.[46] It was an era of colossal development for the world-economy, an era of industry and capitalism, "of the social order it represented, of the ideas and beliefs which seemed to legitimate and ratify it: reason, science, progress, and liberalism."[47] These four ideas were frequently alluded to justify the introduction of the metric system in non-European territories. The period between 1850 and 1875 was critical for the metric system, and for similar international initiatives. Precisely in 1875 the Convention of the Metre took place in Paris, with seventeen nations signing the treaty. Other important global initiatives, like international standard time, the International Telegraph Union, the Universal Postal Union, and the International Meteorological Organization, were created in those years. The metric system was part of a series of projects that helped to build international and interlinguistic mechanisms of standardization and coordination when the world was becoming increasingly unified.[48]

There is a caveat in this characterization of the relationship between capitalism and the rationalization of measurement. That this process happened at large and in the long run does not mean that it was necessary, inevitable, or even easy.[49] It is not possible to equate the expansion of the metric system with the development of capitalism in a simplistic or mechanistic way. The rest of this chapter sheds light on why one of the most industrial and mercantile nations in the world—the United States—rejected the metric system, even though its use would have caused great convenience; and how those interested in inserting the United States economy in free international markets failed to make America a metric nation.

America: One Nation, One Dollar, Multiple Measures

The creation of national currencies and the establishment of standardized weights and measures are, usually, complementary processes. They are driven by the same necessities, maintained by similar institutions, and produce equivalent effects. A single national currency and a single system of measurement are both institutional answers to the needs brought by the unification of national markets and the growth of centralized governments. The state monopoly to coin money and the state monopoly on the means of measurement made monetary and metrological unifications possible; and these processes of homogenization enhanced national identity and state capabilities.

During the nineteenth century, nation builders saw territorial currency as an important instrument to promote national identity—something parallel to the creation of standardized national languages.[50] National currencies have the potential to be a unifying medium of communication. This function can be better appreciated by comparing the present monetary situation in most countries with their situation two hundred years ago, when two distinct problems hindered the communication in the marketplace. First, there was a dislocation between cities, regions, and provinces. Several economic entities coined their own money that circulated locally, and the value of coins and metals varied from one location to the next. Second, there was a marked dissociation between the money employed by upper and lower classes. Gold and silver were almost exclusively used by the most affluent, while the poor used copper coins and low-denomination tokens. The interchangeability between these two sets of coinage was not transparent. The establishment of national currencies, the manufacturing of large amounts of small change (that replaced privately issued tokens), and the banning of private, local, and foreign currencies helped to unify countries across geographic and class lines.[51] That was a direct result of states exercising their monopoly over the production and regulation of money.

Money has had a "qualitatively communistic character," derived from its ability to be a denominator for all values. Money reduces everything in the nation to a common pecuniary language. By making distinctions disappear, the national community could be imagined as leveled on "a kind of horizontal comradeship." With national currencies, the leveling function of money stops at the border, beyond which other currencies operate. In this way, the "leveling" and "communistic" characteristics of national currencies contribute to create a sense of national affiliation and a feeling of distinction from others nations.[52]

The establishment of homogenous systems of measurement has similar leveling, communistic, and unifying properties as the creation of national currencies. When all social classes and regions of a country share the same form of money and the same units of measurement, they become connected by common and homogeneous mediums of communication. In most countries, the metric system and a national homogenous currency were perceived as instruments to unify the country, eliminate local particularities, fashion a homogeneous national population, and create a single national space with no local differences or idiosyncrasies.

The similarities between these processes have some limitations, though. The massive reproduction of a "national imagery" in coins and bills—one of the main nationalistic properties of money—is difficult to replicate in weights and measures. Also, while national currencies underline differences between nations, the global character of the metric system does not allow that in the metrological realm (one can cross borders and even continents and keep using the same meters and kilograms). Nevertheless, it is interesting to notice how in countries like England, nationalistic sentiments nourished the opposition to both the euro and the metric system—a case that helps to confirm that national identity is entangled with both money and measures.[53] In the United States, England, and other places nationalist sentiments were strongly attached to customary measures, and for some social groups, the metric system was seen as a menace to American traditions and values, a sentiment that galvanized the groups that opposed metrication.

The practical difficulties in establishing a national currency and standardizing weights and measures were very similar. In the United States the situation of the monetary system was as knotty as that of weights and measures. The necessity to unify the country under a single currency was imperative. As John Quincy Adams recalled, "At the close of our war for independence, we found ourselves with four English words, pound, shilling, penny, and farthing, to signify all our moneys of account. But, though English words, they were not English things. They were nowhere sterling: and scarcely in any two states of the Union were they representatives of the same sums. It was a Babel of confusion by the use of four words."[54] These conditions persisted decades after independence.

The ultimate success of the United States federal government in establishing a territorial currency during the nineteenth century is evidence that it could muster resources—political, legal, and administrative—to carry out an effective, large-scale centralizing plan. The federal government adopted Jefferson's decimal scheme, created the United States Mint (1792), banned the production and circulation of locally and privately issued money, established in 1865 an office to suppress counterfeiting (the Secret Service Division of the Department of the Treasury), and founded a central banking system.[55] A monopoly on the regulation of the monetary system and on the creation and coining of money was effectively instituted—but that was not the case for the means of measurement.

The process to create a single, standardized national money in America was effective, but it was not easy. It was a stiff challenge. As Viviana Zelizer shows, in the nineteenth century there were more than 5,000 varieties of state banks circulating; several states produced their own gold coins; tradesmen employed privately issued tokens; stores, businesses, and churches issued coins and paper notes. The proliferation of private forms of money was partly a consequence of the shortages of coins (particularly "small change"). Most private monies, though, were local and slowed down the arrival of a truly national economy. An additional problem was that some of the non-official tokens and notes did not even follow the decimal subdivisions of the dollar; some of these retained binary denominations (like tokens with a value of 6¼¢, or what is the same one-sixteenth of a dollar). For the American dollar to become what it is today, the federal government needed to make all private production of monies illegal. It also broadened the definition of counterfeiting to crack down on the production of private currencies. It went as far as prohibiting the practice of inscribing coins with personal or sentimental messages. They also needed to secure a stable amount of coins to avoid the lack of small-denomination coins. It was a large enterprise of prohibition and production. It took decades, but the American state achieved "a significant degree of standardization and monopolization in the physical form of legal tender."[56]

All countries with a functional national currency had to go through similar pains. What was peculiar in the case of the United States is that its regulation of money did not run parallel to a reform of weights and measures. In most national cases, the unification of currency and the standardization of weights and measures (usually through the introduction of the metric system) were part of a two-fold process. In America, monetary homogenization happened earlier and more effectively than national metrological regularization.

What most people ignore is that at the same time as the United States was creating a real national money, some plans were being drafted to create a "universal currency."

The Metric System and Universal Currency

At the heart of the age of capital, during the 1850s and 1860s, the idea of "universal coinage unification" seemed promising and was widely discussed in Europe and the United States. In the Universal Expositions and the International Statistical Congresses—meetings that helped the cause of metrication around the world—numerous thinkers and policy makers from various countries and professional backgrounds pushed forward the idea of an international *metric* gold coinage.[57] These discussions took place in the Fifth International Statistical Congress (Berlin, 1863), the Paris Universal Exposition of 1867, and the International Monetary Conference of 1867 (which gathered government representatives from nineteen European countries and the United States).[58]

Numerous experts held the conviction that a worldwide standard coinage was possible and needed. The ideological underpinning of this project was the ambition to secure a uniform standard of value and exchange, create the conditions for a larger international market, and facilitate the free circulation of goods and capital. The discussions on the international coin, as those of the metric system, were framed in terms of a "universal language"—in this case a language of "measure of value."[59]

The rhetoric used to defend the plans for international coinage was almost identical to the one used to favor the metric system as an international measurement system. For some, the two projects were intertwined. An early example of this marriage of metric and monetary internationalism was the 1858 pamphlet *Universal Currency*, by T. A. Tefft (state commissioner of Industrial Art and Education for Rhode Island), which was both a plan for a common decimal currency in the United States, England, and France, and a guide to rendering the metric system of weights and measures "more simple and popular." Like other advocates of these plans, Tefft's was ultimately a plea for liberalization, commerce, and "civilization." The changing of money, he considered, was burdensome and a "senseless tax" on commerce. The material trading of the age demanded that the "feudal relic of many units of money" should be abandoned and replaced by one standard of coinage used in every country "traversed by the steam-engine."[60]

A decade later, amid the diplomatic negotiations in France on international currency, US senator John Sherman (chairman of the Committee on Finance and future Secretary of Finance and Secretary of State) wrote, in a similar vein, to Samuel Ruggles, the US Commissioner to the Paris Exposition, to endorse his efforts to secure the adoption of the metric system. The tendency of the age, he said, was to break down needless restrictions upon "commercial intercourse." People of different nations learned to respect each other as they found that their differences were the effect of social and local customs. "I trust," he concluded, "that the industrial commission [of the universal exposition] will enable the world to compute the value of all productions by the same standard, to measure by the same yard or meter, and weigh by the same scales."[61]

Shortly later, in a report to the Senate on international coinage, Sherman made the case for uniformity in the coinage of the United States and other countries framing the issue in the same liberal and economic laissez-faire spirit:

> The inconvenience of different standards of value arises mainly in foreign commerce, in the exchange of commodities among nations. The intercourse between modern Christian nations is now more intimate and exchange more rapid than it was between provinces of the same

country two hundred years ago.... Every advance towards a free exchange of commodities is an advance in civilization. Every obstruction to a free exchange is born of the same narrow despotic spirit which planted castles upon the Rhine to plunder peaceful commerce. Every obstruction to commerce is a tax upon consumption; every facility to a free exchange cheapens commodities, increases trade and production, and promotes civilization. Nothing is worse than sectionalism within a nation, and nothing is better for the peace of nations than unrestricted freedom of intercourse and commerce with each other. No single measure will tend in this direction more than the adoption of a fixed international standard of value by which all products may be measured, and in conformity with which the coin of a country may go with its flag into every sea and buy the products of every nation without being disconcerted by the money changes.[62]

However, even if the majority of the participants in the conversations on universal currency were inspired by a similar vision, they did not agree on the means to fulfill it. They faced a similar predicament than those interested in the establishment of a global system of measurement: would it be desirable to employ an already existing currency as the exclusive international standard or would it be better to design a new system and start anew? They were divided on the issue. Some preferred using as standard a currency already in use, mainly the gold franc; others favored a start-afresh solution, arguing that it was better to create a new coin with a fixed amount of pure metal defined with a round metric quantity.

Michael Chevalier, from France, and others pushed for using the metric system as the base for the new international currency. Their plan consisted of a currency with decimalized subdivisions and gold coins weighed in a round number of grams. They wanted to fix the gold content of the proposed new coin at exactly ten grams, or a decagram—thus this group was known as the "decagramists." Their plan was seconded in the United States by congressman William D. Kelley, chairman of the House Committee on Coinage, Weights and Measures, who was a radical northern Republican, like many other influential political figures in Washington that favored the metric system.

Decagramists insisted that the metric system of weights and measures offered the opportunity to unify coinage on a "scientific basis." Scientists, politicians, academics, and journalists who backed "metric currency" were seduced by the "theoretical neatness of an international metric gold coinage," and saw it as a way to interlink universal standards of value, weights, and measures.[63]

However, most of the people in the United States who got involved in the debate preferred employing an existing system as the international standard. Ruggles was the leading voice in this camp, and the federal government gave him support to push his agenda in the international meetings. For this group, the scientific elegance sought by decagramists was too idealistic and the option of a new currency, different from all existing monetary systems, was unrealistic. The practical thing to do, they thought, was to adopt an existing currency as the base for the international standard. Ruggles wanted to equate the American dollar and the British pound sterling to twenty-five gold francs, which would have required England and the United States to slightly modify the weight of their currencies, and for France to abandon silver. This turned the discussions to the problem of bimetallism versus gold monometallism.[64]

Ruggles's plan of using the French franc as the basis for uniform coinage for all nations and the abolition of silver as a monetary standard (plus universal finesses of .900 and the decimal subdivisions), killed the decagramists' hopes. The problem for the metric camp with this initiative was that contrary to France's silver coins, which were measured by exact gram units, the gold franc was not; thus, the demand to follow the French gold currency ruined the possibility of an international metric-based monetary system.

Ultimately, conversations in the Monetary Conference of 1867 did not amount to much. Delegates of many countries were reluctant to commit to any binding agreement. Good intentions were expressed and future conversations scheduled. The topic of universal currency was kept alive in the press, specialized literature, and other international meetings; but it never regained its full impetus.[65] The Franco-Prussian war of 1870 destroyed the spirit of internationalism that animated the Paris meetings. The demise of Napoleon III, who was very interested in exporting the franc and the meter, and the changing balance of power in Europe stripped France of its ability to set conditions in international agreements, which damaged the prospectus of metric currency even further.

An interesting legacy of the intersection between currency and the metric system in the United States was the design of the five-cent nickel coin. Introduced in 1866 (the same year as the Metric Act), nickels were specified to weigh exactly five grams, one gram per every cent. Today the 5-cent nickel coin still weights five grams. In 1873, silver coins of the United States of smaller denominations than one dollar were also given metric weights, with the 10-, 25- and 50-cent silver pieces weighing one gram for each four cents. The potential convenience of this arrangement for everyday life was noted by some commentators who observed that with the nickels "each man carries his own letter-weights in his pocket," and that "two 10-cent pieces will balance one nickel, all of these coins may be conveniently used as weights to check metric scales."[66]

The connection between an international currency and the desire to introduce metric weights and measures in the United States was evident in the affiliations of some key actors involved in the universal currency debates. For example, congressman John Kasson and senator Charles Sumner—the driving forces behind the 1866 metric legislation—supported international coinage unification and the participation of the United States in the 1867 Paris Monetary Conference. More telling, in the 1870s Kasson, along with Samuel Ruggles, and E. B. Elliott (a government actuary in the Treasury Department who prepared a report on metric currency) were members of the first openly pro-metric association in the country, Melvil Dewey's American Metric Bureau—with Elliott working in the Bureau's Committee on International Coinage.[67]

Frederick Barnard, president of the American Metrological Society (the other significant nineteenth-century pro-metric body of the time), published at least four pamphlets and papers on international coinage and the possibility of an invariable standard of value (at the same time as he was leading the cause of the metric system and working with other members of the AMS to develop the plan that culminated in international standard time zones).[68] This interest of the Metrological Society was so explicit that its first objective—as stated in its constitution—was "To Improve existing system of weights, measures, and moneys, and to bring them into relations of simple commensurability with each other."[69]

Later some politicians insisted on using the metric system to push the United States to take the lead in the issue of international currency. In 1896 congressman Charles W. Stone (who had sponsored a bill for the adoption of the metric system) published in *The North American Review* an article on "A Common Coinage for All Nations." He described how the probable adoption, in the near future, of the metric system by the United States, England, and Russia will make it universal and save much of the "friction and loss which have heretofore been the outcome of diverse systems." After suggesting that the dollar may serve as the basis for new international money, he pleaded for a currency that would "change value at no national frontier, that would defy the exactions of brokers and money-changers, that would carry the badge of civilized life into every clime."[70]

There is a lesson from this episode. The case of international currency illuminates an important aspect of the metric system's global triumph. Since there was no other "modern" or "scientific" plan able to seriously challenge the metric system, people in different countries and from different professions interested in a radical metrological reform were galvanized around a single project, instead of being divided by feuds about the virtues and defects of multiple hypothetical plans. For better or worse, the metric system was what they had, and this lack of realistic options gave them greater cohesion. When the suggestive but controversial idea of an international currency was discussed, experts and politicians were divided among many possible ways to materialize that idea, which weakened their position significantly.

But despite the failure to bring to fruition metric coinage (or any other form of universal currency), nineteenth-century metric devotees in America had another chance to espouse metric measures with liberal economic interests, not in a "universal" scale but a continental one.

The Americas and the Fight for a Language of Commerce

The First International Conference of American States of 1889–1890 launched a period of intense pro-metric activity in the United States and Latin America. The plan for a conference of American countries started to take form in 1881 when secretary of State James Blaine conceived the plan to invite representatives from Latin American countries to a hemispheric conference. Invitations were dispatched that year, but the assassination of President James Garfield and the resignation of Blaine halted the project, which was not revived until 1888, and the First International Conference of American States took place from November 1889 to April 1890, in Washington D.C. It was the United States's first experience hosting an international assembly of governments.[71]

The ultimate interest of the United States to arrange this meeting was to strengthen the competitive position of American businesses in Latin America. This interest was partly prompted by the steady increase in commerce between the United States and Latin America. From 1870 to 1900 that trade doubled in volume. In the 1870s Latin America consumed more than 8 percent of US exports and supplied more than 30 percent of the imports; by the early 1890s US exports to the southern part of the continent reached more than 10 percent. However, European rivals had outperformed their North American counterparts in their trade with Latin America during those years. With the Pan-American conference, Blaine wanted to displace European competitors by building a stronger commercial bond with the rest of the continent.[72]

The objectives of the Conference, which gathered a total of eighteen countries, were "preserving peace on the continent," promoting prosperity, establishing a customs union to facilitate trade, improving communication between ports, protecting patents and copyrights, sharing forms of classification and valuation of merchandise, building an inter-American train, creating an international American bank, establishing an office to collect and distribute information of interest to all the countries, adopting a common silver currency (to be issued by each government but accepted as legal tender in commercial transactions across nations), and a uniform system of weights and measures.[73]

The idea of a common currency and common weights and measures resembled, in a continental context, the 1867 Paris monetary conference. Since the plans of the conference were announced, observers saw the opportunity to finish there what was left incomplete in the Paris meetings. As the *Evening Transcript* reported:

> The exigencies of commerce, the adaptation of scientific discoveries to meet the wants of an advancing civilization among peoples newly made neighbors . . . clearly indicate the need of uniformity in the modes of doing business, in estimating the quantities and values of exchangeable

commodities, those prime factors in the intercourse of mankind. The calling of the [Americas] conference evidently means for the people of the United States the early introduction and general use of the metric system. The time is propitious. The relative values of gold and silver have so changed everywhere as to call for a revision of standard units of value. It will be in the power of this conference to do what the several monetary conferences assembled in European capitals, from time to time, have failed to accomplish; namely, the unification of money for the civilized world.[74]

The conference, however, did not have the power to set any resolution on monetary, or any other, policy. It could only make recommendations that would not bind any participant.[75] But the editorial was correct in assuming that when the plan for the conference called for a common system of weights and measures in the Americas, it meant the adoption of the metric system.

The Conference appointed a Committee of Weights and Measures, headed by Jacinto Castellanos, a diplomat from El Salvador, and Clement Studebaker, an industrialist from Indiana known as the world's largest maker of carriages and wagons.[76] The Mexican delegate, Matías Romero, played a significant role in the discussions of the committee.[77] Romero, an experienced Ambassador in Washington and a two-time Minister of Finance, was a firm believer in the principles of economic liberalism.[78] He was akin to the overall principles of the Conference and supported, like other representatives, the idea that metric measures should become the exclusive system in the continent. On January 24, 1890, the Committee concluded its works with this single recommendation: "The international American Conference recommends the adoption of the metrical decimal system to the nations here represented which have not already accepted it."[79]

The resolution had two objectives. First, to pressure the countries that have not fully adopted the metric system yet—mainly the United States—to make metric the only legal system. Second, to compel the countries that had formally adopted the system to enforce it, because in most Latin American countries the metric legislation was merely nominal. It was this ambitious plan of continental free markets and economic integration that gave several Latin American governments the determination to finally introduce the metric system in practice. They did it with the expectation that the United States, as the Conference host and signer of the agreements, would also move forward with its plans for metrication. We know that the first objective of the recommendation of the Committee of Weights and Measures—the adoption of the metric system to the countries in the Americas that had not already accepted it—did not prosper. In a tragic irony for metric supporters, the country that organized a conference that did so much for metrication in the Western hemisphere became the sole nation in the continent that failed to go metric.

This fiasco was not for lack of trying by American pro-metric groups. The conference was the starting point of a large campaign in favor of the metric system in the United States, and Latin America was at the center of these plans. The weights and measures report of the Pan-American conference kickstarted a period in which the adoption of the metric system as the exclusive system of measurement in the United States looked, if not imminent, at least highly probable. In the years after 1890, there was considerable movement in favor of metrication, inside and outside Congress.

After the conference was finished, Blaine sent a letter to President Benjamin Harrison, with the proceedings of the conference, a proposal to adopt the metric system in the United States Customs Service, and the draft of a bill to be considered by Congress.[80] This was the first of many bills regarding metrication discussed by legislators in the following years.

In 1896 a metric bill received considerable support in the House of Representatives. Key members of the Committee on Coinage, Weights, and Measures, like Charles W. Stone, supported metrication unambiguously (he had also sponsored the idea of an international metric currency).[81] The report prepared by the Committee highlighted that Latin America had already gone metric and asked "Why should the United States alone of all the republics of the Western Continent persist in its adherence to a cumbrous and antiquated system, if it may be called a system, of weights and measures, and thus let much of the commerce of its sister republics which it should attract and enjoy drift to the metric-using nations of Europe?"[82] The law for the compulsory and exclusive use of metric measures proposed by the Committee got very close to being accepted.[83]

Sensing that victory was within reach, pro-metric organizations launched numerous efforts to secure favorable legislation. In January 1898, in the Third Annual Convention of the National Association of Manufacturers (NAM), industrialist Albert Herbert read the report of the Association's Committee on Language, Weights and Measures, a lengthy and spirited document that asked for Congress to make the metric system the only legal measures in the United States.[84] The report maintained that the metric system was an instrument to foster universality and compared it to other globally accepted codes: "a very important part of the language of commerce is what is properly called the 'language of quantity,' or the terms and divisions employed in determining weight and measure.... The Arabic numerals are a labor-saving, trade-extending tool, and as a language of commerce it is today the only language that is absolutely universal." The report finalized portraying an enthusiastic future of progress in which the metric system would be the norm and old ideas—as customary measures—would be thrown away:

> We live in an age of marvelous progress. We have now machines which annihilate space and enable us to see and hear and talk over the distance of half a continent, which reproduces the speech of yesterday in its exact tones, and transmit to us with the actual photograph of the speaker if it is desired. Man is entering the twentieth century, equipped with inventions which will compel him to overcome his present ultra conservatism. [...] There is the necessity to find new tools for our new and greater needs—eliminating all waste and discarding the old framework of past customs as we discard the skins and blankets of the patriarch for the modern clothes of the citizen.[85]

Attached to the report was a letter by steel tycoon Andrew Carnegie, who financed the activity of pro-metric groups, like the AMB.[86] The metric system, he declared, "is one of the steps forward which the Anglo-Saxon race is bound to take sooner or later. Our present system inherited from Britain is unworthy of an intelligent nation of today.

The advantage we possess over Britain in our decimal dollar system as compared with their pounds, shillings, and pence, would be fully equaled by the adoption of a metric system of weights and measures."[87] Despite some objections to the idea of industrialists binding themselves to the metric system by the new law and risking to remake "every tool in the hands of American manufacturers," the report was accepted by the NAM.[88]

The federal government was also showing an increased interest in the issues of standardization. In 1901 the government established the National Bureau of Standards (NBS). Its director, Samuel Stratton, was an unapologetic supporter of the idea of the United States going metric. He wrote articles, participated in congressional hearings, and collaborated with pro-metric organizations.[89] The same year the National Board of Trade—an organization interested in securing uniform trade practices—passed a resolution backing a pro-metric bill before Congress, arguing the use of weights and measures is "universal in all civilization," and uniformity was important for economizing time and calculations in "every branch and line of industry," adding that "the decimal plan, which is the basis of the metric system, has become the accepted and established form of notation throughout the entire civilized world."[90]

In 1902 the House Committee on Coinage, Weights, and Measures conducted hearings on the metric system. The inventor Elihu Thomson, from General Electric, was one of many supporters of the metric system[91]—and it was in a supplemental hearing where Lord Kelvin pleaded American lawmakers to pass a favorable metric legislation.[92]

The pro-metric coalition (scientists, inventors, educators, public officials, legislators, and traders) portrayed a vision of universality in which humanity, in a modern world, would share common codes and languages. The metric system was compared to Hindu-Arabic numerals, the alphabet, a universal calendar, musical notation, international auxiliary languages like Esperanto, and sometimes even with a single universal language and a single currency. The more utilitarian objective of propelling international trade was adorned with universalistic rhetoric. At this particular point, the pro-metric movement had considerable momentum in its favor and few people were showing resistance. In newspapers, journals, and congressional hearings metric advocates visibly outnumbered their counterparts.

ASSEMBLING THE ANTI-METRIC CAMP

Despite appearances, not everybody was warm to the idea of metrication in America. Experts who opposed a change in measurement legislation started to build their own platform. In 1902 the tide started to turn. Groups whose immediate economic interests could have been affected by a mandatory adoption of the metric system mobilized their resources to oppose any such resolution. In the meeting of the American Society of Mechanical Engineers some members raised alarm about a government-mandated change in the measurement system.[93] Frederick A. Halsey (1856–1935), a mechanical engineer from New York, and Samuel S. Dale (1859–1940), a textile industrialist from Boston, presented strong opinions warning about the dangers of adopting the metric system. These two men became the voice of the anti-metric movement in America for decades to come.[94]

Halsey was an instrumental piece in the anti-metric movement due to his professional reputation and ability to articulate an opposition to metrication on practical and technical grounds (rather than in nationalistic or religious terms). Trained as a mechanical engineer at Cornell University, he gained fame for his inventions (like the "slugger" rock drill) and for his premium plan of paying for labor, also known as the "Halsey Premium Plan" (a widely recognized method, contemporary of Frederick Taylor's work—another mechanical engineer—on industrial efficiency).[95] He was the first recipient of the gold medal of the American Society of Mechanical Engineers (in 1923) for his contributions to industrial economics. Halsey dedicated a good part of his life to oppose metrication, wrote profusely about the topic (both in technical journals and the popular press), and became the most visible anti-metric figure in twentieth-century America.[96] A lesser-known figure, Dale was editor of *Textile World*, and later owner of the magazine *Textiles*. As Halsey, he spent a great deal of time and energy during half of his life to the defense of customary weights and measures. He penned a deluge of articles, pamphlets, and letters, and was a constant presence in conferences, meetings, and public discussions.[97] Along with his feverish activity against the metric system, he participated in the "tariff question," advocating for the protection of the American textile industry and working with producers, manufacturers, and politicians to impose duties on imported wool.

In a private letter to the president of the Bank of North America, Dale articulated his social and economic vision with a clarity and cohesiveness that was not always present in his publications; but summarizes the position of many metric opponents:

> the defense of our weights and measures has become part of that greater problem involved in the defense of all of our established institutions. That time has come for all of us to stand by the United States of America. That is a duty we owe to those who come after us in return for what others did for us before we saw the light of day.... Protection for our form of government, our language, law, weights and measures and our right to consume what we produce and produce what we consume. That protection is our first duty.[98]

The main positions in the debate were clear. Pro-metric groups wanted a strict law that would mandate metrication; they also sought to remove trade barriers and operate in free international markets—these interests were sublimated in a vision of civilization, progress, and cosmopolitanism. Anti-metric groups rejected compulsory metrication, denounced unjustified state intervention in citizens' lives, opposed the concentration of authority in bureaucratic bodies, and sponsored economic protectionism—they wrapped these interests in a nationalistic discourse.

Following Herbert Spencer's example, Halsey and Dale elaborated their main arguments questioning the alleged practical advantages of metrication and the technical superiority of metric over customary units. Halsey and Dale certainly draw on nationalistic motifs to attack the metric system, but crude nationalism was not at the core of their arguments. They were sophisticated "practical men" with the necessary expertise to challenge pro-metric scientists in their own terrain.

Spencer had started a "rational opposition" to the metric system and Halsey and Dale followed through.[99] They articulated their main ideas in what may be the single most influential book in the history of metrication in America, *The Metric Fallacy*.[100] The book revolved around the argument that, based on practical and economic grounds, the transition to the metric system was undesirable. Halsey argued that changing a system of weights and measures was enormously difficult and the transition would never be fully completed. He emphasized that a change of system represented the destruction of the existing mechanical standards; that foreign commerce does not require the adoption of a new system in manufacture; that for industrial processes the English units were better suited than the metric ones; and that England and the United States have "the simplest and the most uniform system of weights and measures of any country the world."[101] *The Metric Fallacy* questioned the mere possibility of implementing a universal system: "The experience of a century has shown that the idea of a universal system of weights and measures is an 'iridescent dream.' We must make up our minds to get along with diverse systems of weights and measures in the world as we do with diverse languages and systems of currency." As part of their reaction against the introduction of the metric system, Halsey and others developed a defense of diversity as a positive value, not only in the somewhat limited field of technical standards, but in broader cultural aspects; they critiqued, for example, what they called "over standardization" and applauded the defense the of traditional national languages.[102]

The Metric Fallacy was widely discussed. Among the commentators on the book was Charles S. Peirce.[103] He agreed with the authors that the leaders of the metric cause were a reduced group of university people and that metric units were only important just for "a small number of scientific men." American legislators, Peirce said, ought not hear those opinions, and should rather listen to "practical men." He prognosticated a gloomy scenario if customary measures were discarded. The abandonment in American manufactures "of a unit of the international importance of the English inch would bring disaster upon the country." If a law prevented manufacturers from making equipment based on the inch, "the position of our machinery abroad would be irrecoverably surrendered." He concluded that the greatest factor for the attainment of the world supremacy of the American machinery, more than American ingenuity and the American ability for simplification, was "the thorough systematization of American machinery . . . by which the dimensions of the different parts were brought into simple relations to the inch, so that those parts could readily be replaced—a feature only tardily copied by Europe."[104]

The position of Halsey, Dale, Peirce, and others was part of a nascent opposition to the metric system based on economic arguments that was not restricted to mechanical engineers. Many of them asked for an active role of the state in imposing trade tariffs to protect American industries from international competition; but they had a laissez-faire attitude toward metrological standardization, rejecting the intervention of the state in the regulation of weights and measures.

Other groups and associations had started to express similar opinions. The National Machine Tool Builders Association passed a resolution protesting the prospect of metrication in America, arguing that "the adoption of the metric system would entail an enormous first cost of new equipment to conform to the new standards and a constant increased cost in the maintenance of a double standard for repairs and renewals, and a consequent increased cost of the product to the consumer."[105] That same year the NAM, forgetting its previous endorsement of the metric system, adopted a mildly negative resolution wanting to stop any immediate change in the country's weights and measures. They made some estimations on how the compulsory adoption of the metric system would affect the manufacturers' interests: "one-third who are exporters to European countries and dependencies would be benefited; one third who do business in this country and all other countries would neither be benefited nor greatly injured; one third who do business in this country and in England and dependencies would be seriously injured. For all this the expense and incontinence would be very great."[106] The NAM concluded that since the metric system was already legalized for the use of those who find it profitable, it was better if no further action was taken at that time.

In 1904 the NAM completed its U-turn from metric enthusiasts to hostile metric adversaries and asked Halsey to represent them in a Congressional hearing, where he delivered a forceful testimony against any change in metrological legislation.[107] For the anti-metric movement at that time, Halsey and Dale articulated the ideas to oppose metrication, while the NAM provided the political muscle to obstruct metric initiatives in Congress.

It is relatively easy to understand why manufacturers opposed mandatory metrication; it was they who faced the financial burden of retooling and acquiring new equipment if metric legislation was passed.[108] Why mechanical engineers joined manufacturers in this battle is less evident.[109] Their opposition should be framed, first, in the context of the defense by

American industries of their right to set themselves their own technical standards without government intromission; and second, in the federal government's reluctance to intervene decisively in the matter, in one direction or the other—avoidance that was continuous even in cases where national coordination seemed highly desirable.

The NAM used its political influences to oust key pro-metric members of the Committee on Coinage, Weights and Measures, starting with its chairman, James Southard, from Ohio, who lost his nomination in the Republican Party.[110] Lucius Littauer and Solomon Dresser, other active metric proponents in the committee, abandoned Congress as well. Referring to these events, Dale wrote to Marshall Cushing, NAM's secretary, "I think I see in these events the fine Italian hand of Cushing." Cushing responded: "It is, as I can say to you only in the strictest confidence, of course, that I am closely in touch with the [Republican] party managers, and I think that we have reason to fear nothing in the future if political pressure can be made to enter into the situation at all. In other words, we are entitled to feel good and yet to keep our ammunition dry."[111] Cushing was also in close contact with Halsey in organizing the opposition against the metric movement. He was instrumental in the anti-metric campaign, guiding where, when, and against whom actions should be taken.

The combined actions of manufacturers and mechanical engineers to oppose metric legislation during the first two decades of the twentieth century were effective. They halted the momentum in favor of metrication and made the topic a contentious political issue, something that dissuaded some politicians and public servants from showing support for metric initiatives.

Stubborn Factions

Since midway through the first decade of the twentieth century, a landscape of forces battling in the metric war was set and it remained stable for the next decades. Exporters, scientists, and educators favored the metric system. Manufacturers and mechanical engineers opposed it. Congress was prone to inaction. Federal agencies were divided, with some (particularly in the NBS) supporting metric legislation; but others were reluctant to support compulsory legislation despite their interest in accomplishing standardization (like the Secretary of Commerce and members of the military). The general public was indifferent and uninformed about what the metric system was, what was useful for, and which were the implications of adopting it.

Eventually, these camps coalesced into formal associations. Halsey and Dale, with financial backing from Brown & Sharpe Manufacturing Co., established in 1916 the American Institute of Weights and Measures (AIWM), aimed to stop the metric advance and to perfect customary measures. Also that year, during the annual meeting of the American Association for the Advancement of Science, a group of scientists and educators held the first meeting of the American Metric Association (AMA)—with Italian pedagogue Maria Montessori as a special speaker.[112] The first president of the AMA was George Kunz, and among its members was Samuel Stratton, director of the NBS.

In 1919 Albert Herbert—who twenty years earlier, with Andrew Carnegie, championed the metric system in the NAM—hired Aubrey Drury, a professional advertiser, to operate the World Trade Club of San Francisco (later called the All-American Standards Council). The Club promoted complete national metrication, with special emphasis on facilitating commerce with countries from Latin America and the Pacific, and making American companies more competitive in those markets vis-à-vis European industrial powers like Germany (Herbert also paid for a full-time representative in Washington to lobby the cause). Even though he had no expertise in metrology, Drury became a tireless worker for the metric cause; besides his role as director of the World Trade Club, he was vice-president of the AMA. Despite the impact of his work, he is little known, and his papers and correspondence have been scarcely studied. It is difficult to estimate how many metric articles and news bulletins he wrote, as many of his pieces were unsigned, but he should be counted among the most copious writers on the history of the metric system, in the United States or otherwise.[113]

ENGINEERING GENEALOGIES AND CARTOGRAPHIES

The AIWM and the World Trade Club wrestled in newspapers, popular magazines, technical journals, political campaigns, Congressional

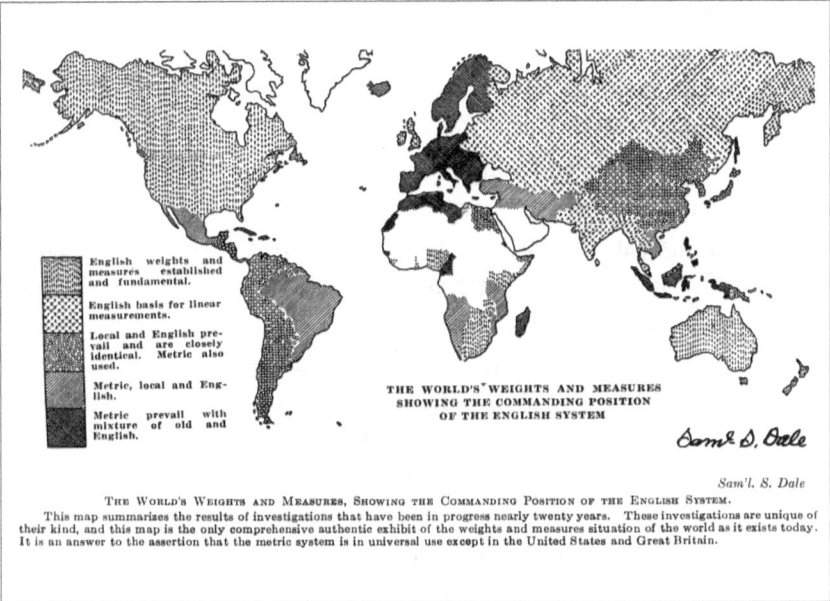

Figure 4.2. World Map by Samuel S. Dale showing the "commanding position of the English system" (1920). Source: Frederick A. Halsey, *The Metric Fallacy* (1920).

hearings, and international conferences.[114] The topics discussed were varied, from the number of countries that were already metricated to the "real" origin of different measurement systems.

One of their most animated subjects of controversy was the extent of the metric diffusion in the world. Both camps conducted investigations to elucidate whether the metric system was actually used around the world or if it was just a formal requirement that people ignored in practice. Their results were presented in contrasting world maps portraying the metrological state of the world, which were distributed in press briefings and other publicity materials. The anti-metric map, made by Dale in 1920 (fig. 4.2), emphasized what he called the "commanding position of the English system." It was divided into five categories. United States, Canada, Great Britain, Ireland, Australia, New Zealand, South Africa, Kenya, and Nigeria were all described as countries where English Weights and Measures were "established and fundamental." Mexico and Brazil were classified as nations where the metric system was mixed with English and local units; the rest of Latin America was marked as a place where "local and

English prevail and are closely identical. Metric also used." Russia, India, and Thailand were presented as "English basis for linear measurements." And finally, a category of countries where "metric prevail with mixture of old and English," which only existed in continental Europe plus a couple of countries in Africa and Southeast Asia.

In 1922 the pro-metric group published its own map, prepared by Drury (fig. 4.3). It intended to show the "well-nigh worldwide use of metric units." It had four categories. The United States and Canada were countries where the use the metric system was used "in many important and practical fields," such as science, medicine, coinage, education, trade, industry, etc. South Africa, India, Australia, and New Zealand were grouped as countries that "have officially petitioned the central British government to adopt the world metric units thruout Britannia." Great Britain and Ireland were described as "countries in transition, now making a declaration of independence against the obsolete German jumble of weights and measures." The rest of the world was depicted as nations that "use meter-liter-kilogram more or less exclusively."

Both maps were a mixture of random facts and half-truths. Dale's map underestimated the momentum and penetration that the metric had already gained worldwide, but it correctly emphasized that the traditional local measures were widely used in metric countries (it should be remembered that it takes several decades for a country to achieve full metrication, that is, when the majority of the people use the metric system in the majority of their measurement operations). Drury's map pointed correctly that the metric system was dominating global metrology, but misjudged how entrenched customary measures were rooted in the United States and England.

A colorful incident of the fight between the AIWM and the World Trade Club was a series of episodes of genealogical imagination (although not as dramatic as pyramid metrology) in which the origins of both metric and customary measures were blatantly reinvented.[115] At the time, during and after World War I, anti-German sentiments were patent in the United States and every camp sought to Germanize its opponent and, conversely, tried to connect some American historical roots with their preferred measures.

Attempting to make the metric units more amicable for the American public and elude the accusations of being part of "foreign attacks on

Searching for a Perfect Language for Commerce   195

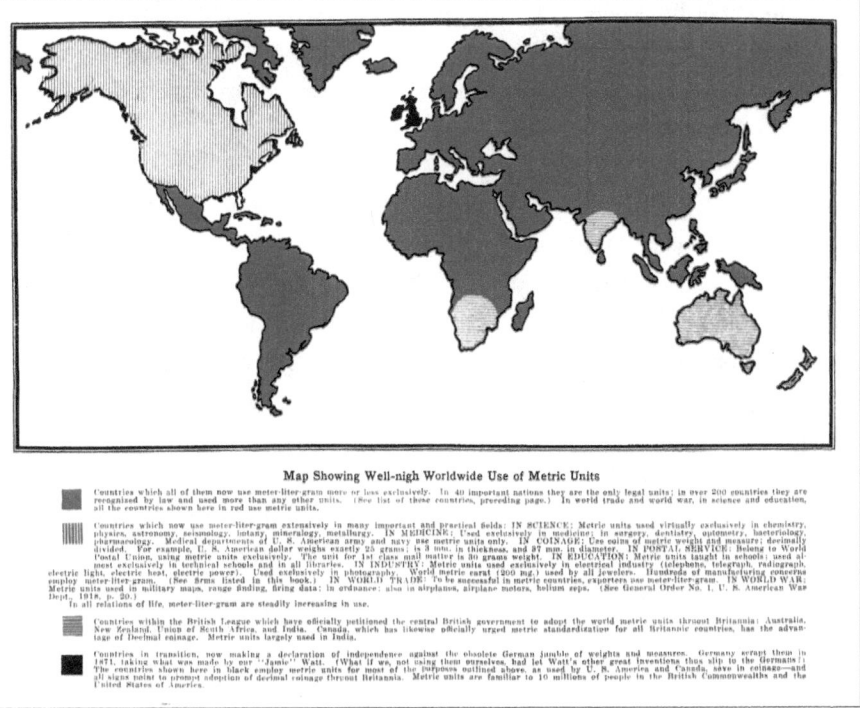

Figure 4.3. World Map by Aubrey Drury, showing the "well-nigh worldwide use of metric units" (1922). Source: Aubrey Drury, *World Metric Standardization* (1922).

American weights and measures," Drury and his colleagues tried to un-French and Anglicize the metric system. To that end, they exalted the figure of Scottish inventor James Watt and attributed him a preponderant role in the inception of the metric system. They argued that the creation of a metric system was first conceived not by French philosophers during the Revolution but by Watt:

> In the United States we commonly think of the decimal metric system standards as "foreign." It is interesting to note that they are of Anglo-Saxon origin. . . . [In 1783 Watt proposed] a simple decimal system whereby the measures of all countries might be standardized on the same basis. The advantages of this suggestion were apparent not only to the scientists of the day, but to the merchants and shippers as well. After Watt had laid his scheme before the leading men of France, in 1786, a world conference on measures was proposed. Britain foolishly

allowed feelings of petty hostility to keep her away from the conference. Thus continental Europe became "metric," while *Anglo-Saxon countries, to whom metric standards were properly a heritage,* retained their old confusion.[116]

In a subsequent move, they assured that English weights and measures were of German origin, and that the Germans forced those standards on the British centuries ago, who in turn landed them in America. In the *New York Times*, W. E. Hague, a member of the World Trade Club, argued that British coinage and American weights and measures "come from the old German Osterling Hanseatic League." More remarkable, Hague considered, "is that America and Britannia continue to use these old German tools after Germany herself has scrapped them. In 1871 she adopted the metric system, the invention of a Briton, James Watt."[117]

To counter Drury's one-two punch of anglicizing the metric system and Germanizing English measures, Halsey and Dale made their own rewriting of metrological history. They tried to show that the real German influence was in the metric camp. Dale sent bulletins to newspapers maintaining that the pro-metric campaign in the United States was "German propaganda." He claimed that Germany hoped to persuade Americans to adopt the metric system to create confusion in their foreign trade, which would enable Germany to seize a part of it.[118] Halsey, in a second edition of *The Metric Fallacy* (published in 1920), claimed erroneously (and probably maliciously) that the transition to the metric system in Latin American countries was done due to "German influence and for German purposes."[119] He proposed that the better way ahead was to unify the weights and measures of North America, Latin America, and the British Empire, as they share a common system "which is no more English than it is Spanish and no more Spanish than it is English, for it is neither. It is Roman."[120] These claims were part of his attempt to counter the idea that the metric system would benefit the United States's international trade. His plan, Halsey promised, would "assure the dominance of the English-speaking peoples over Germany in industry and commerce. To promote the English system is to work for the interest of the English speaking peoples; to promote the metric system is to work for the interest of Germany."[121]

Latin America was, once again, a topic of great relevance in the metric debate. How to harmonize the measurement standards of the United States with the rest of the continent had become a major point of emphasis for both camps.

EXPORTING THE FIGHT: THE BATTLE FOR PAN-AMERICAN UNIFORMITY

The clash between pro- and anti-metric organizations stepped quickly to the international arena, primarily around the question of the economic integration of the Americas and the measures used in Latin America. The disputes focused on what system of measurement was better suited to bring unity to the western hemisphere, what measures were employed in practice in the Spanish- and Portuguese-speaking countries, and whether the metric system was a suitable instrument to bring the continent closer together.

Halsey, Dale, Kunz, Drury, and the rest of the American disputants brought their fight south of the Rio Grande River, repeating their old arguments but attuned to continental circumstances. In the Pan-American conferences dedicated to science, commerce, and standardization during the 1910s and 1920s, members of the AMA, the World Trade Club, and the AIWM made numerous appearances to present papers and lobby representatives from other countries.[122]

This confrontation was substantial because there was a genuine interest in the United States to have a bigger influence in Latin America. Reports of America loosing ground in the area were frequent. For example, the personal representative in Mexico of president Woodrow Wilson reported:

> I asked an intelligent German merchant in Vera Cruz one day to explain to me how it had come about that Germany had absorbed so much of the trade that at one time went to England. He reached into a drawer, pulled out an invoice from England, and said, "Do you see those denominations or yards, feet and inches, gallons and pints, two kinds of ounces, grains and pennyweights, the whole summed up in pounds, shillings and pence? Well," he continued, "a Mexican, even if he can read a little English, needs an interpreter and an accountant to

put this into the language of civilization." ... The American manufacturer and merchant must learn to understand that a foreign market is always a "buyers market."[123]

Since most Latin American countries adopted the metric system during the second half of the nineteenth century, the juncture was an opportunity for the metric camp to score some points showing how international commerce demanded the establishment of an international system of weights and measures. They had the support of Americans working in the Pan-American Union, like chief statistician William C. Wells, who insisted that the world was gradually becoming one big market where the sale of manufactured products was more important than the sale of gross materials, and asked for a common system of measures to produce standardized goods whose dimensions were measured the same in manufacturing and consuming countries. That common system, he thought, was the metric system.[124]

The interest in Pan-Americanism was a door to circumvent metric legislation in the United States, as Halsey quickly recognized.[125] Faithful to his initial arguments, Halsey questioned whether the metric system was actually used by regular people in Latin America and suggested that several systems of measurement coexisted in those countries. To document this opinion, Halsey sent questionnaires to selected people in South America asking what units of measurement were used for international and domestic commerce in their place of residence. With the responses Halsey prepared a report trying to demonstrate the failure of the introduction of the meter, liter, and kilogram as exclusive units of measurement in Latin America.[126] To counter that, metric proponents were forced to prepare their own study to refute Halsey.[127]

In 1919, during the Second Pan-American Commercial Conference, the president of the AMA presented a paper on "The Metric System as a Factor in Pan American Unity," in which he advocated for the metric system and the introduction of a common continental currency. He maintained that nothing could pave the way for a good understanding among the American nations than uniformity in measures and currency, "for this will obviate many causes of misunderstanding and dispute, and will aid powerfully in developing trade among these nations."[128] As he later put it, "Has it ever occurred to you that musicians in every part of the world can read and play the music written by anybody in the civilized world? Why not the meter-liter-gram?"[129]

In that same conference, Halsey reinstated his plan to standardize the units of the customary English measures with the Spanish colonial measures (using the former as the base to alter the latter, of course) and to use that system in all the continent as a Pan-American system. The plan was based on the similarities between some units used in both systems, like inch and *pulgada*, foot and *pie*, etc. He also repeated the claim of a common metrological past of all nations in the continent: "Let us unify the weights and measures of the two Americas and the British Empire on the basis of the system which came to us from the mother of us all—the Roman Empire."[130] This move served Halsey and Dale as a counterpunch to the internationalistic pretensions of the metric camp. As they saw it, three centuries of Spanish dominance in the new world created something close to a Hispanic-American system of measurement. Virtually all countries from Chile to Mexico share a considerable number of units (like *vara, carga, fanega, libra, onza*, and *cuartillo*). It was, however, an imperfect "system" from the standpoint of uniformity, due to the great variability in the magnitudes of those units between and within countries.

Halsey, Dale, and their anti-metric associates used that apparent Ibero-American metrological unity, and its similarities with the English system, to recommend an alternative Pan-American system of measures—which basically meant the adoption in Latin America of the English units, but using Spanish terms: pound-*libra*, ounce-*onza*, yard-*vara* and so forth. But despite its internationalistic façade, their plan was, for all practical matters, intended solely to influence the metric debate in the United States. Their objective was to stop the penetration of the metric system in the United States. The situation in Latin America was not a serious concern for them. Halsey and Dale never engaged in significant conversations with Latin American experts and governments. And even if their Pan-American plan was sincere, it arrived too late. By the 1920s, Latin-American countries had already invested too much time and money in introducing the metric system to seriously consider new standards. "New" because the plan was not just to "get back to where you once belonged," as it may looked. It implied unifying all units of measure according to physical standards that, besides the names, were only randomly used. It also represented wasting all the efforts invested in the metric system. Not surprisingly, the plan was not taken seriously by any Latin American government or interest group. What the representatives from those countries sought was *metric* continental unification, They were, thus, more receptive to Drury's plans.[131]

Despite Dale's and Halsey's candor, Latin American countries did not receive well the idea of a non-metric Pan-American unity. In response to Dale's paper "Uniformity or Confusion in Pan America?" presented in the First Pan American Standardization Conference in Lima, a member of Peru's delegation replied that Latin American countries were well on their way to metric uniformity and that it would be preferable for the United States to join them.[132]

Other Latin American countries voiced their disappointment over the United States not making good on its pledge, made in 1890, to join the other American nations in the use of the metric system, something that hurt their common business interests. In 1927, the Secretary of the Ministry of Industry of Colombia wrote to the Drury:

> The United States are proud of the large amount of out-of-date machinery that they discard to be replaced with more efficient equipment as soon as a practical innovation appears—without considerations of expense. A scientific and uniform system of weights and measures is a vital equipment for domestic and international commerce. Abandoning the antiquated and incoherent system inherited from colonial times is worth as much as throwing away dated machinery. Adopting the French metric system would be a convenient and lucrative achievement, worth of such a great people; it would create a bridge to the other American republics and enhance their commerce with them.[133]

American merchants, bankers, and diplomats with interests in Mexico received information that contradicted Halsey's and Dale's claims on the prevalence of customary measures and the subsequent metrological confusion. Regarding foreign commerce, Mexico had become a fully metric nation. This is what the US vice-consul at Ciudad Porfirio Díaz (today Piedras Negras), on the Mexican-American border, reported: "Our English weights and measures are stumbling-blocks to the Latin countries whose trade we seek. The persistent use by our merchants in their catalogues, circulars, etc., of our old English weights and measures causes us to lose considerable trade with Mexico and other Latin-American countries, as they are unintelligible to many foreign buyers."[134]

In 1926 a more emphatic call of attention came directly from Mexican businessmen, when the Mexican Confederation of Chambers of Commerce made a call for the "urgent necessity" to use the metric system in the United States. They expressed their disappointment with the United States' refusal to adopt the metric system, noting that the adoption "would help them tremendously in their domestic dealings, it would bring immensely positive results to improve their relationship with clients all over the world, and they would also collaborate with the establishment of universality in weights and measures."[135] The document recalled the United States participation in the 1890 agreement for Pan-American metrication. Finally, the Confederation asked for its American and British counterparts to take steps toward adopting the metric system.

The reluctance of the United States to adopt the metric system shocked and annoyed Latin American countries where the United States was seen by many as an advanced nation and as their hoped-to-be future. America's unwillingness to implement the metric system was considered a weird anomaly—it did not match with their image of a prosperous nation. In the end, all exhortations—within and outside the United States—to persuade the US federal government and major economic actors to embrace the metric system were not enough to prompt any decisive action.

Time and again, the American government recognized the importance of the metric system to enhance the commercial relations of the country, but that was insufficient to convince them to pursue a forceful metric policy. Illustrative of this is a 1943 letter by Lyman J. Briggs (then director of the NBS) to J. T. Johnson (president of the Metric Association):

> we should cooperate to the fullest extent with the Latin American countries and should promote trade with them. But there is a serious question as to whether it is either necessary or desirable for this country to adopt the metric system in order to use it in foreign trade. A knowledge and understanding of the metric system and a willingness to use it properly are essential. No one proposes that we should change our language to Spanish or Portuguese in order to promote trade relations with Latin

America because in the matter of language it is recognized that knowledge and understanding are the essential items.... If the metric system is a sufficient step forward, then its use in our trade with our neighbors to the South may give it the impulse for use among ourselves.[136]

That impulse never arrived and for the rest of the twentieth century the obstacles on the road to metrication in the United States proved insurmountable.

NORTH AMERICAN FREE TRADE AND METROLOGICAL COOPERATION

In the late twentieth century, international cooperation and the prospects of free trade in the Americas influenced, once more, the development of metrology and the management of the metric system. The old goal of the Pan-American Conference of 1890 of unifying weights and measures in the continent was partially materialized decades later with the help of the Organization of American States (OAS)—which itself has its origins in the 1890 meeting. The OAS's aim was not to establish a single system of weights and measures for all the countries in the Americas, but the creation of a coordinating body. Interested in solving technical and legal problems related to production, in 1974 the OAS unveiled a special project on a regional system for metrology and calibration, the Sistema Interamericano de Metrología y Calibración (SIMYC), which included ten Latin American countries and was supported by the National Bureau of Standards.[137] The SIMYC served as the basis for the Inter-American Metrology System (SIM), created in 1979. The objectives of the SIM are the definition of the national measurement system in each country, compatibility in measurement results, training of technical personnel, distribution of scientific information, and collaboration with other international bodies (mainly with the General Conference on Weights and Measures, in Paris). The SIM aims, ultimately, to help international economic integration and free trade by reducing technical barriers among countries and the promotion of programs of technical cooperation in metrology and standardization.[138]

The creation of the OAS and renewed plans of free trade in the Americas also played a role in the development of new technical institutions devoted to measurement in Latin America. During the negotiations of the North American Free Trade Agreement (NAFTA) in the 1990s, it was stipulated that participating countries should have frameworks for intellectual property protection and effective systems of standards and metrology. The coincidence in these objectives with the stated goals of the First International Conference of American States is no accident.

It has been a recurring irony that Pan-American efforts to unify weights and measures result in the advancement and improvement of the metric system in Latin America, but are never sufficient to compel the United States to ditch customary measures. Several Mexican scientists and engineers had worked for many years to establish a metrological center in Mexico, but never received assistance from the government to finance the project. It was with the negotiation of NAFTA that they found a favorable situation to get that support. The creation of a new and improved metrological center was one of the results of that Treaty.[139] Coming full circle, in Mexico the plan for a Pan-American commerce agreement gave prominence to a federal office of weights and measures at the end of the nineteenth century, and it was a North American free trade treaty that reinvented that agency at the end of the twentieth century.

The Price of Isolation and the Power to Travel Solo

It is difficult to quantify the economic consequences for the United States of not adopting the metric system. Some have estimated more than two billion dollars of lost opportunities for non-metric American industries.[140] Having two legal systems of measurement operating simultaneously creates a dual economy. Manufacturing for foreign markets requires considering the metric specifications of other countries and the duplication of production processes—like making slight changes in the same car models, one for the United States and another with metric specifications for Europe. Big companies dependent on international supply chains need to synchronize the measurement specifications of all the participants to avoid costly misunderstandings. Importing products from abroad forces the acquisition of metric parts and tools for spare parts and reparations (as anyone in the United States who has tried to assemble an Ikea desk with their own set of Allen wrenches can attest).

Is this a validation of the pro-metric claims about the necessity to metricate America? Just partially. At the dawn of the twentieth century, Frederick Halsey argued that foreign commerce does not require the adoption of a new system of measurement in manufacturing. He was partially right—at least in the case of the United States. American industries have been a global force for more than a century now. Standing in a position of influence they can expect others to adopt their standards, instead of the other way around. It does not work like that every single time, but is a luxury that only very powerful countries can afford (as England did in the nineteenth century). For most other countries that is not a feasible option. Some countries dictate the rules, others adapt to them.

Could an eventual decline in American economic hegemony change these conditions? Absolutely. After the British lost their empire and they needed to participate more actively in the European markets in the 1970s, their anti-metric pride was eroded—material conditions trumped cultural resistance. In the future something similar may happen in United States.

Does Capitalism Begets Metrication?

Despite its lure, we should be careful with capitalism-centric narratives and broad generalizations that equate global metrication with global capitalism. Almost a century ago George Sarton posted a challenging question, "Why did the most industrial and mercantile nation in Europe reject the metric system, while its use would have caused great economies in time and money?" His question came with a warning: "Suppose the situation had been reversed, how tempting it would have been to explain the creation of the metric system as a necessary result of the superior mercantilism of England."[141]

As we saw in the first chapters of this book, other factors beyond the economy should be accounted for to explain the worldwide success of the metric system. Ultimately, capitalism only explains a part of that process. Other macro-historical forces were acting as well, like ruptures in sovereignty, national unifications, revolutions, colonization, and independence from colonial rule. The transformations of the states and shifts in political structures were as important for international metrication—it is not an accident that the Bolsheviks introduced the metric system in all the Soviet republics, something impossible to explain with the "capitalist hypothesis."

An internationally uniform system of measurement is helpful for global capitalism, no question about it; but as the case of the United States shows, the building of such a system cannot be left in the hands of self-interested economic actors. A more forceful intervention is required.

The conflicting interests of different industries regarding standardization in the United States are revealing in this context. Anti-metrics wanted the state to be active in controlling the market, but to be passive in the control of standards—a regulated economy with laissez-faire politics. Pro-metrics wanted a strong state that could homogenize standards, but a weak one in the regulation of the market—regulated politics with laissez-faire economy. The relationship between states and free markets is not simple and unilateral; they can complement each other, or they can get in each other's way—the metric question in America got stuck in the middle.

CONCLUSION

Why Is There No Metric System in the United States?

Five Reasons

> Bring out number, weight, and measure in a year of dearth.
> —WILLIAM BLAKE, *The Marriage of Heaven and Hell (1790)*

When I began my readings on the history of weights and measures in the United States, I checked out from a university library in New York City a copy of Richard Deming's *Metric Power: Why and How We Are Going Metric*, published in 1974.[1] There I found, inscribed in blue ink, an anonymous handwritten annotation: "The publication of this book proves that America is a backward country!" I found the comment amusing. I made a photocopy and kept it in my files.

The note expressed succinctly one of the common self-images of the United States. Many adhere to the legend of "exceptionalism," the idea of a special country, unique and exemplary, destined for greatness; its people, values, and achievements are superior and incomparable. If there is no metric system in the United States it is because the American way of measuring is better—or as some put it, "There are two kinds of country, those that use the metric system and those that have landed on the moon."[2] Others, like the spontaneous scribbler, do not see an American exceptionalism but an American "backwardism." They see their country as shamefully odd: widespread possession of weapons, persistence of the death penalty, reluctance to ratify international treaties like the Kyoto Protocol and the International Criminal Court . . . and no metric system.

Self-aggrandizing and self-derogating Americans play both into the myth of the uniqueness of the United States, a position that obscures the understanding of American history and society. When exceptionalists are open to comparing the United States with other countries, they do it with the expectation that comparisons should be made based on the American example, aiming to show how other nations are progressing in approximating America's institutions and way of living. "Backwardists," on the other hand, tend to see Americans as people with an accentuated propensity to selfishness, stupidity, and ignorance. If they have not adopted the metric system, that is just additional evidence of the stubbornness and self-centeredness of their fellow countrymen.

Both sides are wrong. Certainly, there is something ironic—and potentially illuminating—in studying a case in which the United States seems to be at the end carriage of the train of history, instead of playing the self-ascribed role of driver. Mischievously, Benedict Anderson pursued such an image in his famous study on the origins of nationalism, *Imagined Communities*. "I decided," he wrote, "to compare the early US with the welter of new nationalisms in Spanish America and put it at the end of the chapter rather than at the start. I enjoyed anticipating the annoyance that would be caused by calling Franking and Jefferson 'Creoles,' as if they were simply an extension of patterns everywhere visible south of the US border, and commenting that Simón Bolívar was a more impressive figure than George Washington."[3] My intentions in writing *Yardstick Nation* were not as wicked as Anderson's, but I feel that the American politicians, scientists, and industrialists who had pushed for metrication were an extension of patterns everywhere visible south of the United States border.

When the differences and similarities between the so-called "underdeveloped" and "developed" countries (or the ill-named "global north" and "global south") are compared, there is a tendency to consider the history, values, and institutions of the latter as "normal," and those of the former as immature forms or deviations from "the norm." With metrication, however, the United States is the country that looks "atypical" and "underdeveloped." This inversion of roles is helpful to denaturalize American history.

I hope no one thinks that the publication of *Yardstick Nation* proves that America is a backward or an exceptional country. It is just another country, with a history formed by intelligible human actions, beliefs, and institutions. Its cultural traits are not mysterious or inexplicable and its self-defining myths are no more entrenched than others.

The lack of a unified metric system in the United States is an interesting object of inquiry not because America is unique in that regard (in the end, other countries remain outside of the metric sphere as well), but because it was something expected to happen considering its social and historical conditions. A century ago, German sociologist Werner Sombart was interested in why no socialist party had been able to reach prominence in America, a country that was considered "capitalism's land of promise" and where the tendency of capital accumulation attained historically unprecedented levels. According to authoritative arguments, the development of commanding socialist political organizations was predictable under such conditions.[4] Likewise, one would think that America was a perfect place for the thriving of the metric system: openness to the future, constant technological innovation, incessant industrialization, international capitalist enterprises, etc. Despite those favorable conditions, it did not happen here. Why?

Five Reasons: Loosely Centralized State, Divided Experts, International Economic Predominance, Aversion to Compulsion, and Unfavorable Historical Timing

To answer this question, it is important to understand the historical and social specificity of actors and circumstances (national and international) directly involved in the use, establishment, and enforcement of weights and measures in the United States. That is to say that any aspect of the so-called "American exceptionalism" should not be solved by resorting to ahistorical, vague, or abstract characteristics of the "American people" or the "American character."

Clichés are remarkably persistent. During the years I conducted the research for this book I had many conversations with intelligent and sophisticated people about the American metric riddle. More often than not, when they heard that I was trying to find out why there is no metric system in the United States, I received essentialist comments. A professor of International Affairs at Columbia University, for example, said to me

"Because we Americans want to be different." A graduate student at the University of Chicago asked me "So, are you studying stubbornness?" A novelist in Mexico City claimed, "Well, that's because *gringos* are very square, isn't it?" These type of reasonings can transmute into attempts at apparently more sophisticated explanations about the United States' resistance to metrication. For instance, commentators have tried to portray Canada, New Zealand, South Africa, the United Kingdom, and the United States as natural barriers to the metric system because—in contrast to countries with authoritarian, top-down forms of government—they are societies based on democracy and private property (disregarding thus that in all those countries, except America, the metric system was introduced decades ago and that most other democracies have gone metric as well).[5]

Instead of these mystifications, it is more constructive to look at the way in which American institutional and cultural idiosyncrasies have been explained in historical and sociological terms. Take for instance Sombart's approach to the enigma of why there is no socialism in the United States. It would have been very easy—and popular—to answer the question by alluding to Americans' "love for freedom" or their "ingrained individualism." Sombart evaded such easy exits. He focused, instead, on the workers and their living conditions (like their style of public life), their economic situation (income and cost of living), and their political position (like the monopoly of two major parties, the failure of third parties, and the place of workers in the state). Analogously, Markovits and Hellerman, while tackling the question of why soccer was highly popular around the world except in the United States, avoided widespread answers like "soccer is boring because there are very few goals," or "baseball is a more beautiful sport." Instead, they explained this absence by looking at the formation of the American sports landscape and how baseball and football had already crowded out other professional sports at the time of the introduction of soccer into the United States.[6] These works stressed the historical specificity, the structural conditions, and the institutional development of the phenomenon at hand, instead of deducing answers from conventional national images.

The metric question in America, similarly, cannot be solved through the issue of the intrinsic virtues of the metric system or through the "character" or "essence" of Americans. The metric system has not been adopted in the United States not because it is better or worse than the English customary system, nor because Americans are "pragmatic," "democratic," or "conservative." The metric question demands a sociological and historical explanation, as I have tried to show in this book.

The main factors that explain the United States' decision not to adopt the metric system as its exclusive system of measurement are a loosely centralized state, a divided field of experts, the United States international economic predominance, an entrenched aversion to compulsion, and unfavorable historical timing.

LOOSELY CENTRALIZED STATE

Political processes in America operate within the framework of a federalized and fragmented state. Since its beginnings in revolutionary France, metrication has been an achievement—and a symbol—of political centralization. In every country, the meter has subjugated hundreds of local and traditional metrological units, legislations, and practices. The American state has not been able to concentrate the regulation of weights and measures in the sole hands of the federal government. The ratio of power of the states and local government vis-a-vis Washington is markedly high, and they have been able to maintain a considerable degree of metrological autonomy. Furthermore, the American state has lacked legitimacy (symbolic capital) to be perceived as a rightful authority to set national standards unilaterally. All this has halted any plans for a federal, mandatory metric policy. The alternative path of state-by-state metrication has not been pursued, and it does not seem like a feasible solution for the problem of standardization. Full metrication is the result of mandatory national regulations. The US government has shied away from mandatory metric policies and has not provided economic and logistic support to those who wanted to explore a voluntary transition. Many names have been given to this approach: "the government's policy is to have no policy," "a country of laissez-faire standards," "governing in the absence of government," "voluntary metrication," "patchwork of standards," etc. Whatever the label, this method of no carrot and no stick failed to produce an effective path to measurement homogeneity.

ANTI-METRIC OPPOSITION AND A DIVIDED FIELD OF EXPERTS

Most scientists in America were in favor of the adoption of the metric system, but other groups of experts, mainly engineers, rejected the idea. A key economic sector in the country, manufacturers, was against the transition to metric measures throughout the twentieth century. Their opposition prevented the pro-metric front from presenting unanimous support for the metric system among experts. Mechanical engineers, an important and well-organized group, mounted a solid case in mass media, scientific publications, and political debates against the convenience of adopting a mandatory metric policy. The antagonism among experts and the opposition by manufacturers contributed to make the compulsory introduction of the metric system a polemic issue and impeded a cohesive front of scientists, specialists, and policy advisors. A unified field of experts and a less vociferous opposition probably would have improved the chances of securing favorable metric legislation—*probably* should be stressed here, as this was not the main or sole condition that prevented the introduction of the metric system in America. In important junctures, like in Reconstruction, compulsory metric legislation was not obtained despite the absence of any organized opposition. Structural conditions in politics and the economy greatly favored the chances of anti-metric groups.

INTERNATIONAL ECONOMIC PREDOMINANCE

The size and might of the United States economy grants it the possibility to impose its conditions in international trade agreements—including defining metrological standards. This has allowed the United States to enforce the use of English customary measures as de facto standards in several areas of production and exchange. For instance, despite growing economic integration and the subsequent necessity for international coordination (like the European Union that pushed England and Ireland to advance in their metric transitions) Mexico and Canada did not have the leverage to demand the exclusive use of the metric system in all the NAFTA-USMCA territory. The technological dominance of the United States has also helped to lessen the problem of keeping customary units; by designing new technologies with customary units like the inch, American companies have maintained those measures operative.

AVERSION TO COMPULSION

People at large have been indifferent or hostile toward the metric system, something common in most countries. Large metrication from below has not existed—spontaneous adoption of the metric system has been confined to a few professional groups. Despite multiple efforts by schools, government agencies, and pro-metric associations, the knowledge among the American population of what the metric system is and how it works has been limited and inadequate. Paradoxically, and contravening the predictions of metric advocates, when polls have been conducted the results show that resistance to metrication grew as familiarity with the system increased.[7] According to Gallup surveys, in 1971 little more than 50 percent of the people said that they were not aware of what the metric system is, while one out of every five persons declared that they opposed its introduction in the United States. In 1991, only 20 percent were not aware of what the metric system was, and one out of two persons opposed its introduction. As more people claimed familiarity with the system "the margin by which opponents outnumbered advocates continued to grow."[8] Additionally, a deep-rooted trait in American political culture, the aversion to compulsion, has greatly undermined the chances of the metric system becoming the exclusive and mandatory system of measurement in the country.[9] Skepticism toward an activist state has persisted outside and inside the federal government. Federal agents frequently advised against compulsory solutions and suggested, instead, a laissez-faire approach.

UNFAVORABLE HISTORICAL TIMING

At the end of the eighteenth century and the first decades of the nineteenth, America had good perspectives for adopting a system of measurement different from English customary measures. It was a context in which the metric system could have been a serviceable answer to problems early Americans were confronting. One of those opportunities was in the 1780s with Thomas Jefferson's report on weights and measures, but at that point, the metric system was not yet finished, so it was not an option. Another moment occurred with John Quincy Adams's report in the 1820s, but at that time the metric system was not in actual use in any other country (even France, in those years, was using a mixed system, the *mesures usuelles*). American metrological reformers were in the right political place at the wrong historical time. In hindsight, the key year in America's metrication history was 1866, after the Civil War. The metric system was already a recognizable international force, and the Reconstruction era was a propitious moment for metrication as plans of national unification, expansion of the administrative functions of the state, and economic modernization were on the rise. However, many in the government did not want a mandatory introduction of the metric system if England was not committed to make the transition as well. Failure to procure definitive actions at that moment generated—unintentionally—the development of numerous industrial and technical processes in the country without the use of metric units, something that made any future change more expensive and burdensome; customary measures locked in. In the second half of the twentieth century, another window opened when the U.K. and its former colonies, including Canada, switched to metric. Now America could have gone along with its closest allies, but this time coordination with the Great Britian was not as urgent as in the past. The United States had become a power to be reckoned with. In the 1860s America was so weak that it could not afford to go metric without England; in the 1970s it was so strong that it could remain non-metric without the United Kingdom.

A Future Metric America?

Will the metric system ever take root in the United States? Historical predictions usually fail, and it is futile to categorically answer the question either positively or negatively. But looking at the persistent historical trends in the two-centuries-old global process of metrication, we can venture into some social forecasting.[10] There are structural conditions that need to change for the metric system to set foot—and oust the foot—in America.

First, there should be an increased level of political centralization, an expansion of the administrative capacities of the state, and a greater accumulation of symbolic capital. This not only requires a transformation in the structure of the federal state, but a change in the balance of power between federal and local governments in favor of the former. There is also a need for the state to be perceived as a legitimate authority to decide those issues. So far, neither Reconstruction nor the New Deal—periods of state expansion—were effective in opening the door for the federal government to take control of metrological matters.

Second, there should be an American decline in the international economy. Many countries have decided to adopt the metric system in times when they feel that they are falling behind, thus entering a "catch-up mode" and start searching for ways to participate more effectively in international markets. There have been some moments in American history when the feeling of "falling behind" created a similar sense of urgency. The space race and the Soviet Union's launching of the *Sputnik* triggered a campaign to revamp math and science education and channeled unrestrained funds to NASA.[11] The absence of the metric system, however, has not prompted a similar response. The pride and anxieties of Americans were more deeply stirred by a picture of Yuri Gagarin in a spacesuit than by a cold and inanimate platinum-iridium meter bar.

Third, a period of large social reorganization could also be propitious for metrication. A noted American historian once told me, in an informal conversation, that the United States did not adopt the metric system in the early 1970s "because Nixon had too many problems to worry about weights and measures." I would rather say that the country in the entire twentieth century (let alone the 1970s) had too few problems to embark on a full-scale metrological transition. Many countries did not adopt the metric system when conditions were calm and stable; they mustered the energy to go from one measurement system to another when their political and economic structures were being deeply reshaped, shortly after revolutions, civil wars, struggles for independence, etc. The prolonged stability of the United States—the absence of ruptures in sovereignty in the last century and a half—have made the project of metrication unplausible.

I have stressed the importance of structural conditions and historical timing. Structural conditions promote continuity and obstruct change. They are assemblages, frameworks, habits of action and thought, fixed series of relationships between social realities that time only erodes very slowly.[12] They limit what societies can do, but they can and do change. Historical timing points to crucial moments when continuity or structural crisis are determined, and they can serve as launching pads for periods of transformation. Thus, we should be careful not to think that what exists today was an unescapable outcome of history; but we should also avoid the idea that we can change whenever we want.

Things with the metric system in America could have been different, but they are the way they are for no trivial reasons. Emphasizing the importance of timing and path dependency, as I have done, does not mean that the absence of the metric system in America can be attributed to a "butterfly effect." This is not a case where a few marginal and accidental changes in the past would have transformed a whole historical outcome. Some people have explored that possibility, unconvincingly. Among journalists, for example, it is popular to tell the tale of how "pirates might be the reason why the United States doesn't use the metric system." The story goes like this: in 1794, French botanist Joseph Dombey was on his way from Le Havre to Philadelphia. One of the objectives of his travel was to deliver copies of the preliminary metric standards of length and mass to Thomas Jefferson. With those standards in their hands, some people believe, Jefferson and the US Congress would have been persuaded to adopt the metric system. Tragically, Dombey never made it to America. His ship was captured by British pirates, who imprisoned him. He died—unable to complete his scientific-diplomatic mission—on the island of Montserrat waiting to be rescued, and "America's hopes for using the metric system died with him." If that meter bar and that copper kilogram had been delivered as planned, the fable says, "today the United States might not be the last country in the world to resist the metric system."[13] This is a captivating but implausible supposition. Dombey's standard eventually got to the United States and produced no effect. A few years later, F. R. Hassler presented a more advanced meter prototype to public functionaries and members of the American Philosophical Society, and the result was identical: no metric policy.

The actions of metric hopefuls have occurred within assemblages of relationships that make change difficult, and continuity has prevailed. The functioning of the American federal government, the interest of powerful economic actors, and the weight of the United States in the international arena have set a scenario where the possibilities of altering the conditions of measurement regulation are limited. Things being as they are, the chances of a sudden change in the status of the metric system in America are slim.

Who Won?

Who won and who lost the American metric battles? Metric proponents lost the most. Despite the best efforts of several generations of metric enthusiasts, the majority of their fellow Americans keep thinking primarily in non-metric terms. They may find solace in the fact that science, medicine, and some other professional areas are primarily metric. Also, customary measures are currently defined relative to the metric system. The scientific infrastructure behind English measures has disappeared. The yard and pound enjoy a robust public life; but behind doors, they are anchored in metrological laboratories where only the meter and kilogram exist.

Those who opposed metrication won something. Despite two centuries of continuous pressure, America has not been engulfed by the metric ocean. The dam trembled but did not crack. The names, magnitudes, and arithmetic of customary measures conceptually and materially shape everyday life and many industries. It is not a small feat. For some, that is good enough, but others aspired to much more. The dream of a strong English system of measures that could prevent metric global supremacy vanished in the second half of the twentieth century.

What about ordinary people who did not care to participate in the battle of the standards and simply kept on living their lives with what they had at hand? They routinely deal with the blessings and curses of living in a country with two legal systems of measurement. Some see it as a richer and freer life, and exalt the virtues of living in a "bimensural nation."[14] It is, however, a reward not free of perils. The cohabitation of two measurement systems generates problems for science, and for the communication between experts and laypeople. Some cases are well known, like NASA's Mars Climate Orbiter, lost in space in 1999. The mission failed due to a data mismatch. The ground navigation software (supplied by a spacecraft builder company) produced results in customary units, while the rest of the system (designed by NASA), treated data from that software as if it were in metric units.[15] Millions of dollars and years of planning were wasted. Public ridicule for the space agency—and the country—ensued.

Other situations are less conspicuous but more consequential for common people. According to the American Academy of Pediatrics each year in the United States more than 70,000 emergency visits occur due to unintentional overdoses. The main cause is that parents sometimes confuse tablespoons and teaspoons and give too much medicine to their children; in other instances, when the prescriptions are written with metric units, people misread the position of the decimal point, creating 10-fold dosing errors.[16] A sole national system of measurement and a population properly trained to use the decimal point would solve these issues, but the United States has neither.

All this is the product of an odd historical trajectory. Long-term processes—like the history of weights and measures in the United States—do not follow a master plan.[17] Many projects were proposed in the 250 years since the Articles of Confederation gave Congress the exclusive right to fix the standards of weights and measures. None of those plans had the goal of arriving at the situation we have today: an isolated nation with two legal systems of measurement working simultaneously. Thomas Jefferson wanted an original decimal system; John Quincy Adams proposed to work with other countries to design something more perfect than the metric system; Charles Latimer wished that no metric units would ever be used at all; Herbert Spencer asked for a novel duodecimal number system; Frederick Halsey sought to unify the measures of the British Empire, the United States, and Latin America; Melvil Dewey, Thomas Mendenhall, and others pursued a policy of mandatory metrication; functionaries in different decades hoped that the country would voluntarily switch to the metric system. All those plans failed. The end result was an unforeseen and unintended scenario that no one really wanted.

The only plan that succeeded was the one proposed in the 1980s, when the final scenario had already materialized: the world had turned metric and the United States was trying to get there with the government coordinating a voluntary transition. When the Reagan administration disbanded the US Metric Board, they decided it was better to do even less, that things were fine as they were, and that nothing else was worth trying. No compulsion and no persuasion. The plan was to have no plan.

The United States flirted with metrication, but never truly got there. It has been attentive to international changes in metrology, but has not taken decisive actions to get in sync with the rest of the world. It was close to England but did not adopt the UK's imperial system. It developed its own customary units but never planned to impose them globally—as

France and England did with their systems. The United States has just drifted in the metrological sea, making small adjustments according to the changing situations or simply doing nothing—that's the American way of measuring.

APPENDIX

Adoption of the Decimal Metric System of Weights and Measures by Country

Sources of information on the progress of metrication in the world, even the more comprehensive, are either outdated or incomplete. For this book, I assembled a complete and accurate dataset on global metrication country by country. I cross-compared, corrected, and updated existing data.[1] This was complemented with new information from official documentation on metric adoption from numerous countries and by consulting national experts, metrological agencies, and diplomatic offices. The result was a dataset that includes 197 countries according to the grid of national states in 2024, that I used for the analysis in Chapter 1.

To define the beginning of metrication in a single country I decided to use the year of the first legislation ordering exclusive and compulsory use of the metric system in every nation-state. This excludes those legislations that made the metric system only optional.

Contrary to other criteria used to trace global metrication, I do not use the moment when the actual enforcement of the metric legislation started taking place, which sometimes happened several decades after the first legislation was approved. There are two reasons for this. First, my research wants to indicate when countries were first committed to fully switching to the metric system, even if they were unsuccessful in the initial implementation (which is a much harder, costlier, and slower affair). Second, it is complicated to track the actual changeover to the metric system due to the messiness of the process, which has many gray areas to determine how thorough the transition has been at any given moment.

There are at least two important limitations when studying the global spread of the metric system in the way it is done here. First, by using independent countries as the units of observation two crucial elements are missed: a) what happened *within* every country, once the adoption was made, to convince or force populations to use the metric units; b) how individuals, associations, and organizations appropriated the metric system even if they were not required to do so by law. In the dissemination of the metric system, the adoption by a nation-state tells just half of the story, as the official adoption does not mean that the population will use it. The second half of the story is what happens within those countries. This point is critical to understanding why many countries are still trying to introduce the metric system, even decades after the official adoption was registered (to mention just one example, South Korea officially became metric in 1949, but as late as 2008 it was still launching campaigns to make the reform tangible).

The following list (Table 1) is a product of my dataset. It shows the 197 countries existing in 2024 and their respective year of official adoption of the decimal metric system as the compulsory and exclusive system of measurement.

The countries at the end of the table, marked with an N/M, are the five remaining non-metric countries.

When countries are followed by a **C** it indicates that the metric system was introduced when that country was a colony; if it is followed by an **S** it means that the metric system was introduced when that country was part of a larger political entity from which it seceded. Both of these cases are considered non-voluntary adoptions.

The column **Opt.** indicates, when appropriate, the year when the use of the metric system became optional in that country.

Table 1. Years of Metric Adoption by Country

COUNTRY	COLONY OR SECESSION	YEAR METRIC ADOPTION	OPTIONAL
France		1795	
Belgium		1816	
Luxembourg		1816	
Netherlands		1816	
Algeria	C	1840	
Senegal	C	1840	
Spain		1849	
Portugal		1852	
Monaco		1853	
Colombia		1853	
Mexico		1857	
Venezuela		1857	
Cuba	C	1858	
Italy		1861	
Brazil		1862	
Peru		1862	
Uruguay		1862	
Romania		1864	
Chile		1865	1848
Ecuador		1865	
Dominican Republic		1867	
Germany		1868	
Bolivia		1868	
Turkey		1869	
Suriname	C	1871	
Croatia	S	1871	
Czech Republic	S	1871	
Liechtenstein	S	1871	
Montenegro	S	1871	
Slovakia	S	1871	
Slovenia	S	1871	
Austria		1871	
Serbia		1873	
Hungary		1874	
Sweden		1874	
Switzerland		1875	1868
Mauritius	C	1876	
Argentina		1877	1863

COUNTRY	COLONY OR SECESSION	YEAR METRIC ADOPTION	OPTIONAL
Seychelles	C	1878	
Bosnia and Herzegovina	S	1878	1910
Costa Rica		1881	
Norway		1882	1875
Benin	C	1884	
Chad	C	1884	
Congo, Republic of the	C	1884	1960
Côte d'Ivoire	C	1884	
Mauritania	C	1884	
Niger	C	1884	
El Salvador		1885	
Finland		1886	
Macedonia	S	1888	
Bulgaria		1888	
Sao Tome and Principe	C	1891	
Tunisia	C	1893	
Nicaragua		1893	
Honduras		1895	
Djibouti	C	1898	
Puerto Rico	C	1899	
Paraguay		1899	
Equatorial Guinea	C	1900	
Iceland		1900	
Guinea	C	1901	1959
Guinea-Bissau	C	1905	
Angola	C	1905	
Cape Verde	C	1905	
Mozambique	C	1905	
Philippines	C	1906	
Denmark		1907	
San Marino		1907	
Congo, Democratic Rep.	C	1910	
Burundi	C	1910	
Rwanda	C	1910	
Malta	C	1910	
Belize	S	1910	
Guatemala		1910	1894
Vietnam	C	1911	
Thailand		1912	
China		1913	

COUNTRY	COLONY OR SECESSION	YEAR METRIC ADOPTION	OPTIONAL
Comoros	C	1914	
Panama		1915	
Mongolia		1916	
Russia		1918	1899
Poland		1919	
Haiti		1920	
Cambodia	C	1922	
Morocco	C	1922	
Western Sahara	C	1922	
Kazakhstan	S	1922	
Kyrgyzstan	S	1922	
Tajikistan	S	1922	
Turkmenistan	S	1922	
Uzbekistan	S	1922	
Armenia	S	1922	
Azerbaijan	S	1922	
Belarus	S	1922	
Estonia	S	1922	
Georgia	S	1922	
Latvia	S	1922	
Lithuania	S	1922	
Ukraine	S	1922	
Libya	C	1923	
Indonesia	C	1923	
Afghanistan		1923	
Togo	C	1924	
Iran		1927	
Iraq		1930	
Syria	C	1934	
Andorra		1934	
Lebanon		1934	
Israel		1947	
Albania		1948	
Korea, North		1948	
Korea, South		1949	
Egypt		1951	1873
Japan		1951	
Bhutan		1951	
Taiwan	S	1952	
Jordan		1952	

COUNTRY	COLONY OR SECESSION	YEAR METRIC ADOPTION	OPTIONAL
India		1954	1920
Sudan		1954	
Timor-Leste	C	1957	
Macau	C	1957	
Greece		1957	1836
Madagascar		1957	
Maldives		1959	
Mali		1960	
Burkina Faso		1960	
Central African Rep		1960	
Gabon		1960	
Somalia		1960	1950
Cameroon		1961	
Kuwait		1961	
United Arab Emirates		1961	
Eritrea	S	1962	
Ethiopia		1962	
Nigeria		1962	
Nepal		1963	
Laos		1963	
Saudi Arabia		1964	
United Kingdom		1965	1864
Namibia	S	1967	
Ireland		1967	1897
South Africa		1967	
Pakistan		1967	
Tanzania		1967	
Uganda		1967	1950
Kenya		1967	1953
Singapore		1968	
Australia		1969	
New Zealand		1969	1925
Bahamas, The		1969	
Zimbabwe		1969	
Botswana		1969	
Swaziland		1969	
Bahrain		1969	
Grenada		1969	
Dominica		1969	

Appendix 227

COUNTRY	COLONY OR SECESSION	YEAR METRIC ADOPTION	OPTIONAL
Saint Vincent and the Grenadines		1969	
Saint Kitts and Nevis		1969	
Canada		1970	1871
Sri Lanka		1970	
Trinidad and Tobago		1970	
Zambia		1970	1937
Lesotho		1970	
Papua New Guinea		1970	
Solomon Islands		1970	
Malaysia		1971	
Guyana		1971	
Ghana		1972	
Cyprus		1972	
Fiji		1972	
Qatar		1972	
Jamaica		1973	
Barbados		1973	
Nauru		1973	
Oman		1974	
Antigua and Barbuda		1974	
Tonga		1975	
Sierra Leone		1976	
Malawi		1976	
Gambia, The		1976	
Tuvalu		1978	
Yemen		1981	
Bangladesh		1982	
Kiribati		1984	
Brunei		1985	
Vanuatu		1988	
Saint Lucia		2000	
South Sudan	S	2011	
Myanmar		2013	1920
Samoa		2015	
Liberia		N/M	2018
Marshall Islands		N/M	1986
Micronesia, Federated States of		N/M	1986
Palau		N/M	1994
United States		N/M	1866

CHRONOLOGY OF THE METRIC SYSTEM IN THE UNITED STATES

1784   Thomas Jefferson (delegate from Virginia to the Congress of the Confederation) prepared a proposal concerning coinage—Notes on the Establishment of a Money Unit, and of a Coinage for the United States—with the intention to base the money system on decimal reckoning.

1785   The American dollar became the first fully decimal currency in the world (1 dollar: 100 cents).

1789   Beginning of the French Revolution. Initial plans got underway to reform weights and measures, which culminated a few years later in the creation of the decimal metric system.

1790   President George Washington called Congress for the creation of a uniform system of currency and weights and measures. Thomas Jefferson (Secretary of State) proposed a decimal-based measurement system in his *Plan for Establishing Uniformity in the Coinage, Weights, and Measures of the United States.*

1793   In France, the republican calendar replaced the Gregorian calendar.

1795   France started using the metric system with preliminary standards. The basic unit of the system, the meter, was originally defined as one ten-millionth of the distance from the equator to the North Pole.

1799   Congress on Definite Metric Standards is held in Paris. Final calculations to determine the length of the meter. Representatives from nine European countries participated in the Conference, but none from England and the United States.

1805 In France, Napoleon abolished the republican Calendar.

1807 Establishment of the United States Survey of the Coast, the first scientific agency of the United States Government. Its first superintendent, in 1816, was Ferdinand R. Hassler, a Swiss scientist who in the 1790s studied in France with some of the inventors of the metric system. Among the tasks was the standardization of weights and measures in the country.

1812 In France, Napoleon suspended the compulsory use of the metric system and adopted the *mesures usuelles*, a system that mixed metric and customary measures: the *toise* was defined as exactly two meters, the *livre* was defined as 500 grams, etc.

1816 President James Madison urged Congress to act on the problem of uniformity of weights and measures.

The United Kingdom of the Netherlands (today Belgium and Netherlands) and Luxembourg adopted the metric system.

1821 John Quincy Adams presented his *Report of the Secretary of State, upon Weights and Measures*. The report suggested a two-part plan: 1) to fix the current standard, excluding innovations; 2) to consult with other countries for the establishment of a future universal system. Congress took no action. Facing inaction by the federal government, several states of the Union enacted their own legislations.

1822 The American Colonization Society began sending free people of color to Liberia to establish a colony. Liberia is today the only African country that has not adopted the metric system.

1836 Joint resolution by the Senate and House of Representatives resolving that the Secretary of the Treasury should deliver to custom-houses and the governor of each state in the Union a complete set of standard weights and measures.

1837 France issued new legislation to abolish the *mesures usuelles* and reinstated the exclusive use of the metric system.

1840 France introduced the metric system in Algeria and Senegal.

1857 The Chamber of Commerce and the American Geographical and Statistical Society presented a proposal for a new measurement system, different from the English and metric systems. It was the first of multiple plans that failed to challenge the metric system as a modern alternative to customary measures.

Mexico adopted the metric system—by 1867 ten Latin American Countries had approved compulsory metric legislation.

1865 The Secret Service Division of the Department of the Treasury was established with the mission of suppressing counterfeiting, as part of a larger effort by the federal government to create a single, standardized national currency. A similar concentration of authority and resources was not invested to create and regulate a nationally standardized system of measurement.

1866 Metric Act of 1866. The use of the metric system became legal, but not mandatory, in the United States. Promoters of the Metric Act justified the lack of compulsory provisions arguing that the country needed to be in tune with England in terms of commercial standards.

1867 International Monetary Conference held in Paris. Representatives from nineteen European countries and the United States discussed the possibility of an international currency based on metric measures.

1868 Germany adopted the metric system.

1873 The American Metrological Society was created, headed by Frederick A. P. Barnard.

1875 The Convention of the Metre was signed in Paris by seventeen nations, including the United States. The convention created the International Bureau of Weights and Measures (BIPM).

1876 The American Metric Bureau, the first organization exclusively dedicated to the goal of introducing the metric system in America, was founded by Melvil Dewey—that same year he patented his decimal classification system for libraries, also known as Dewey Decimal Classification.

1878 The US Senate ratified the Metre Convention.

1879 The International Institute for Preserving and Perfecting Weights and Measures was founded by Charles Latimer, chief engineer of the Atlantic and Great Western Railway. Scottish astronomer Charles

Piazzi-Smyth was named counselor of the organization. They opposed the metric system and promoted "pyramid metrology" in the US.

1880 Coleman Sellers (chief engineer of William Sellers & Co. and president of the American Society of Mechanical Engineers) questioned the pertinence of metrication and published *The Metric System: Is It Wise to Introduce It into Our Machine Shops?*

1884 The International Meridian Conference was held in Washington, DC. Representatives from twenty-five countries agreed on a common prime meridian (Greenwich). The idea of the conference was promoted by the American Metrological Society. In the negotiations, the French delegation petitioned, unsuccessfully, that if France was to accept the British meridian, then England should adopt the metric system.

1889 Copies of the international kilogram and meter were assigned among the signers of the Metre Convention. The United States received kilograms number 4 and 20, and meters 21 and 27. The prototypes were later received, unsealed, and exhibited in the White House in a ceremony with president Benjamin Harrison.

The First International Conference of American States is held in Washington, DC. The Conference recommended "the adoption of the metrical decimal system to the nations here represented which have not already accepted it." As a result, new efforts to complete metrication in Latin American countries were carried out, and discussions to adopt the metric system intensified in the US.

1893 *Mendenhall Order*. T. C. Mendenhall, Superintendent of Weights and Measures, with the approval of the Secretary of the Treasury, defined officially the yard and pound in terms of the metric system. The metric prototypes were declared the fundamental standards of length and mass in the US.

1896 Herbert Spencer published in England and the US a series of influential articles opposing metrication. In response, William Thompson (Lord Kelvin) urged both countries to adopt the metric system.

1901 The National Bureau of Standards (NBS) was created by the federal government; its director, Samuel Stratton, was a strong supporter of metrication.

1902 Kelvin appeared before the US Congress to endorse the adoption of the metric system in America.

1904   *The Metric Fallacy*, by Frederick A. Halsey, and *The Metric Failure in the Textile Industry*, by Samuel S. Dale, were published. The National Association of Manufacturers (NAM) strongly opposed the adoption of the metric system. During the coming years, Halsey and Dale articulated the ideas to oppose metrication and the NAM provided the political muscle to obstruct metric initiatives in Congress.

1905   The First National Conference on Weights and Measures was hosted by the NBS. The Conference has met annually since then. As an unofficial organization, it makes recommendations on specifications and model laws but has no legal authority to enforce them.

1906   The Philippines, under an interim US military government, adopted the metric system.

1913   China adopted the metric system.

1916   The American Metric Association is founded during the annual meeting of the American Association for the Advancement of Science (in 1974 it changed its name to US Metric Association).

Halsey and Dale, with financial backing from Brown & Sharpe Manufacturing Co., established the American Institute of Weights and Measures, aimed to perfect customary measures and stop the metric advance.

1918   Russia adopted the metric system.

1919   The World Trade Club, a pro-metric association, is founded in San Francisco, with the financial backing of industrialist Albert Herbert. The organization is run a by professional advertiser, Aubrey Drury.

1921   Herbert Hoover, Secretary of Commerce, established the Division of Simplified Practice in the NBS. Hoover insisted that the government would only help in the voluntary adoption of national specifications in commerce and rejected state-imposed standards.

1925   The "Britten Metric Bill" is introduced in Congress. It called for the use of metric units in the buying and selling of commodities. The legislation proposed that the meter should be called "world yard;" the liter, "world quart;" and the half kilogram, "world pound."

1932   The Amateur Athletic Union and the sports programs of several universities converted to the metric system for field events to be in sync with the Olympic Games.

1947 The Trust Territory of the Pacific Islands was established by the United Nations and administered by the United States. The Trust included Palau, Micronesia, and the Marshall Islands. These three nations are the only countries in Oceania that have not adopted the metric system.

1951 In Japan. the metric system is introduced during the Allied occupation with US General Douglas MacArthur as supreme commander.

1954 India adopted the metric system.

1959 The United States, United Kingdom, Australia, Canada, New Zealand, and South Africa agreed to a common yard and pound, called "international yard" and "international pound," which were defined in metric terms (international yard = 0.9144 meters; international pound = 0.45359237 kg). No agreement was reached on a common pint. The "US survey foot," slightly longer than the "international foot," was not changed; this created problems and confusion due to the simultaneous use of two nearly identical versions of the same units.

1960 The International System of Units (SI), established by the BIPM, became the modern form of the metric system (but in regular use it is still called "metric system"). The SI has six base units: meter, kilogram, second, ampere, kelvin, and candela. The meter was redefined as the "length equal to 1 650 763.73 wavelengths in vacuum of the radiation corresponding to the transition between the levels $2p_{10}$ and $5d_5$ of the krypton 86 atom."

1965 England committed to the metric system.

1967 Ireland and South Africa adopted the metric system.

1968 The US Congress authorized the Secretary of Commerce to make a three-year study (the Metric Study Act) to determine the advantages of increased use of the metric system.

1969 New Zealand adopted the metric system.

1970 Canada adopted the metric system.

1971 The NBS conducted the report to Congress and presented it with the title *A Metric America, a Decision Whose Time Has Come*.

In United Kingdom and Ireland, "Decimal Day" (February 15th). British currency switched over from the system of pounds, shillings, and pence (1:20:12) to a decimal system (1 pound: 100 pence).

1972	Lewis M. Branscomb, director of the NBS, proposed a non-mandatory policy toward metrication.
1973	The State Department of Transportation in Ohio placed two pairs of road signs along an interstate highway showing distances in kilometers and miles. The try was unsystematic and never expanded.
1974	Congress passed the "Education Amendments," which included a section stating that it is the policy of the United States to encourage (but not force) educational institutions to prepare students "to use the metric system with ease."
1975	Metric Conversion Act of 1975 is passed. The US Metric Board was established to coordinate a voluntary conversion to the metric system.
1979	Some gas stations, mainly in California, changed their fuel pumps to sell gasoline by the liter.
1982	The US Metric Board was disbanded by the Reagan administration.
1983	The meter was redefined as "the length of the path travelled by light in vacuum during a time interval of 1/299,792,458 of a second."
1988	The National Bureau of Standards changed its name to the National Institute of Standards and Technology (NIST). The institution was reorganized, but it remained in the Department of Commerce.
1999	NASA lost the Mars Climate Orbiter due to a measurement mismatch in software supplied by a spacecraft builder company that produced results in customary units while a second system, supplied by NASA, used metric units.
2013	Bill introduced in the Hawaii House of Representatives that would have made the metric system mandatory within that state.
2015	Samoa adopted the metric system, the latest national adoption. This completed the transition to the metric system by all member countries of the Commonwealth of Nations.
	A bill was introduced in the Oregon State Senate that would have established the metric system as the official unit of measurement within that state.

2019   The BIPM redefied four of the seven base units of the International System (including the kilogram) in terms of physical constants.

2023   NIST, the National Geodetic Survey, and other federal agencies indicated that beginning on January 1, the US survey foot should be avoided, "except for historic and legacy applications." It was superseded by the international foot definition of 1 foot = 0.3048 meters exactly. The US survey foot was phased out to facilitate the modernization of the National Spatial Reference System.

2024   Five countries in the world remain non-committed to adopt the metric system as their exclusive system of measurement: the United States, Liberia, Palau, Micronesia, and the Marshall Islands.

NOTES

Introduction

1. Melvil Dewey Papers, Columbia University, Box 66, File "Metric System—Associations—American Metric Bureau—Publicity."
2. "Progress and Metrics," *Daily News*, November 15, 1945, 9C.
3. Leo O'Mealia, "The Family," cartoon, *Daily News*, November 15, 1945, 31.
4. On the importance of the things that did not happen for historical and sociological analysis, see Rebecca Jean Emigh, "The Power of Negative Thinking: The Use of Negative Case Methodology in the Development of Sociological Theory," *Theory and Society* 26 (1997): 649–84.
5. On how this idea can be overstated, see Peter Berger, "The Cultural Dynamics of Globalization," in *Many Globalizations: Cultural Diversity in the Contemporary World*, ed. Peter Berger (Oxford: Oxford University Press, 2002), 2.
6. "Measurement," *Oxford English Dictionary*, 3rd ed., http://www.oed.com/view/Entry/115513.
7. A. Hunter Dupree, "The Measuring Behavior of Americans," in *Nineteenth-Century American Science: A Reappraisal*, ed. (Evanston: Northwestern University Press, 1972), 26; A. Hunter Dupree, "Metrication as Cultural Adaptation," *Science* 185 (1974): 208.
8. A study by the US Department of Commerce estimated that the average adult makes use of at least fifty measurements each day (more than 20,000 measurements per year). See Robert Frederick Smith, "Cognitive Effects of the Introduction of a New Measurement Language into American Culture" (PhD diss., New York University, 1978), 2.
9. John Quincy Adams, *Report of the Secretary of State, upon Weights and Measures* (Washington, DC: Gales & Seaton, 1821), 119–20.
10. Roger Chartier, *The Cultural Uses of Print in Early Modern France* (Princeton, NJ: Princeton University Press, 1988), 183; Hector Vera, "Quantitative Measurement and the Production of Meaning," in *The Oxford Handbook of Symbolic Interactionism*, ed. W. H. Brekhus, T. DeGloma, and W. R. Force (New York: Oxford University Press, 2021), 104–21.

11. In the present our basic measuring standards and instruments work so (apparently) effortlessly, that some of the tools that for centuries were central to ensure some degree of correctness and permanence in measurement, like public standards of length displayed in public building, are today completely ignored and are considered extinct and obsolete objects: Emanuele Lugli, *The Making of Measure and the Promise of Sameness* (Chicago: University of Chicago Press, 2022), xi–xvii; David Rooney, "Public Standards of Length," in: *Extinct: A Compendium of Obsolete Objects*, ed. B. Penner, A. Forty, O. H. Turner, and M. Critchley (London: Reaktion Books, 2021), 260–63.
12. Ken Alder, *The Measure of all Things* (New York: The Free Press, 2002), 1–2.
13. John Perry, *The Story of Standards* (New York: Funk & Wagnalls, 1955), 5.
14. Personal communication, June 22, 2010.
15. Carmen J. Giunta, *A Brief History of the Metric System: From Revolutionary France to the Constant-Based SI* (Cham: Springer Nature, 2023), 69–78; James Vincent, *Beyond Measure: The Hidden History of Measurement from Cubits to Quantum Constants* (London: Faber & Faber, 2022); Claire Cock-Starkey, *The Curious History of Weights & Measures* (Chicago: University of Chicago Press, 2023). For recent studies on quantification: Steffen Mau, *The Metric Society: On the Quantification of the Social* (London: Polity, 2019); Jerry Z. Muller, *The Tyranny of Metrics* (Princeton, NJ: Princeton University Press, 2019); Marion Fourcade and Kieran Healy, *The Ordinal Society* (Cambridge: Harvard University Press, 2024).
16. Dan Bouk, *Democracy's Data: The Hidden Stories in the U.S. Census and How to Read Them* (New York: Farrar, Straus and Giroux, 2022); Caitlin Rosenthal's *Accounting for Slavery: Masters and Management* (Cambridge: Harvard University Press, 2018). Lugli, *The Making of Measure*; Aashish Velkar, *Markets and Measurements in Nineteenth-Century Britain* (Cambridge: Cambridge University Press, 2012); Aashish Velkar, "'Imperial Folly': Metrication, Euroskepticism, and Popular Politics in Britain, 1965–1980," *The Journal of Modern History* 92 (2020): 561–601.
17. Witold Kula, *Measures and Men* (Princeton, NJ: Princeton University Press, 1986), 94.
18. Otis Dudley Duncan, *Notes on Social Measurement* (New York, Russell Sage Foundation, 1984), 12–38.
19. Duncan, *Notes on Social Measurement*, xiii.
20. Duncan, *Notes on Social Measurement*, vii, 26–27.
21. Kula, *Measures and Men*, 185–227.
22. Ken Alder, "A Revolution to Measure," in *The Values of Precision*, ed. Norton Wise (Princeton, NJ: Princeton University Press, 1994), 43; Ronald E. Zupko, *French Weights and Measures before the Revolution: A Dictionary of Provincial and Local Units* (Bloomington: Indiana University Press, 1968).
23. Karl Marx, *Capital* (New York: Penguin, 1990), 163–64.
24. Alexis de Tocqueville, *The Ancien Régime and the French Revolution* (Cambridge: Cambridge University Press, 2011), 183.
25. Tocqueville, *The Ancien Régime*, 39.

26. Yann Fauchois, "Centralization," in Furet and Ozouf, *Critical Dictionary of the French Revolution*, 629–30.
27. Edmund Burke, *Reflections on the Revolution in France* (New Haven, CT: Yale University Press, 2003), 167.
28. Alder, "A Revolution to Measure," 39. On language unification in revolutionary France, see Brigitte Schlieben-Lange, *Ideologie, revolution et uniformité de la langue* (Hayen: Mardaga, 1996), 99–134.
29. Charles Coulston Gillispie, Science and Polity in France,: The Revolutionary and Napoleonic Years (Princeton, NJ: Princeton University Press, 2004), 458–94; Denis Guedj, *The Measure of the World* (Chicago: University of Chicago Press, 2001).
30. On the gritty effort needed to fulling some of these tasks, see Alder, *The Measure of All Things*, 74; Lugli, *The Making of Measure*, 25–31; Hector Vera, "Counting Measures: The Decimal Metric System, Metrological Census, and State Formation in Revolutionary Mexico, 1895–1940," *Histoire & Mesure* 32 (2017): 121–40.
31. Mona Ozouf, "Regeneration," in Furet and Ozouf, *A Critical Dictionary of the French Revolution*, 781–91.
32. Tocqueville, *The Ancien Régime*, 1.
33. Tocqueville, *The Ancien Régime*, 21.
34. Kula, *Measures and Men*, 185–227; John Markoff, *The Abolition of Feudalism* (University Park: Pennsylvania State University Press, 1996), 30–36.
35. On the different proposals discussed, see Charles Coulston Gillispie, *Science and Polity in France*, 235–49.
36. Isabel F. Knight, *The Geometric Spirit: The Abbé de Condillac and the French Enlightenment* (New Haven, CT: Yale University Press, 1968), vii–viii, 18–20.
37. Jean Starobinski, *1789, The Emblems of Reason* (Charlottesville: University Press of Virginia, 1982), 69.
38. Emmet Kennedy, *A Cultural History of the French Revolution* (New Haven, CT: Yale University Press, 1989), 77–81.
39. José Ortega y Gasset, *The Modern Theme* (New York: Harper, 1961), 131.
40. Octavio Paz, *The Labyrinth of Solitude* (New York: Grove Press, 1961), 125.
41. On the optical telegraph in the context of the creation of the metric system, see Denis Guedj, *Le mètre du monde* (Paris: Editions du Seuil, 2000), 118–19, 141–42.
42. Victor Hugo, *Les Misérables* (New York: Kelmscott Society, 1887), 18–19.
43. Alejo Carpentier, *Explosion in a Cathedral* (Boston: Little, Brown and Company, 1963), 7.
44. Jonathan H. Grossman, "Standardization (Standardisation)," *Critical Inquiry* 44 (2018): 450.
45. Burke, *Reflections on the Revolution in France*, 147; also 46–47, 155 and 167–68.
46. Guedj, *Le mètre du monde*, 269.
47. Condorcet, "Observations on the Twenty-Ninth Book of *The Spirit of Laws*," annexed in Antoine Louis Claude Destutt de Tracy, *A Commentary and Review of Montesquieu's Spirit of Laws* (Philadelphia: William Duane, 1811), 273.
48. John L. Heilbron, *Weighing Imponderables and Other Quantitative Science Around 1800* (Berkeley: University of California Press, 1993), 249.

49. On decimalization as a general tendency during the revolution, see Alder, *The Measure of all Things*, 125–159; Kula, *Measures and Men*, 250–51.
50. Hector Vera, "Decimal Time: Misadventures of a Revolutionary Idea, 1793–2008," *KronoScope: Journal for the Study of Time* 9 (2009): 33–37.
51. These words came from a member of the Commission of Weights and Measures of the Cisalpine Republic in 1801. Quoted in Kula, *Men and Measures*, 83; see also 250.
52. Rebecca L. Spang, *Stuff and Money in the Time of the French Revolution* (Cambridge: Harvard University Press, 2015), 249–59.
53. In 1704 Russia had implemented a partially decimal currency, with ruble divided into 100 kopeks; but other units did not follow a decimal progression. See Adrian Tschoegl, "The International Diffusion of an Innovation: The Spread of Decimal Currency," *Journal of Socio-Economics* 39 (2010): 104.
54. Deirdre Mask, *The Address Book: What Street Addresses Reveal about Identity, Race, Wealth, and Power* (New York: St. Martin Press, 2020), 110–28.
55. Dupree, "The Measuring Behavior of Americans," 30–31.
56. American Metric Bureau, "International Measures," *Bulletin of the American Metric Bureau* 20 (1878): 316.
57. Thomas Paine, *Common Sense: Addressed to the Inhabitants of America* (London: D. Jordan, 1792), 59.
58. "Kilometers and Miles on Ohio Road Signs," *New York Times*, February 18 (1973): 51.
59. Charles Latimer, *The French Metric System or the Battle of the Standards* (Chicago: Thomas Wilson, 1880), 7.
60. Samuel S. Dale, *The Foreign Attack on Our Weights and Measures* (Boston: self-published, 1926), 1; see also "The Itch of Change," *The Wall*, February 18 (1920): 9.
61. Jean D. Steed, "Cowboy Hall Director Assails Courts after Losing Metric System Challenge," *The Oklahoman*, December 1, 1981; "Is the Metric System a Communist Plot?" *CBC News Canada*, https://www.cbc.ca/player/play/1775836699.
62. On how populists frame metrication and how it is used to exploit anxieties about globalization, see Velkar, "'Imperial Folly': Metrication, Euroskepticism, and Popular Politics in Britain," 561–601.
63. Dan Satherley, "Metric System 'Creepy', Symbol of 'Tyranny': Fox News Host Tucker Carlson," *Newshub*, July 6, 2019, https://www.newshub.co.nz/home/world/2019/06/metric-system-the-yoke-of-tyranny-fox-news-host-tucker-carlson.html.
64. John Marciano, *Whatever Happened to the Metric System? How America Kept Its Feet* (New York: Bloomsbury, 2014), 244.
65. Douglas V. Frost, "Logical Steps to Metric Conversion," *Poultry Science* 44 (1965): 1227.
66. Stephen Mihm, "Inching toward Modernity: Industrial Standards and the Fate of the Metric System in the United States," *Business History Review* 96 (2022): 47–76; Edward Franklin Cox, "The International Institute: First Organized Opposition to the Metric System," *Ohio Historical Quarterly* 58 (1959): 54–83.
67. Ian Bartky, "The Adoption of Standard Time," *Technology and Culture* 30 (1989): 34–48.
68. Perry, *The Story of Standards*, 13.

69. Anne Hanley, "Men of Science and Standards: Introducing the Metric System in Nineteenth-Century Brazil," *Business History Review* 96 (2022): 18.
70. Héctor Vera, "Counting Measures," 130–33; Andrew Coyne, "50 Years Later, Metric Still Hasn't Won the Day in Canada," *The Globe and Mail*, January 22 (2020), https://www.theglobeandmail.com/opinion/article-50-years-later-metric-still-hasnt-won-the-day-in-canada; Eugen Weber, *Peasants into Frenchmen* (Stanford, CA: Stanford University Press, 1976): 30–33.
71. Doug McAdam, Sidney Tarrow, and Charles Tilly, *Dynamics of Contention* (New York: Cambridge University Press, 2001), 78.
72. Michael Mann, "The Autonomous Power of the State: Its Origins, Mechanisms and Results," *European Journal of Sociology* 25 (1984): 192.
73. On the concept of "symbolic power," a term originally coined by Pierre Bourdieu, see Mara Loveman, "The Modern State and the Primitive Accumulation of Symbolic Power" *American Journal of Sociology* 110 (2005): 1651–83.
74. Andrew L. Russell, *Open Standards and the Digital Age: History, Ideology, and Networks* (Cambridge: Cambridge University Press, 2014), 42.
75. While in other countries metrological legislation is brief and inclusive, in the US compendiums of local laws regulating weights and measures amounted to more than nine hundred pages. See National Bureau of Standards, *Federal and State Laws Relating to Weights and Measures* (Washington, DC: Bureau of Standards, 1926); Jessie V. Coles, *The Consumer-Buyer and the Market* (New York: John Wiley & Sons, 1938), 522–28.
76. Robert R. Jenks, "Governing in the Absence of Government: The Birth and Development of the United States Industrial Standards System" (PhD diss., University of California, Santa Barbara, 1999).
77. Officials are so frightened by the controversy caused by any form of banning, that they coin exotic formulae to curb the use of subpar technological products. In the case of incandescent lightbulbs, they tried to convince the public that it is not a ban, and they call the lightbulb-limiting guidelines "restrictions" and "efficiency standards." Nevertheless, a lot of space has been used in mass media to demonize compact fluorescent light—the maligned "pigtails"—and other energy-saving innovations. Since 2007 the Energy Department has issued and then rolled back several energy efficiency regulations; regulations that were not even strict banning of all incandescent light bulbs. Adam Sternbergh, "The Fight Over Plastic Bags Is About a Lot More Than How to Get Groceries Home," *New York Magazine* July 15, 2015, https://nymag.com/intelligencer/2015/07/plastic-bag-bans.html; Rachel DuRose, "What's Going on with Your Lightbulbs?" *Vox*, Aug 12, 2023, https://www.vox.com/science/2023/8/12/23827110/light-bulb-incandescent-led-energy-efficiency-ban-explained.
78. On the Conference of the Metre of 1875, see Terry Quinn, *From Artifacts to Atoms: The BIPM and the Search for Ultimate Measurement Standards* (New York: Oxford University Press, 2012), 74–87.
79. Rexmond C. Cochrane, *Measures for Progress* (Washington, DC: US Dept. of Commerce, 1966), 23.

80. Charles S. Peirce, "Review of Noel's *The Science of Metrology*," in *Writings of Charles S. Peirce* (Bloomington: Indiana University Press, 1982), 378.
81. Similar arguments have been articulated in different moments in time. For instance, an article in the *Pittsburgh Post* ("Pros and Cons of Metric System," June 26, 1927) expressed that "If we were without any system of weights and measures and were called upon to choose between the system now in vogue and the metric system, there is scarcely any doubt that we would favor the latter. A sudden change under existing conditions would, however, result in intolerable confusion." See also Jonathan Hogeback, "Why Doesn't the U.S. Use the Metric System?" *Encyclopedia Britannica*, November 17 (2016): https://www.britannica.com/story/why-doesnt-the-us-use-the-metric-system.
82. Philipp Genschel, "Path-Dependence," in *International Encyclopedia of Economic Sociology*, J. Beckert and M. Zafirocski eds. (London: Routledge, 2005), 507–8; Paul David, "Clio and the Economics of QWERTY," *American Economic Review* 75 (1995): 332–37.
83. Daniel Immerwahr, *How to Hide an Empire: A History of the Greater United States* (New York: Farrar, Straus and Giroux, 2020), 339.
84. Richard L. Hopkins, *Origin of the American Point System for Printers' Type Measurement* (Terra Alta, WV: Hill & Dale Press, 1976).
85. Manuel Castells, *The Information Age, vol. III End of Millennium* (Oxford: Blackwell, 2010), 32.
86. I owe this apt formulation to one the anonymous reviewers.

Chapter 1

1. "Mesure," in *Encyclopédie, ou Dictionnaire raisonné des sciences, des arts et des métiers, par une société des gens de letters* (Paris: De l'imprimerie de Le Breton, imprimeur ordinaire du Roy, 1765), 10: 424.
2. Maurice Crosland, "The Congress on Definitive Metric Standards, 1798–1799: The First International Scientific Conference?" *Isis* 60 (1969): 226–231.
3. These were the Batavian Republic, Cisalpine Republic, Denmark, Helvetian Republic, Ligurian Republic, Kingdom of Sardini, Spain, Roman Republic, and Tuscany.
4. Hector Vera, "Weights and Measures," in *Blackwell Companion to the History of Science*, ed. Bernard Lightman (Oxford: Blackwell Publishers, 2016), 463–64.
5. Vera, "Weights and Measures," 463–64.
6. George Sarton, "The Spread of Understanding," in *The Life of Science: Essays in the History of Civilization* (Bloomington: Indiana University Press, 1960), 7.
7. Sarton, "The Spread of Understanding," 11.
8. Sarton, "The Spread of Understanding," 11–12.
9. Karl Menninger, *Number Words and Number Symbols* (Cambridge: MIT Press, 1969), 389–445.
10. D. J. Struik, "The Prohibition of the Use of Arabic Numerals in Florence," *Archives Internationales d'Histoire des Sciences* 21 (1968): 291–94; D. J. Struik, *A Concise History of Mathematics* (New York: Dover Publications, 1987), 81.

11. Eviatar Zerubavel, "The Standardization of Time: A Sociological Perspective," *American Journal of Sociology* 88 (1982): 1–23.
12. Magdi Abdelhadi, "Muslim Call to Adopt Mecca Time," BBC, April 21, 2008; "Saudis hoping giant clock will set 'Mecca Time'" *AFP*, August 10, 2010.
13. Martin H. Greyer, "One Language for the World: The Metric System, International Coinage, Gold Standard, and the Rise of Internationalism, 1850–1900," in *The Mechanics of Internationalism*, ed. Martin Geyer and Johannes Paulmann (New York: Oxford University Press, 2001), 55–92.
14. On the concept of "social distribution of knowledge," see Peter Berger and Thomas Luckmann, *The Social Construction of Reality* (New York: Doubleday, 1966), 14–15, 127.
15. "Cálculo del costo que tiene una serie de los pesos y medidas del nuevo sistema métrico decimal," Archivo General de la Nación (AGN), Pesas y medidas, box 1, file 22.
16. There is a large body of literature on the republican calendar: Eviatar Zerubavel, "The French Republican Calendar: A Case Study in the Sociology of Time," *American Sociological Review* 42 (1977): 868–77; Noah Shusterman, *Religion and the Politics of Time: Holidays in France from Louis XIV through Napoleon*. (Washington, DC: The Catholic University of America Press, 2010), 98–236; Sanja Perovic, *The Calendar in Revolutionary France: Perceptions of Time in Literature, Culture, Politics* (Cambridge: Cambridge University Press, 2015); Bronislaw Baczko, "Le calendrier républicain," in *Les lieux de mémoire*, ed. Pierre Nora (Paris: Gallimard, 1997), 67–106; James Friguglietti, "Gilbert Romme and the Making of the French Republican Calendar," in *The French Revolution in Culture and Society*, ed. N. Andrews (New York: Greenwood Press, 1991), 13–22; Matthew John Shaw, *Time and the French Revolution: The Republican Calendar, 1789-year XIV* (Woodbridge: Royal Historical Society, 2011); Michael Dudzik, "The Decimalisation of Republican Time. The French Revolution Which Failed (1793–1795)," *Acta Polytechnica* 64 (2024), 182–93.
17. For works focused on the decimalization of time in the French revolution: Paul Smith, "La division décimale du jour," in *Genèse et diffusion du système métrique*, ed. J.-C. Hocquet and B. Garnier (Caen: Editions-diffusion du Lys, 1990), 123–34; Louis Marquet, "24 heures ou 10 heures? Un essai de division décimale du jour (1793–1795)," *L'Astronomie* 103 (June 1989): 285–290; Shaw, *Time and the French Revolution*, 131–137. On some subsequent attempts see Hector Vera, "Decimal Time," 29–48; Peter Galison, *Einstein's Clocks, Poincaré's Maps: Empires of Time* (New York: Norton, 2003), 153–73.
18. "Decree Establishing the French Era, November 25, 1793 (4 Frimaire, Year II)," in *A Documentary Survey of the French Revolution*, ed. John Hall Stewart (New York: Macmillan, 1951), 509.
19. "Decree Establishing the French Era," 509, 512.
20. Numerous pictures and descriptions of decimal clocks can be seen in *Les heures révolutionnaires*, eds. Yves Droz and Joseph Flores (Besançon: Association Française des Amateurs d'Horlogerie Ancienne, 1989).

21. On police regulation of time see Shusterman, *Religion and the Politics of Time*, 161–205; on the force of habit, Marquet, "24 heures ou 10 heures?," 287.
22. On the social consequences of eliminating the seven-day week (and Sundays) see: Eviatar Zerubavel, *The Seven Day Circle: History and Meaning of the Week* (New York: The Free Press, 1985), 27–35.
23. Vera, "Decimal Time," 36.
24. Adrian Tschoegl, "The International Diffusion of an Innovation," 101.
25. For different interpretations of the breakdown of decimal time and the republican calendar see: Zerubavel, "The French Republican Calendar," 874–76; Perovic, *The Calendar in Revolutionary France*, 238.
26. In the discussions about metrication it is widely assumed that there are only three non-metric countries (Liberia, Myanmar, and the United States), an unfounded assertion that has taken a life of its own and has been repeated thousands of times for more than two decades.
27. This theory was condensed in Everett Rogers, *Diffusion of Innovations* (New York: The Free Press, 1995).
28. David Easley and Jon Kleinberg, *Networks, Crowds, and Markets* (New York: Cambridge University Press, 2010), 498.
29. Easley and Kleinberg, *Networks, Crowds, and Markets*, 497–536.
30. These isolated voluntary adopters are France, Colombia, Mexico, Romania, Costa Rica, El Salvador, China, Bhutan, Saint Kitts and Nevis, Nauru, and Canada.
31. Easley and Kleinberg, *Networks, Crowds, and Markets*, 505, 509.
32. Theda Skocpol, *States and Social Revolutions* (New York: Cambridge University Press, 1979), 4.
33. See for example Bronislaw Baczko, "Rationaliser revolutionnairement," in *Les mesures et l'histoire*, ed. Institut D'Histoire Moderne et Contemporaine Centre National de la Recherche Scientifique (Paris: Éditions du Centre National de la Recherche Scientifique, 1984); Ken Alder, "A Revolution to Measure"; Kula, *Measures and Men*, 228–64; Yannick Marec, "L'ambition revolutionnaire: Mesurer toutes choses rationalment," in *La révolution Française et les processus de socialisation de l'homme moderne* (Paris: Éditions Messidor, 1989), 691–700.
34. Vera, "Weights and Measures," 462.
35. Endymion Wilkinson, *Chinese History: A Manual* (Cambridge: Harvard University Asia Center, 2000), 239–42, and in Kula, *Measurement and Men*, 284–86.
36. Michael D. Gordin, "Measure of All the Russias: Metrology and Governance in the Russian Empire," *Kritika* 4 (2003): 783–815; N. A. Shost'in, "History of Russian Metrology: D. I. Mendeleev and the Metric System of Measures," *Measurement Techniques* 4 (1968): 429–31; K. P. Shirokov, "Fifty Years of the Metric System in the USSR," *Measurement Techniques* 9 (1968): 1141–46.
37. Frederick Engels, *The Peasant War in Germany* (New York: Routledge, 2015), 24–25.
38. Karl Marx and Friedrich Engels, *The Communist Manifesto* (London: Penguin, 2002), 224.

39. W. G. Tinckom-Fernandez, "Turks Are Getting Many New Habits," *New York Times*, March 11, 1928; "Angora Bewilders By Swift Reforms," *New York Times*, June 3, 1928.
40. "Myanmar to Adopt Metric System," *Eleven Myanmar*, October 10, 2013.
41. Zertuche, *Estudio sobre pesas y medidas en los países centroamericanos* (United Nations, 1958).
42. G. D. Burdun, "Worldwide Dissemination of the Metric System," *Measurement Techniques* (1968): 1151.
43. Hector Vera, "The Social Life of Measures: Metrication in the United States and Mexico, 1789-2004" (PhD diss., New School for Social Research, 2012), 68.
44. Mexico is, by far, the country that had to wait the most to have a metric neighbor. France, the creator of the system, waited twenty-one years; Costa Rica had to wait twelve years; the other isolated voluntary adopters waited for ten years or less.
45. It has to be said, though, that after Mexico made metric compulsory in 1857, the United States accepted the metric as a legal, optional system in 1866, and Canada did the same in 1871, which means the region got fully integrated as a "soft metric" zone in less than fifteen years.
46. For a comparison between Canada and the United States, see Grace Ellen Watkins and Joel Best, "Successful and Unsuccessful Diffusion of Social Policy: The United States, Canada, and the Metric System," in *How Claims Spread: Cross-National Diffusion of Social Problems*, ed. J. Best (New York: Aldine de Gruyter, 2001), 267-81; and for a comparison between Mexico and the United States see Vera, "The Social Life of Measures."
47. Benjamin Constant, *Political Writings* (New York: Cambridge University Press, 1988), 73-74. This was an echo of older philosophical ideas in France against centralization and uniformity. See, for example, Montesquieu's thoughts on the idea of uniformity: "There are certain ideas of uniformity that sometimes seize great spirits (for they touched Charlemagne), but that infallibly strike small ones. They find in it a kind of perfection they recognize because it is impossible not to discover it: in the police the same weights, in commerce the same measures, in the state the same laws and the same religion in every part of it. But is this always and without exception appropriate? Is the ill of changing always less than the ill of suffering? And does not the greatness of genius consist rather in knowing in which cases there must be uniformity and in which differences? . . . When the citizens observe the laws, what does it matter if they observe the same ones?" *The Spirit of Laws* (Cambridge: Cambridge University Press, 1989), 617.
48. Kula, *Measurement and Men*, 268-69.
49. Kula, *Measurement and Men*, 275-77.
50. The French exportation of the metric measures through military power—at least within Europe—halted when Napoleon's soldiers marched through Prussia, Austria, and Poland, places where Paris did not try to implant the revolutionary system.
51. Arthur Kennelly, *Vestiges of Pre-Metric Weights and Measures Persisting in Metric-System Europe, 1926-1927* (New York: Macmillan, 1928), viii.

52. These 38 countries are Algeria, Angola, Benin, Burundi, Cambodia, Cape Verde, Chad, Comoros, Côte d'Ivoire, Cuba, Democratic Republic of Congo, Djibouti, Equatorial Guinea, Guinea, Guinea-Bissau, Indonesia, Libya, Macau, Malta, Mauritania, Mauritius, Morocco, Mozambique, Niger, Philippines, Puerto Rico, Republic of the Congo, Rwanda, Sao Tome and Principe, Senegal, Seychelles, Suriname, Syria, Timor-Leste, Togo, Tunisia, Vietnam, and Western Sahara.

53. This is 37 out of 150 countries.

54. Not directly about the metric system, but illustrative of the problems to standardize measurement units in a colony is Debdas Banerjee, *Colonialism in Action* (New Delhi: Orient Longman Limited, 1999), 49–52.

55. *Rapport du jury centrale sur les produits de l'agriculture et de l'industrie exposés en 1849*, vol. I (Paris: Imprimerie Nationale, 1850), lxiv, quoted by Arthur Chandler, "Exposition of the Second Republic: Paris 1849," http://www.arthurchandler.com/seond-republic-exposition-1849. For a sense of the context in which the metric system was introduced in the French colonies see Tom M. Hill, "Imperial Nomads: Settling Paupers, Proletariats, and Pastoralists in Colonial France and Algeria, 1830–1863" (PhD diss., University of Chicago, 2006): 222.

56. These are the year of the introduction of the metric system in other French territories and departments: Guadeloupe, 1844; Martinique, 1844; Réunion, 1839; New Caledonia, 1862; Polynesia, 1847. See Moreau, *Le système métrique* (Paris: Chiron, 1975), 107.

57. Toby Lester, "New-Alphabet Disease?," *Atlantic Monthly*, July, 1997, 20.

58. After Niyazov's death, and due to popular demand, the old names of the calendar were restituted in 2008. See Paul Theroux, "The Golden Man," *The New Yorker*, May 28, 2007, 57; "Turkmen go Back to Old Calendar," *BBC News*, April 24, 2008, http://news.bbc.co.uk/2/hi/asia-pacific/7365346.stm.

59. Neil MacFarquhar, "Libya under Qaddafi: Disarray Is the Norm," *New York Times*, February 14, 2001.

60. The only exception to this trend that I have heard about is China, where allegedly Mao restituted traditional measures (at least in name, because the magnitudes of those traditional measures were standardized in metric equivalences) as a sign of anti-imperialism and cultural pride. But I have not been able to find solid references to confirm it. In any case this was a short-lived restoration; in 1984 the State Council ordered that the metric system would be the sole legal system in the country after 1990. See Endymion Wilkinson, *Chinese History*, 240.

61. Nadeem Badshah, "Boris Johnson to Reportedly Bring Back Imperial Measurements to Mark Platinum Jubilee," *The Guardian*, May 28, 2022.

62. Department for Business and Trade, "Choice on Units of Measurement: Consultation Response," December 27, 2023, https://www.gov.uk/government/consultations/choice-on-units-of-measurement-markings-and-sales/outcome/choice-on-units-of-measurement-consultation-response.

63. Feza Günergun, "Introduction of the Metric System to the Ottoman State," in *Transfer of Modern Science and Technology to the Muslim World*, ed. Ekmeleddin Ihsanoglu (Istanbul: Research Centre for Islamic History, Art, and Culture, 1992), 297–316.

64. In Eastern Asia, Japan had passed a law making the metric optional in 1893, but it brought very little penetration, see Iwata, "Weights and Measures in Japan," 1022.
65. Egypt had actually a long metric past, as the system had been accepted on a permissive basis since 1873, a status ratified in further legislations in 1892 and 1914.
66. On Australia's to metrication see, Jan Todd, *For Good Measure: The Making of Australia's Measurement System* (Crows Nest: Allen & Unwin, 2004).
67. "The Metric System," *New York Tribune*, December 23, 1902; Vera, "The Social Construction of Units of Measurement," 183.
68. Robert K. Merton, "Self-Fulfilling Prophecy," in *Social Theory and Social Structure* (New York: The Free Press, 1968), 477.
69. See, for example, Andro Linklater, *Measuring America*, 242.
70. On the metric system in Spain see Juan Gutiérrez Cuadrado, *Metro y kilo: el sistema métrico decimal en España* (Madrid: Akal, 1997); José Ramón Gómez Martínez, María Teresa Sánchez Trujillano, and José Antonio Tirado Martínez, *¿Y esto en onzas cuánto es?: 1853–2003* (Logroño: La Rioja, 2003); Gustavo Puente Feliz, "El Sistema métrico decimal. Su importancia e implantación en España," *Cuadernos de Historia Moderna y Contemporánea* 3 (1982): 95–125; Antonio E. Ten, "El sistema métrico decimal y España," *Arbor. Ciencia, Pensamiento y Cultura* 527–528 (1989): 101–122.
71. What should be said about Spain's impact in Latin America, however, is that publications by Spanish scientists were highly influential in intellectual circles in the Americas; in particular, the books by the military engineer and mathematician Gabriel Ciscar, one of Spain's representatives to the Paris conference of 1799, served as entrance door for men of knowledge who sought technical literature, like Gabriel Ciscar's *Memoria elemental sobre los nuevos pesos y medidas decimales fundados en la naturaleza* (Madrid: Imprenta Real, 1800).
72. On metrication in Austria see Kennelly, *Vestiges of Pre-Metric Weights and Measures*, 96–101; William Hallock and Herbert Treadwell Wade, *Outlines of the Evolution of Weights and Measures and the Metric System* (New York: Macmillan, 1906), 86.
73. Quoted in Alfred Crosby, *The Measure of Reality* (New York: Cambridge University Press, 1997), 88.
74. Robert Poole, "'Give Us Our Eleven Days!': Calendar Reform in Eighteenth-Century England," *Past and Present* 149 (1995): 95–139.
75. H. Peyton Young, "The Economics of Convention," *The Journal of Economic Perspectives* 10 (1996), 105–22.
76. For example, Linklater, *Measuring America* (New York: Walker and Company, 2002), 248–58.
77. On the creation of the imperial system see Ronald E. Zupko, *Revolution in Measurement: Western European Weights and Measures since the Age of Science* (Philadelphia: American Philosophical Society, 1990), 176–207.
78. See the articles on "The Colonial Conference and the Metric System" published in London's *Times*, August 23, 28, 29, September 3, 4, 11, and October 13, 1902. Also, House of Commons. Papers by Command, volume 77 (London: Darlyn & Son, 1906), 54; "Weights and Measures (Metric System) Bill," in *The Parliamentary*

Debates: Fourth Series, Second Session of the Twenty-Eighth Parliament of the United Kingdom of Great Britain and Ireland, volume 171 (London: Wyman and Sons, 1907), 1313.
79. Vera, "Weights and Measures," 466.
80. Federico Beigbeder Atienza, *Manual de pesos, medidas y monedas del mundo con equivalencias al sistema métrico decimal* (Madrid: Castilla, 1959).
81. Statistical Office of the United Nations, *World Weights and Measures* (New York: United Nations, 1966), 103–30.
82. On how measures disappear, see Héctor Vera, *A peso el kilo: Historia del sistema métrico decimal en México* (Mexico: Libros del Escarabajo, 2007), 145–50.
83. Shigeo Iwata, "Weights and Measures in Japan," in *Encyclopaedia of the History of Science, Technology, and Medicine in Non-Western Cultures*, ed. Helaine Selin(Boston: Kluwer Academic, 1997), 1021.
84. Andreas Wimmer and Brian Min, "From Empire to Nation-State: Explaining Wars in the Modern World, 1816–2001," *American Sociological Review* 71 (2006): 872.
85. These 24 former colonies are Egypt, Sudan, Madagascar, Burkina Faso, Central African Republic, Gabon, Mali, Somalia, Cameroon, Nigeria, Ethiopia, Namibia, Kenya, Tanzania, Uganda, South Africa, Botswana, Swaziland, Zimbabwe, Lesotho, Ghana, Gambia, Malawi, and Sierra Leone.
86. Anil Kumar Acharya, *History of Decimalisation Movement in India* (Calcutta: Indian Decimal Society, 1958); Lal C. Verman and Jainath Kaul, eds., *Metric Change in India* (New Delhi: Indian Standards Institution, 1970).
87. Sujay Rao Mandavilli, "Metrication in India," July 22 (2009), https://usma.org/metrication-in-other-countries#india.
88. Marie Tyler-McGraw, *An African Republic Black and White Virginians in the Making of Liberia* (Chapel Hill: University of North Carolina Press, 2009), 7.
89. Robin Dopoe, "Gov't Pledges Commitment to Adopt Metric System," *Liberian Observer*, May 25, 2018.
90. *The Europa World Year 2004* (London: Europa Publications, 2004), 2835, 2903, 3304.
91. Bureau of Standards, *War Work of the Bureau of Standards* (Washington, DC: Government Printing Office, 1921), 220–21; Bureau of Standards, *Metric Manual for Soldiers* (Washington, DC: Government Printing Office, 1918).
92. Sheldon S. Cline, "Metric System Urged for U.S. and England. Only Two Civilized Nations Retaining Ancient and Cumbersome Weights and Measures," *Evening Star*, June 15, 1919.
93. "A.A.U. Convention Adopts Metric Plan," *New York Times*, November 23, 1932, 23.
94. This important observation was made by one of the anonymous reviewers.
95. Ian Whitelaw, *A Measure of All Things: The Story of Man and Measurement* (New York: St Martin's Press, 2007), 50–51.
96. Ken Hoke, "Aviation's Crazy, Mixed Up Units of Measure," *Aerosavvy: Aviation Insight*, September 5 (2014), https://aerosavvy.com/metric-imperial.
97. Aristide R. Zolberg, *A Nation by Design: Immigration Policy in the Fashioning of America* (New York: Russell Sage Foundation, 2006), 461–75.

98. Pew Research Center, "Immigrant Share of the U.S. Population, 1850 to 2022," https://www.pewresearch.org/short-reads/2024/07/22/key-findings-about-us-immigrants/sr_24-07-22_immigrant-facts_1.

Chapter 2

1. It was not rare in history for authorities to publicly destroy measures in displays that had the aura of a ceremonial exhibition of justice. In England, the walls of the Exchequer showed the image of Henry VII's "trial of weights and measures," in which the king is superintending the proof of standards in the Exchequer chamber; the picture shows how some clerks verify the proper dimensions and functioning of bushels, barrels, and balances, while another one is burning short measures. In general, the public destructions of short measures served to reduce widespread suspicions about the inefficiency of inspectors to guarantee that storekeepers' scales would not be fraudulent or noncompliant. At a deeper level, these actions functioned as ritualistic purification in which the authorities destroyed the instruments of corrupt practices and restored proper moral conduct in the market. See Charles Knight, *The Popular History of England* vol. II (London: Bradbury and Evans, 1856), 251–52; Stephen Mihm, "Whole Foods and the Old Thumb on the Scale," *Bloomberg* July 2 (2015), https://www.bloomberg.com/view/articles/2015-07-02/whole-foods-and-the-old-thumb-on-the-scale.
2. Department of Consumer Affairs, Mayor's Bureau of Weights and Measures, city of New York, *What the Purchasing Public Should Know* (New York: J. W. Pratt Company, 1911), 5.
3. "Says Housewives Lose Money by Short Measures," *Evening Journal* (Wilmington, DE.), June 17, 1914, 1.
4. National Geographic Society, "A Wonderland of Science," *National Geographic Magazine* 27 (1915), 168.
5. Correspondence, A. W. Epright to Samuel Dale, June 19, 1924, CU, Samuel S. Dale Papers, box 2.
6. National Geographic Society, "A Wonderland of Science," 169.
7. On the coproduction of science and the state, see Patrick Carroll, *Science, Culture, and Modern State Formation* (Berkeley: University of California Press, 2006).
8. Theodore Porter, *Trust in Numbers* (Princeton, NJ: Princeton University Press, 1995), 26.
9. Meketre was a chancellor during the reign of Mentuhotep II in ancient Egypt, twentieth century B.C.
10. Metropolitan Museum of Art, "Model of a Granary with Scribes," accessed January 15, 2024, https://www.metmuseum.org/art/collection/search/545281.
11. Gordon Childe, *Man Makes Himself* (Wiltshire: Moonraker, 1981), 153–54; see also his *What Happened in History* (Baltimore: Penguin, 1961), 107–8. Denise Schmandt-Besserat proposed the theory that the oldest systems of writing derived from archaic forms of counting and measuring, see her *How Writing Came About* (Austin:

University of Texas Press, 1996). For more recent archeology work on measurement, math, writing, and state-making, see D. Alex Walthall, *Sicily and the Hellenistic Mediterranean World: Economy and Administration during the Reign of Hieron II* (Cambridge: Cambridge University Press, 2024), and Eleanor Robson, *Mathematics in Ancient Iraq: A Social History* (Princeton, NJ: Princeton University Press, 2009).

12. Childe, *Man Makes Himself*, 154.
13. Norbert Elias, *Time: An Essay* (Oxford: Blackwell, 1992), 50–53; Arnold Toynbee, *A Study of History: Abridgement of Volumes VII-X* (New York: Oxford University Press, 1987), 59.
14. Toynbee, *A Study of History*, 54–55.
15. Kula, *Men and Measures*, 18. As William Hallock and Herbert Treadwell Wade, observed, "As the fixing of weights and measures is manifestly an attribute of government, so any successful reforms must depend upon the character and strength of a particular government, and in order to influence neighboring countries the territory affected should be comparatively large and the number of its inhabitants considerable." William Hallock and Herbert Treadwell Wade, *Outlines of the Evolution of Weights and Measures* (New York: Macmillan, 1906), 81.
16. Kula, *Men and Measures*, 18–19.
17. Jean Bodin, *On Sovereignty* (New York: Cambridge University Press, 1992), 80–81.
18. Bodin, *On Sovereignty*, 81.
19. The *Oxford English Dictionary* actually reports that both meanings date from the fourteenth century.
20. A. E. Berriman, *Historical Metrology: A New Analysis of the Archaeological and the Historical Evidence Relating to Weights and Measures* (New York: Greenwood Press, 1953), 4 (see the statue of Gudea, ruler of Lagash, on 54). On King Henry see *Walton's Set of Kings*, British Museum, https://www.britishmuseum.org/collection/object/P_1870-0514-2901.
21. Kula, *Men and Measures*, 23.
22. Max Weber, "Politics as a Vocation," in *For Max Weber: Essays in Sociology* (New York: Oxford University Press, 1958), 78.
23. Elias developed this idea on what he called the "monopoly mechanism." See *The Civilizing Process* (Oxford: Blackwell, 2000), 268–77; Stephen Mennell, *The American Civilizing Process* (Cambridge: Polity, 2007), 11–17.
24. Elias, *Time: An Essay*, 53.
25. Pierre Bourdieu, "Rethinking the State: Genesis and Structure of the Bureaucratic Field," in *State/Culture,* ed. George Steinmetz (Ithaca: Cornell University Press, 1999), 59.
26. Max Weber, *Economy and Society* (Berkeley: University of California Press, 1978), 166.
27. Bourdieu, "Rethinking the State," 59.
28. John C. Torpey, *The Invention of the Passport: Surveillance, Citizenship, and the State* (New York: Cambridge University Press, 2000), 4–5.
29. On "multiple sovereignty" as a political phenomenon, see Charles Tilly, *European Revolutions, 1492–1992* (Cambridge, MA: Blackwell, 1993), 10; on the concept used

for the administration of weights and measures see Hector Vera, "Weights and Measures," 462–63; Heong Hong and Tan Miau Ing, "Contested Colonial Metrological Sovereignty: The *Daching* Riot and the Regulation of Weights and Measures in British Malaya," *Modern Asian Studies* 56.1 (2022): 407–26.
30. Kula, *Measures and Men*, 79.
31. Kula, *Measures and Men*, 80–81; Emanuele Lugli, *The Making of Measure*, 75-94.
32. Edward Cox, "The Metric System: A Quarter-Century of Acceptance (1851–1876)," *Isis* 13 (1958): 77; Vera, "Weights and Measures," 467.
33. Georges Gurvitch, *The Social Frameworks of Knowledge* (New York: Harper, 1972), 74; Norbert Elias, "Knowledge and Power," in *Knowledge and Society*, ed. N. Stehr and V. Meja (New Brunswick: Transaction Books, 1984), 251–91.
34. Gurvitch, *The Social Frameworks of Knowledge*, 73–78.
35. Bourdieu, "Rethinking the State," 56–57.
36. Bourdieu, "Rethinking the State," 61.
37. Bourdieu, "Rethinking the State," 67.
38. Benedict Anderson, *Imagined Communities* (London: Verso, 1991), 163–85.
39. Anderson, *Imagined Communities*, 184–85.
40. James C. Scott, *Seeing Like a State* (New Haven, CT: Yale University Press, 1998), 24–33.
41. Scott, *Seeing Like a State*, 24. As Porter also stresses, the problem in metrological matters of "separating knowledge from its local context" is shared by the political, economic, and scientific spheres. *Trust in Numbers*, 22.
42. Scott, *Seeing Like a State*, 27. This is also matter for the "administrative power" of the state based on the "storage of information," see Anthony Giddens, *The Nation-State and Violence* (Berkeley: University of California Press, 1987), 172–81.
43. Scott, *Seeing Like a State*, 30.
44. William J. Ashworth, "Metrology and the State: Science, Revenue, and Commerce," *Science* 19 (2004): 1314.
45. Ashworth, "Metrology and the State," 1315.
46. Ashworth, "Metrology and the State," 1316.
47. On the tight link of morals with measurement and standardization see Graeme J. N. Gooday, *The Morals of Measurement: Accuracy, Irony, and Trust in Late Victorian Electrical Practice* (New York: Cambridge University Press, 2004), 23–29; Lawrence Busch, "The Moral Economy of Grades and Standards," *Journal of Rural Studies* 16 (2000), 273–83.
48. See, for example, these verses from an ancient Egyptian *Tale of Woe*, which describes a time of decay and hunger in a place governed by a local tyrant: "His tax had been burdensome to me more than can be imagined. / [. . . The people said of his false grain measure:] / "How wicked is the carpenter that made it! / One sack becomes in it a sack and half." Quoted in Jan Assmann, *The Mind of Egypt: History and Meaning in the Time of the Pharaohs* (New York: Metropolitan Books, 1996), 292–93.
49. See also Deuteronomy 25:13–16; Leviticus 19:35–36.
50. Flavius Josephus, *Josephus: The Essential Writings* (New York: Kregel, 1990), 21.

51. Richard Sheldon, Adrian Randall, Andrew Charlesworth, and David Walsh, "Popular Protest and the Persistence of Customary Corn Measures: Resistance to the Winchester Bushel in the English West," in *Markets, Market Culture and Popular Protest in Eighteenth-Century Britain and Ireland*, ed. A. Randall and A. Charlesworth (Liverpool: Liverpool University Press, 1996), 27; Charles Piazzi Smyth, "Weights and Measures," in Edward Hine, *Flashes of Light. Part 2 of Forty-Seven Identifications of the British Nation with the Lost Ten tribes of Israel* (London: W. H. Guest, 1876), 77–78.
52. E. P. Thompson, "The Moral Economy of the English Crowd in the Eighteenth Century," in *Customs in Common: Studies in Traditional Popular Culture* (New York: The New Press, 1993), 193, 218.
53. Sheldon et al., "Popular Protest and the Persistence of Customary Corn Measures," 37.
54. Quoted in Ken Alder, *The Measure of all Things*, 329, 393. For other cases of metric opposition in France, see Gustave Tallent, *Histoire du Système Métrique* (Paris: Libraire H. Le Soudier, 1910), 88–95; Yannick Marec, "Autor des résistances au sysème métrique," in *Genèse et diffusion du système métrique*, ed. J.-C. Hocquet and B. Garnier, 135–144 (Caen: Editions-Diffusion du Lys, 1990). Around that time, Savoy and Piedmont were ready for the metric reform ordered by the king in 1845. In Sardinia peasants obstinately adhered to their customary measures, "either out of traditional attachment or because they could not afford—in the wake of a poor harvest—to acquire new [standards]," and they rioted against the introduction of the metric system. See Kula, *Measures and Men*, 275.
55. On the Quebra-Quilo revolt see Roderick J. Barman, "The Brazilian Peasantry Reexamined: The Implications of the Quebra-Quilo Revolt, 1874–1875," *Hispanic American Historical Review* 53 (1977): 401–424; Kim Richardson, *Quebra-Quilos and Peasant Resistance: Peasants, Religion, and Politics in Nineteenth-Century Brazil* (Lanham: University Press of America, 2012).
56. Héctor Vera, "Medidas de resistencia: grupos y movimientos sociales en contra del sistema métrico," in *Metros, leguas y mecates: Historia de los sistemas de medición en México*, eds. H. Vera and V. García Acosta (Mexico: CIESAS, 2011), 181–99; Francie R. Chassen López, *From Liberal to Revolutionary Oaxaca: The View from the South, Mexico 1867–1911* (University Park: Pennsylvania State University Press, 2004), 370–77.
57. James C. Scott, *Weapons of the Weak: Everyday Forms of Peasant Resistance* (New Haven, CT: Yale University Press, 1985), 31.
58. Jennifer Steinhauer, "Constitution Has Its Day (More or Less) in House," *New York Times*, January 7, 2011.
59. Theda Skocpol, *Social Policy in the United States* (Princeton, NJ: Princeton University Press, 1994), 33.
60. Constitution of the United States, article 1, section 8, clause 5. See also Jay Wexler, *The Odd Clauses: Understanding the Constitution through Ten of Its Most Curious Provisions* (Boston: Beacon Press, 2011), 21–38.

61. Among the many accounts of the history of weights and measures in the United States, see Edward Franklin Cox, "A History of the Metric System of Weights and Measures: With Emphasis on Campaigns for its Adoption in Great Britain and in the United States prior to 1914" (PhD diss., Indiana University, 1956); Charles F. Treat, *A History of the Metric System Controversy in the United States* (Washington, DC: National Bureau of Standards, 1971); Ellen Watkins, "Measures of Change: A Constructionist Analysis of Metrication in the United States" (PhD diss., Southern Illinois University, 1998); Lewis Judson and Louis E. Barbrow, *Weights and Measures Standards of the United States: A Brief History* (Washington, DC: National Bureau of Standards, 1976); Arthur Frazier, *United States Standards of Weights and Measures, their Creation and Creators* (Washington, DC: Smithsonian Institution Press, 1978); Arthur McCourbrey, "Measures and Measuring Systems," in *The Encyclopedia Americana. International Edition* (Danbury: Grolier, 1993), 18: 584–97.
62. Herbert Wade, "Weights and Measures," in *Encyclopaedia of the Social Sciences* (New York: The MacMillan Company, 1930), 15: 390. There are several directories of ancient measures, one of the more detailed is Zupko, *French Weights and Measures*.
63. The number of measures in the American colonies is mentioned by Andro Linklater, *Measuring America*, 104; unfortunately, he does not cite any authoritative reference. A. Hunter Dupree suggested that the colonies "had less diversity in their measurement system than any country in Europe," but he also does not offer evidence to support that claim—see his article "Measurement," in *The History of Science in the United States: An Encyclopedia*, ed. Marc Rothenberg (New York: Garland Publishing, 2001), 339. On the number of France's measures see Alder, "A Revolution to Measure," 43.
64. Ralph W. Smith, *The Federal Basis for Weights and Measures: A Historical Review of Federal Legislative Effort, Statutes, and Administrative Action in the Field of Weights and Measures in the United States* (National Bureau of Standards Circular 593. Washington, DC: National Bureau of Standards, 1958), 3.
65. *The Colonial Laws of New York, from the Year 1664 to the Revolution* (Albany: James B. Lyon, 1894), I: 64. A large variety of documents on this topic can be found in *American State Papers: Documents, Legislative and Executive of the Congress of the United States. Class X, Miscellaneous: Volume II* (Washington, Gales and Seaton, 1834), 538–750.
66. *Articles of Confederation*, article 9, paragraph 4.
67. *The Addresses and Messages of the Presidents of the United States, Inaugural, Annual, and Special, from 1789 to 1846*, ed. Edwin Williams (New York: Edward Walker, 1846), I: 34.
68. *The Addresses and Messages of the Presidents*, I: 37.
69. *The Addresses and Messages of the Presidents*, I: 42.
70. The Public Land Survey System was used for the cadaster and it literally shaped the land during the westward territorial expansion. On apportionment of Representatives in Congress, see Michel L. Balinski and H. Peyton Young, *Fair Representation: Meeting the Ideal of One Man, One Vote* (New Haven, CT: Yale University Press,

1982); on Jefferson's long-lasting interest on measurement, see I. B. Cohen, *The Triumph of Numbers* (New York: Norton, 2005), 73–80.
71. Thomas Jefferson, "Notes on the Establishment of a Money Unit, and of a Coinage for the United States," in *The Papers of Thomas Jefferson,* ed. Julian P. Boyd (Princeton, NJ: Princeton University Press, 1953), VII: 175–88.
72. Jefferson, "Notes on the Establishment of a Money Unit," 176.
73. *The Papers of Thomas Jefferson,* 602–75.
74. On Jefferson's work for these reports, see C. Doris Hellman, "Jefferson's Efforts towards Decimalization of United States Weights and Measures," *Isis* 16 (1931): 266–314; I. B. Cohen, *Science and the Founding Fathers* (New York: Norton, 1995), 102–8; Linklater, *Measuring America,* 103–16.
75. *The Papers of Thomas Jefferson,* XVI: 663–64.
76. The similarities between Jefferson's plan with that of the French scientists appointed by the French National Assembly (what would become the metric system) is not accidental. While in Paris, Jefferson met Condorcet and other French mathematicians and astronomers who participated in the invention of the metric system.
77. Keit Ellis, *Man and Money* (London: Priory Press, 1973), 96–97.
78. Deming, *Metric Power,* 21–23; McCourbrey, "Measures and Measuring Systems," 593–94.
79. For a compass to navigate the sea of documents produced by the federal government regarding weights and measures, see Jeanette Smith, "Take Me to Your Liter: A History of Metrication in the United States," *Journal of Government Information* 25 (1998): 419–38.
80. *The Addresses and Messages of the Presidents,* I: 335.
81. John Quincy Adams, *Report of the Secretary of State, upon Weights and Measures* (Washington, DC: Gales & Seaton, 1821), 91.
82. Adams, *Report of the Secretary of State,* 133.
83. Adams, *Report of the Secretary of State,* 134. To Adams's credit, when he became president, he advocated cooperation with other nations to find a definitive system. In his first Annual Message (1825), he said to Congress that "The establishment of a uniform standard of weights and measures was one of the specific objects contemplated in the formation of our constitution; and to fix that standard was one of the powers delegated by express terms, in that instrument to Congress. The governments of Great Britain and France have scarcely ceased to be occupied with inquiries and speculations on the same subject since the existence of our constitution; and with them it has expanded into profound, laborious, and expensive researches into the figure of the earth, and the comparative length of the pendulum vibrating seconds in various latitudes, from the equator to the pole. These researches have resulted in the composition and publication of several works highly interesting to the cause of science. The experiments are yet in the process of performance. Some of them have recently been made on our own shores, within the walls of one of our own colleges, and partly by one of our own fellow-citizens. It would be honorable to our country if the sequel of the same experiments should be countenanced by the patronage of our government, as

they have hitherto been by those of France and Great Britain." *The Addresses and Messages of the Presidents*, I: 591.
84. Smith, *The Federal Basis for Weights and Measures*, 7.
85. Sarah Ann Jones, "Weights and Measures in Congress: A Historical Summary of Events Culminating in the Weights and Measures Act of 1836" (M.A. Diss., George Washington University, 1935), 25. See also *Register of Debates in Congress: Comprising the Leading debates and Incidents of the Second Session of the Eighteenth Congress*, volume II (Washington, DC: Gales and Seaton, 1826), 2648–54.
86. Jones, "Weights and Measures in Congress," 32; Smith, *The Federal Basis for Weights and Measures*, 8–10; *Report of the Secretary of the Treasury of the Construction and Distribution of Weights and Measures* (Washington, DC: A. O. P. Nicholson, 1857).
87. A. Hunter Dupree, *Science in the Federal Government* (Baltimore: The Johns Hopkins University Press, 1986), 52.
88. S. W. Stratton, "Address to the Conference," *Weights and Measures: Thirteenth Annual Conference of Representatives from Various States held at the Bureau of Standards* (Washington, DC: Government Printing Office, 1921), 13.
89. See, as an illustration, "The Metric System," *Hudson Star*, December 5, 1927.
90. *Texas Agriculture Code* (chapter 13, § 13.022); "Arpent in Louisiana," accessed February 24, 2020, http://www.sizes.com/units/arpent_louisiana.htm. For information about Spanish measures in California and New Mexico see, for example, J. N. Bowman, "Weights and Measures of Provincial California," *California Historical Society Quarterly* 30 (1951): 315–38; John Baxter, "Measuring New Mexico's Irrigation Water: How Big is a Surco?," *New Mexico Historical Review* 75 (2000): 397–413; and Eric Perramond, "How Metrics Shape Water Politics in New Mexico: From Quantifying Governance to Active Monitoring," *Water Alternatives* 17 (2024): 455–68.
91. Eric Foner, *Reconstruction: America's Unfinished Revolution, 1863–1877* (New York: Harper, 2014), 308.
92. "An Act to Authorize the Use of the Metric System of Weights and Measures," in *The Statutes at Large, Treaties, and Proclamations of the United States of America from December 1865, to March 1867*, ed. George P. Sanger (Boston: Little, Brown, and Company, 1886), 339. For some reactions and polemics around this law, see "The Proposed Metric System," *Scientific American* 15 (July 21, 1866), 50; "Metric System of Weights and Measures," *Chicago Tribune*, May 25, 1866.
93. Charles Sumner, "The Metric System of Weights and Measures," in *The Works of Charles Sumner* (Boston: Lee and Shepard, 1876), X: 525, 539.
94. Sumner, "The Metric System," 539.
95. "Report of the Committee on Coinage, Weights, and Measures," *The Reports of the Committees of the House of Representatives, Made During the First Session Thirty-Ninth Congress, 1865–'66* (Washington, DC: Government Printing Office, 1866), 20; see also *The Metric System: A Compilation, Consisting of Extracts from the Report of the Committee of the House Of Representatives, and Law of Congress Adopting the System* (Philadelphia: J. B. Lippincott & Co., 1867).
96. Foner, *Reconstruction*, 365. On the role of the state at that time, see Richard Franklin Bensel, *Yankee Leviathan: The Origins of Central State Authority in America*,

1859–1877 (New York: Cambridge University Press, 1990), ix; for a different interpretation see: Stephen Skowronek, *Building a New American State: the Expansion of National Administrative Capacities, 1877–1920* (New York: Cambridge University Press, 1982), 30–31.
97. Skocpol, *Social Policy in the United States*, 21.
98. Foner, *Reconstruction*, 453–54.
99. American Enterprise Institute for Public Policy Research, *Metric Conversion Bills* (Washington, DC: American Enterprise Institute, 1974), 2.
100. On the works of the Bureau, see Chester Hall Page and Paul Vigoureux, eds. *The International Bureau of Weights and Measures 1875–1975* (Washington, DC: National Bureau of Standards, 1975).
101. Louis Albert Fischer, "History of the Standard Weights and Measures of the United States," *Bulletin of the Bureau of Standards* 1 (1905), 379; Victor F. Lenzen, "The Contributions of Charles S. Peirce to Metrology," *Proceedings of the American Philosophical Society* 109 (1965): 41; *Report of the Superintendent of the United States Coast Survey, 1890* (Washington, DC: Government Printing Office, 1891), 750- 751: "Meter and Kilogram," *Evening Star*, January 2 (1890), 5.
102. Judson, *Weights and Measures Standards of the United States*, 16–20.
103. According to the Mendenhall Order "1 avoirdupois pound = 453.5924277 grammes" and the "yard in use in the United States is equal to $\frac{3600}{3937}$ of the metre."
104. Smith, *The Federal Basis for Weights and Measures*, 17.
105. Dupree, *Science in the Federal Government*, 272.
106. "The Action of the House of Representatives on the Metric Bill," *Science* 3, April 24 (1896), 626–27; Treat, *A History of the Metric System Controversy*, 106.
107. Samuel Stratton to Melvil Dewey, December 13, 1902, CU, Melvil Dewey Papers, box 66.
108. Rexmond C. Cochrane, *Measures for Progress*, 33–38.
109. National Bureau of Standards, *Laws Concerning the Weights and Measures of the United States* (Washington, DC: Government Printing Office, 1904. For single state example see *Laws of the State of New York in Relation to Weights and Measures and Uniform Rules and Regulations* (Albany: J. B. Lyon Company, 1916).
110. On the origin and functions of the Bureau, see Cochrane, *Measures for Progress*; P. G. Agnew, "The Work of the Bureau of Standards," *Annals of the American Academy of Political and Social Science* 82 (1919): 278–88; Louis Albert Fischer, "Recent Developments in Weights and Measures in the United States," *Popular Science Monthly* 84 (1914): 345–69; Dupree, *Science in the Federal Government*, 271–77.
111. See Samuel Dale to George M. Bond, December 12, 1922, CU, Samuel S. Dale Papers, box 2.
112. Ralph Weir Smith, *Weights and Measures Administration* (Washington, DC: National Bureau of Standards, 1962), 79.
113. Smith, *Weights and Measures Administration*, 82.
114. Joan Koenig, "Uniformity in Weights and Measures Laws and Regulations," in *A Century of Excellence in Measurements, Standards, and Technology*, ed. David Lide (New York: CRC Press, 2002), 368–70.

115. On the Division of Simplified Practice see Peri E. Arnold, "The 'Great Engineer' as Administrator: Herbert Hoover and Modern Bureaucracy," *The Review of Politics* 42 (1980): 342–43; Lawrence Busch, "Herbert Hoover and the Construction of Modernity," *Journal of Innovation Economics & Management* 22 (2017): 29–55.
116. Newspaper clipping, in Aubrey Drury Papers, box 14. For more examples on the activities of the Division of Simplified Practice see "Mr. Hoover's Dictionary," *Time*, June 18, 1923.
117. Pan American Standardization Conference, *Primera Conferencia Panamericana para la Uniformidad de Especificaciones, reunida en Lima, Perú, del 23 de diciembre de 1924 al 6 de enero de 1Actas y demás documentación, publicación oficial* (Lima: E. Moreno, 1927), 93.
118. See, for example, "Address by the Secretary of Commerce, Hon. Herbert Hoover," *Weights and Measures: Sixteenth Annual Conference of Representatives from Various States held at the Bureau of Standards* (Washington, DC: Government Printing Office, 1923), 76–81.
119. David Hart, "Herbert Hoover's Last Laugh: The Enduring Significance of the 'Associative State,'" *Journal of Policy History* 10 (1998): 419–44.
120. On some characteristics of this style of setting standards, see Andrew Russell, "The American System: A Schumpeterian History of Standardization," Progress on Point Paper 12 (Washington, DC: Progress & Freedom Foundation, 2005).
121. A pint in the United Kingdom is equal to 20 imperial fluid ounces (568.261 milliliters); the US pint is equal to 16 US fluid ounces (473.176473 milliliters), and the US dry pint is equal to 550.6104713575 milliliters.
122. "International Yard," in *Sizes*, accessed February 20, 2022, http://www.sizes.com/units/yard_international.htm.
123. National Bureau of Standards, *A Metric America*; National Bureau of Standards, *U.S. Metric Study Report* (Washington, DC: Government Printing Office, 1971). The twelve volumes of the report were dedicated to International Standards; Federal Government: Civilian Agencies; Commercial Weights and Measures; The Manufacturing Industry; Nonmanufacturing Business; Education; The Consumer; International Trade; Department of Defense; A History of the Metric System Controversy in the United States; Engineering Standards; and Testimony of Nationally Representative Groups.
124. Treat, *A History of the Metric System Controversy*.
125. *A Metric America*, iii.
126. Lewis M. Branscomb, "The Metric System in the United States," *Proceedings of the American Philosophical Society* 116 (1972), 295.
127. Judson, *Weights and Measures Standards of the United States*, 26.
128. "Title 15 U.S.C. Chapter 6 §(204) 205a - 205l Metric Conversion Law (Pub. L. 94–168, §2, Metric Conversion Act, Dec. 23, 1975)," https://uscode.house.gov/view.xhtml?path=/prelim@title15/chapter6/subchapter2&edition=prelim, accessed December 18, 2022.
129. Constance Holden, "Metrication: Craft Unions Seek to Block Conversion Bill," *Science* 184 (1974): 48–50, 94; Grace Ellen Watkins, "Measures of Change," 190–288;

Malcolm Browne, "Kinder, Gentler Push for Metric Inches Along," *New York Times*, June 4, 1996.
130. *A Metric America*, iii.
131. Watkins and Best, "Successful and Unsuccessful Diffusion," 280.
132. Ronald Reagan to Louis Polk, March 9, 1A copy of the letter can be seen at https://usma.org/laws-and-bills/history-of-the-united-states-metric-board#locale-notification, accessed April 12, 2024.
133. Jason Zengerle, "Waits and Measures," *Mother Jones* 24 (1999): 70.
134. U.S. General Accounting Office, *Metric Conversion: Future Progress Depends Upon Private Sector and Public Support* (Washington, DC: Government Printing Office, 1994).
135. Hank Jenkins-Smith, Carol Silva, and Geoboo Song, *Health Policy Survey 2010: A National Survey on Public Perceptions of Vaccination Risks and Policy Preferences* (Norman, OK: University of Oklahoma, 2010), 5–7.
136. "National School Standards, at Last," *New York Times*, March 13, 2010.
137. Thomas Jefferson to John Quincy Adams, November 1, 1817, in *The Writings of Thomas Jefferson*, ed. H. A. Washington (Washington, DC: Taylor & Maury, 1854), VII: 87.
138. Frederick R. Hutton to Melvil Dewey, December 9, 1CU, Melvil Dewey Papers, box 66.
139. See National Institute of Standards and Technology, "Metrication in Law," https://www.nist.gov/pml/owm/metrication-law.
140. Gregory Higby and Glenn Sonnedecker, "Adoption of the Metric System by the U.S. Pharmacopoeia," *Journal of the History of Medicine and Allied Sciences* 40 (1985): 207.
141. Guy M. Wilson, Review of *The Metric System of Weights and Measures*, by National Council of Teachers of Mathematics, *The Journal of Educational Research* 43 (1949): 75.
142. Medical Group Circular of the Metric Bureau. Boston: American Metric Bureau, 1878; Edward Wigglesworth, "Why Should We Use the Metric System?" *The Western Lancet: A Journal of Medicine, Surgery, and Reviews of Medical Literature* 7 (1878): 424–27.
143. Samuel Dale to James B. Gardner, March 21, 1923, CU, Samuel S. Dale Papers, box 4.
144. Samuel Dale to William C. French, September 6, 1923, CU, Samuel S. Dale Papers, box 4.
145. Samuel Dale to John W. Gaines, December 22, 1903, CU, Samuel S. Dale Papers, box 4. By the same token, Frederick Halsey, one of the major public figures against the metric system in American history, wrote in a *The New York Times* article, during Prohibition: "Whatever the reader's view of prohibition, our experience with it should convince all that laws that attempt to change the habits and customs of the people by creating crimes of acts that many do not consider wrong are difficult of enforcement. Fancy fining . . . for selling dry good by the yard and butter by the pound—things they have done since the Pilgrim Fathers have done at Plymouth Rock! And yet that is the meaning of compulsory laws, and all history shows that nothing less will accomplish the purpose." Frederick Halsey, "Disputes Metric Success," *New York Times*, August 23, 1925; also of interest for this issue are Gertrude

Cushing Yorke, "Three Studies on the Effect of Compulsory Metric Usage," *Journal of Educational Research* (1944): 343–51; Fred A. Geiger, "Why the Metric System Must Not be Made Compulsory," *Machinery* May (1920); and W. Le Conte Stevens, "The Metric System: Shall It Be Compulsory?," *The Popular Science Monthly* 64 (1904): 394–405.
146. "Advocates the Adoption of Metric System," *San Francisco Chronicle*, Feb 13, 1927; "The Metric System," *Tribune* (Cheyenne, Wyoming.), December 12, 1927.
147. "Freedom 2 Measure," https://web.archive.org/web/20051230152441/http://www.freedom2measure.org/, accessed January 4, 2024.
148. For a brief but thoughtful reflection in these lines see A. J. Stubbs, "The Necessity of Compulsion," *Decimal Educator* 2 (1919): 155.
149. "On a Highway in Arizona, the Mile Makes Way for Metric Standard," *The New York Times*, September 10, 1979.

Chapter 3

1. Clyde Haberman and Albin Krebs, "Foe of Metric System Won't Give an Inch," *New York Times*, May 10, 1979, C16; Jeff Jarvis, "The Meter and Liter Don't Measure Up, Says Seaver Leslie, Who Wants to Give the Foot a Hand," *People*, June 22 (1981); John Marciano, *Whatever Happened to the Metric System?* 250–53; James Vincent, *Beyond Measure*, 272–73.
2. Paul L. Montgomery, "800, Putting Best Foot Forward, Attend a Gala Against Metrics," *New York Times*, June 1, 1981, B4.
3. André Balazs, "Artist Stands Up for the Foot," *New York Magazine*, May 14, (1979), 7; Malcolm W. Browne, "Kinder, Gentler Push for Metric Inches Along," *New York Times*, June 4, 1996, C1.
4. Frank Donovan, *Prepare Now for a Metric Future* (New York: Weybright and Talley, 1970); Richard Deming, *Metric Power*.
5. Calvin Trillin, "Anti-Metrics," *The New Yorker*, August 31 (1981), 82–85; Stewart Brand, "Stopping Metric Madness!," *New Scientist* 88, October 30 (1980); Randy Bancroft, "Brand-ing In 30 Seconds?" *Metric Maven*, October 10 (2017), https://themetricmaven.com/brand-ing-in-30-seconds/.
6. Douglas Martin, "John Michell, Counterculture Author Who Cherished Idiosyncrasy, Dies at 76," *New York Times*, May 2, 2009; John Michell, "A Defense of Sacred Measures," *The CoEvolution Quarterly* 17, Spring (1978): 2–6.
7. Besides Asimov there are a number of popular science communicators who have become the face of metric advocacy, including Bill Nye (aka "The Science Guy") and Neil deGrasse Tyson. See Isaac Asimov, "How Many Inches in a Mile?," in *Today and Tomorrow and...* (London: Abelard-Schuman, 1974), 147–154; Bill Nye, "Walk a Kilometer in My Shoes and Go Metric," *Wall Street Journal*, July 19, 2022, https://www.wsj.com/articles/bill-nye-the-science-guy-metric-system-imperial-measurement-education-competitiveness-11658180474; "Neil deGrasse Tyson Explains the Metric System," *Star Talk*, May 25 (2021), https://www.youtube.com/watch?v=gYa1bvFOouk.

8. Keith Michael Baker, "Science and Politics at the end of the Old Regime," in *Inventing the French Revolution* (Cambridge: Cambridge University Press, 1990), 156–59; Louis Marquet, "Condorcet et la creation du système métrique decimal," in *Condorcet, mathématicien, économiste, philosophe, homme politique*, ed. Pierre Crépel and Christian Gilain (Paris: Minerve, 1989), 52–62; Jean-Pierre Poirier, *Lavoisier: Chemist, Biologist, Economist* (Philadelphia: University of Pennsylvania Press, 1998), 319–24; Charles Coulston Gillispie, *Pierre-Simon Laplace, 1749–1827: A Life in Exact Science* (Princeton, NJ: Princeton University Press, 1997), 149–55; Ruth Inez Champagne, "The Role of Five Eighteenth Century French Mathematicians in the Development of the Metric System" (PhD diss., Columbia University, 1979); Paul Tunbridge, "James Watt and the Metric System," in *Lord Kelvin, His Influence on Electrical Measurements and Units* (London: Institution of Electrical Engineers, 1992), 86–87; Juan Gutiérrez Cuadrado, *Metro y kilo: el sistema métrico decimal en España* (Madrid: Akal, 1997), 15–22; Héctor Vera, *A peso el kilo*, 95–96; N. A. Shost'in, "History of Russian Metrology," 429–31; Maria Montessori, "The House of Children. Lecture, Kodaikanal, 1944," *NAMTA Journal* 38 (2013): 19; Céline Fellag Ariouet, "Marie Curie, the International Radium Standard and the BIPM," *Applied Radiation and Isotopes* 168 (2021): 109528.

9. When Thomas Jefferson's best friend, Dabney Carr, died of bilious fever in May of 1773, he channeled his grief by making lists and conducting measures. His personal documents show that shortly after Carr's passing, he spent hours doing calculations, like how much time it would take for one of his slaves to prepare the burial ground for the coffin: "two hands grubbed the grave yard 80. f. sq. = $1/_7$ of an acre in 3½ hours, so that one would have done it in 7. hours, and would grub an acre in 49. hours = 4. days." Thomas Jefferson, "Notes for Epitaph and Grave of Dabney Carr," https://founders.archives.gov/documents/Jefferson/01-27-02-0595; Geoffrey C. Ward, *Thomas Jefferson*, documentary, Ken Burns (dir.), Public Broadcasting Service, 1997.

10. Robert Tavernor, *Smtoot's Ear: The Measure of Humanity* (New Haven, CT: Yale University Press, 2007), xi–xvi.

11. Elizabeth Durant, "Smoot's Legacy," *MIT Technology Review*, June 23, 2008, https://www.technologyreview.com/2008/06/23/219971/smoots-legacy

12. Incidentally, and more significant for the history of measurement in the United States, in 2001 Oliver Smoot served as chairman of the American National Standards Institute, a private nonprofit organization in charge of overseeing the development of voluntary consensus standards in the country and coordinating American and international standards to ensure that domestic products can be used and sold worldwide (as I underlined in the previous chapter, the key word here is *voluntary*, the Achilles heel of all metric standardization campaigns).

13. On the origins of the growing distance between common metrological knowledge and scientific metrology, see Armand Machabey, "Techniques of Measurement," in *A History of Technology & Invention: II. The First Stages of Mechanization*, ed. Maurice Daumas (New York: Crown, 1969), 342.

14. G. F. Kunz, "New International Metric Diamond Carat of 200 Milligrams," *Science* 38 (1913): 523.
15. Kunz, "New International Metric Diamond Carat," 523–24.
16. Hector Vera, "The Social Construction of Units of Measurement: Institutionalization, Legitimation and Maintenance in Metrology," in L. Huber and O. Schlaudt eds., *Standardization in Measurement: Philosophical, Historical and Sociological Issues* (London: Pickering and Chatto, 2015), 184.
17. G. F. Kunz, "The New International Metric Diamond Carat of 200 Milligrams (Adopted July 1, 1913, in the United States)," *Bulletin of the American Institute of Mining and Metallurgical Engineers* 79 (1913): 1225–45.
18. On the concepts of general knowledge and special knowledge, see Thomas Luckmann, "Common Sense, Science and the Specialization of Knowledge," *Phenomenology + Pedagogy* 1 (1983): 59–73; Berger and Luckmann, *The Social Construction of Reality*, 70–73.
19. Benjamin Lee Whorf, *Language, Thought, and Reality* (Cambridge: Massachusetts Institute of Technology, 1956), 139–42; Caleb Everett, *Numbers and the Making of Us* (Cambridge: Harvard University Press, 2017), 82–102.
20. Some cases of this process can be seen in Witold Kula, *Men and Measures*, 267–88.
21. John L. Heilbron, "The Politics of the Meter Stick," *American Journal of Physics* 57 (1989): 992.
22. Bronislaw Malinowski and Julio de la Fuente, *Malinowski in Mexico: The Economics of a Mexican Market System* (Boston: Routledge, 1982), 176–77.
23. Manuel Moreno Fraginals, *The Sugarmill: The Socioeconomic Complex of Sugar in Cuba 1760–1860* (New York: Monthly Review Press, 1976), 153.
24. Vera, "Weights and Measures," 467.
25. One of the few accounts of competitors to the metric system are George Adam, "Alternatives to the Metric System, Based on British Unit," *Decimal Educator* (1935), 17: 43–46, 18: 11–12, 20–21, 28–29; Ronald Zupko, *Revolution in Measurement*, 209–25.
26. Linklater, *Measuring America*, 260.
27. See Marcello Maestro, "Going Metric: How It All Started," *Journal of the History of Ideas* 3 (1980): 479–86; *The Papers of Thomas Jefferson* (Princeton, NJ: Princeton University Press, 1971), 16: 619–623; "James Watt: Pioneer of Decimal Systems," *Decimal Educator* 19 (1936): 9–10.
28. In Spain, for example, Vicente Pujals de la Bastida, *Sistema métrico perfecto ó docial: y demostración de sus inmensas ventajas sobre el decimal y sobre todo otro sistema de medidas, pesos y monedas* (Madrid: Imprenta de la Esperanza, 1862).
29. C. E. Macqueen. *The Advantages of a Complete Decimal System of Money, Weights and Measures* (Liverpool: Financial Reform Association, 1855); *Report of the Joint Special Committees of the Chamber of Commerce and American Geographical and Statistical Society on the Extension of the Decimal System to Weights and Measures of the United States* (New York: W. C. Bryant & Co. Printers, 1857). Other decimal proposals were J. S. H. Aslit, *Decimal Coinage: With a Proposal for Decimalizing our Weights, and Measures of Length and Capacity* (London: Silverlock, 1854); William

Henry Jessop, *A Complete Decimal System of Money and Measures* (Cambridge, 1855); John Felton, *The Decimal System: An Argument for American Consistency* (New York: G. P. Putnam & Co., 1857). On the opposite direction (rejecting decimalization), R. E. W. Goodridge, *On the Proposed Change of Time Marking to a Decimal System: A Plea that the Duodecimal System be Retained* (Winnipeg: Manitoba Daily Free Press, 1886).

30. John William Nystrom, *Project of a New System of Arithmetic, Weight, Measure and Coins: Proposed to Be Called the Tonal System, with Sixteen to the Base* (Philadelphia: J. B. Lippincott & Co., 1862); Lisha Pace, "Nystrom," July 25, 2023, https://history-computer.com/nystrom; Alfred B. Taylor. "Octonary Numeration, and its Application to a System of Weights and Measures," *Proceedings of the American Philosophical Society* 24 (1887): 296–366.

31. W. Wilberforce Mann, *A Decimal Metric New System Founded on the Earth's Polar Diameter, and Designed for the Adoption of all Civilized Nations as the One Common System* (New York: University Publishing Company, 1872).

32. Edward Noel, *Science of Metrology or Natural Weights and Measures: A Challenge to the Metric System* (London: Edward Stanford, 1889). On Peirce and metrology, see Victor F. Lenzen, "The Contributions of Charles S. Peirce to Metrology," *Proceedings of the American Philosophical Society* 109 (1965): 29–46; Charles S. Peirce, "Testimony on the Organization of the Coast Survey," in *Writings of Charles S. Peirce* (Bloomington: Indiana University Press, 1982), 3: 149–61.

33. Charles S. Peirce, "Review of Noel's *The Science of Metrology*," in *Writings of Charles S. Peirce, 1886–1890* (Bloomington: Indiana University Press, 1982), 6: 378–79.

34. Zupko, *Revolution in Measurement*, 176.

35. Zupko, *Revolution in Measurement*, 200.

36. The adage is attributed to Max Weinreich, quoted in John C. Maher, *Multilingualism: A Very Short Introduction* (Oxford: Oxford University Press, 2017), 8.

37. Kasson was not a senator but a congressman.

38. Cards titled "Biog metric," in CU, Melvil Dewey Papers, box 67; also reproduced in Grosvenor Dawe, *Melvil Dewey, Seer: Inspirer: Doer, 1851–1931* (Lake Placid Club, NY: Melvil Dewey Biografy, 1932), 277–80. This is a "translation" of the text with regular spelling: "In school in Adams Center I rebelled against compound numbers. I told the teacher that geometry taught us a straight line was the shortest distance between two points and that it was absurd to have long measure, surveyor's measure, and cloth measure; also absurd to have quarts and bushels of different sizes and to have avoirdupois, troy, and apothecary weights, with a pound of feathers heavier than a pound of gold. I spread out on my attic room table sheets of foolscap and decided that the world needed just one measure for length, one for capacity, and one for weight, and that they should all be in simple decimals like our money. I was puzzling over the names to give the new measures when I read that John A. Kasson of Iowa had passed in Congress a bill legalizing the metric system. I looked it up at once, found that it met my plan ideally and the next week went to our village lyceum and gave a talk on the great merit of international weights and measures. From that day I became a metric apostle."

Notes to Pages 131–138   263

39. Wayne Wiegand, *Irrepressible Reformer: A Biography of Melvil Dewey* (Chicago: American Library Association, 1996), 10–11.
40. Quoted in Wiegand, *Irrepressible Reformer*, 20.
41. On Dewey's classification project: Markus Krajewski, *Paper Machines: About Cards & Catalogs, 1548–1929* (Cambridge: MIT Press, 2011), 87–106.
42. Melvil Dewey, *A Classification and Subject Index, for Cataloguing and Arranging the Books and Pamphlets of a Library* (Amherst: n. p., 1876), 4. In further editions the book was titled *Decimal Classification and Relative Index for Libraries*. The system has been revised through 23 major editions.
43. *Metric Advocate* 38 (1881): 513.
44. Postcard of the American Metric Bureau, CU, Melvil Dewey Papers, box 66.
45. CU, Melvil Dewey Papers, box 66.
46. Wiegand, *Irrepressible Reformer*, 43.
47. Melvil Dewey to James H. Southard, April 13, 1904, Melvil Dewey Papers, box 66.
48. "Metric Advantages: Melvil Dewey Scores Opponents of Decimal Measures," *New York Herald Tribune*, February 16, 1925.
49. Dewey Melvil, "Efficiency Society," *The Encyclopedia Americana* (New York: The Encyclopedia Americana Corporation, 1918), 9: 719–20.
50. "Apostle of Simplification," *Journal of Calendar Reform* 2 (1932): 1: 18.
51. Isaac Asimov, "Forget It!," in *Asimov on Numbers* (New York: Pocket Books, 1978), 131–46.
52. From 1887 to 1889, he served as secretary of the American Metrological Society; years later he was a member of the Advisory Board of the All-America Standards Council (a California-based organization that promoted metric unification for all countries in the Americas); and he functioned as member of the Advisory Board and chairman of the Metric Education Committee in the American Metric Association.
53. Edward Franklin Cox, "The International Institute: First Organized Opposition to the Metric System," *Ohio Historical Quarterly* 58 (1959): 59.
54. "Charles Latimer," *Virtual American Biographies*, http://famousamericans.net/charleslatimer/.
55. Charles Latimer, *The Divining Rod: Virgula Divina-Baculus Divinatorius (Water-Witching)* (Cleveland: Fairbanks, Benedict & Co., 1876).
56. Cox, "The International Institute," 63.
57. Simon Schaffer, "Metrology, Metrication, and Victorian Values," in *Victorian Science in Context*, Bernard Lightman ed. (Chicago: University of Chicago Press, 1997), 450.
58. H. A. Brück, and M. T. Brück, *The Peripatetic Astronomer: The Life of Charles Piazzi Smyth* (Philadelphia: Adam Hilger, 1988): 95–97.
59. Eric Michael Reisenauer, "'The Battle of the Standards': Great Pyramid Metrology and British Identity, 1859–1890," *The Historian* 65 (2003): 932–46.
60. John Taylor, *The Battle of the Standards: The Ancient, of Four Thousand Years, against the Modern, of the Last Fifty Years—the Less Perfect of the Two* (London: Longman, Green, Longman, Roberts, & Green, 1864). On Herschel's contributions

to the invention of traditions in metrology, see Schaffer, "Metrology, Metrication, and Victorian Values," 443–49.
61. Daniel J. Boorstin, "Afterlives of the Great Pyramid," *Wilson Quarterly* 16 (1992): 132.
62. Isaac Newton, "A Dissertation upon the Sacred Cubit of the Jews and the Cubits of the Several Nations," in *Miscellaneous Works of Mr. John Greaves* vol. 2 (London: J. Hughs, 1737), 405–433; Tessa Morrison, *Isaac Newton's Temple of Solomon and his Reconstruction of Sacred Architecture* (Cham: Springer Science, 2010), 63–71.
63. Brück, and Brück, *The Peripatetic Astronomer*, 98–99.
64. Charles Piazzi Smyth, *Life and Work at The Great Pyramid. Vol. III* (Edinburgh: Edmonston and Douglas, 1867), 572.
65. Charles Piazzi Smyth, *Our Inheritance in the Great Pyramid* (London: Alexander Strahan & Co., 1864). Some later editions of the book appeared under the title of *The Great Pyramid: Its Secrets and Mysteries Revealed*.
66. Reisenauer, "The Battle of the Standards," 958.
67. Charles A. L. Totten, "The Metrology of the Great Pyramid," *Van Nostrand's Engineering Magazine* 31 (1884): 226–34; Charles Piazzi Smyth, "Weights and Measures," in Edward Hine, *Flashes of Light. Part 2 of Forty-Seven Identifications of the British Nation with the Lost Ten tribes of Israel* (London: W. H. Guest, 1876), 77.
68. Reisenauer, "The Battle of the Standards," 964–65.
69. Charles Piazzi Smyth, "Weights and Measures," in Edward Hine, *Flashes of Light. Part 2 of Forty-Seven Identifications of the British Nation with the Lost Ten tribes of Israel* (London: W. H. Guest, 1876), 78.
70. Piazzi Smyth, *Life and Work at The Great Pyramid*, 580.
71. Piazzi Smyth, "Weights and Measures," 77–78. He was invoking here the old idea from the Jewish tradition that Cain was the inventor of weights and measures and used them for cheating and avarice.
72. Piazzi Smyth, *Life and Work at The Great Pyramid*, 581.
73. Piazzi Smyth, "Weights and Measures," 79.
74. Charles Latimer, *The French Metric System or the Battle of the Standards: A Discussion of the Comparative Merits of the Metric System and the Standards of the Great Pyramid* (Chicago: Thomas Wilson Publisher, 1880).
75. *International Conference Held at Washington for the Purpose of Fixing a Prime Meridian and a Universal Day, October, 1884: Protocols of the Proceedings.* (Washington, DC: Gibson Bros., 1884), 207.
76. Robert P. Crease, *World in the Balance: The Historic Quest for an Absolute System of Measurement* (New York: W. W. Norton, 2011), 157–59.
77. *The International Standard* 1 (1833), 18; Treat, *A History of the Metric System Controversy*, 90.
78. C. A. L. Totten, *An Important Question in Metrology, Based upon Recent and Original Discoveries: A Challenge to "the Metric System" and an Earnest Word with the English-Speaking Peoples on Their Ancient Weights and Measures* (New York: John Wiley & Sons, 1884), v.
79. Treat, *A History of the Metric System Controversy*, 89. The expression "a pint's a pound the world around" comes from the fact that a US liquid pint, at room temperature, holds 1.042 pounds of water.

80. Cox, "The International Institute," 67, 75.
81. Frederick A. P. Barnard, *The Imaginary Metrological System of the Great Pyramid of Gizeh* (New York: John Wiley & Sons, 1884).
82. Stephen Mihm, "Inching toward Modernity: Industrial Standards and the Fate of the Metric System in the United States," *Business History Review* 96 (2022): 54–61.
83. Schaffer, "Metrology, Metrication, and Victorian Values," 443–49. The formula "invention of traditions" was made famous by social historians Eric Hobsbawm and Terence Ranger.
84. Eviatar Zerubavel, *Time Maps: Collective Memory and the Social Shape of the Past* (Chicago: Chicago University Press, 2003), 55–81; Eviatar Zerubavel, *Ancestors and Relatives: Genealogy, Identity, and Community* (New York: Oxford University Press, 2013).
85. Zerubavel, *Time Maps*, 62.
86. On the debates on metrication in Parliament see Bernard Semmel, "Parliament and the Metric System," *Isis* 54 (1963): 125–33 (esp. 130–31); Joseph Mayer, "Parliament and the Metric System. Comments," *Isis* 57 (1966): 117–19.
87. *The Times*, April 4, 7, 9 and 25, 1896.
88. Herbert Spencer, "The Metric System," *Popular Science Monthly* 49 (1896): 186–99. Alongside Spencer's paper appeared a letter by Frederick Bramwell, a British engineer, fellow of the Royal Society of London for the Improvement of Natural Knowledge and former president of the Institution of Civil Engineers, which echoed Spencer's attack on ten-base systems by saying that decimals are "absolutely incompatible with mental arithmetic" and are prone to errors in placing the decimal point.
89. This was also announced in "Herbert Spencer Opposes Metric System," *The New York Times*, June 22, 1896.
90. *The Times*, March 28, April 4, 8, and 13, 1899. The whole set of eight articles from 1896 and 1899 was later reprinted as "Against the Metric System" and "The Metric System Again," in Spencer's volume of essay's *Various Fragments* (New York: D. Appleton and Company, 1914), 142–70, 225–39. I quote from the latter version.
91. Herbert Spencer, *Autobiography* (New York: D. Appleton, 1904), 617–623.
92. And another affinity between Spencer and the metric advocates is described by David Duncan in *The Life and Letters of Herbert Spencer* (London: Methuen & Co., 1908), 22: "It is interesting to note how, after experience in the measurement of brickwork at [a] bridge, the future opponent of the metric system resolved 'to have a foot-rule made divided into decimals instead of into inches.' 'I am trying to bring decimal arithmetic into use as much as possible.'"
93. From 1812 to 1837 there was an impasse in the use of the metric system in France, starting when Napoleon implemented a compromise between the metric and customary measures with the so called *mesures usuelles*. The metric system was reintroduced as the exclusive and mandatory system during the reign of Louis Philippe.
94. Auguste Comte, *The Positive Philosophy of Auguste Comte: Freely Translated and Condensed by Harriet Martineau*, vol. III (London: Geroge Bell & Sons, 1896), 302; Auguste Comte, *Cours de philosophie positive 6: Le complément de la philosophie sociale et les conclusions générales* (Paris: Bachelier Impimeur-Libraire,1842),

450–51; Luce Langevin, "The Introduction of the Metric System: The First Example of Scientific Rationalization by Society," *Impact of Science on Society* 11 (1961): 95.

95. William Graham Sumner, *Folkways: A Study of the Sociological Importance of Usages, Manners, Customs, Mores, and Morals* (New York: Dover, 1906), 155–56.
96. Spencer, *Various Fragments*, 231, 157, 165, 168.
97. Spencer, *Various Fragments*, 143, 148, 152–53. As I mentioned in a previous chapter, it was on this ground that Spender questioned the pertinence of using the decimal principle for time reckoning.
98. Spencer, *Various Fragments*, 155–56.
99. Spencer, *Various Fragments*, 154.
100. Spencer, *Various Fragments*, 159.
101. For a detailed explanation of how base-12 number systems work, see the "Manual of the Dozenal System," compiled by the Dozenal Society of America: https://dozenal.org/drupal/sites_bck/default/files/DSA_mods_rev.pdf.
102. Spencer, *Various Fragments*, 233.
103. Spencer, *Various Fragments*, 160.
104. Spencer, *Various Fragments*, 168–69, 234. On this topic, a biographer of Spencer describes this scene "He wrote his well-known articles to *The Times* against the metric system, and brought out a pamphlet soon afterward on the same subject. His pleasure at its success both at home and in America was like that of a novice over his first victory. He was in such spirits about it that we felt constrained to inquire if he had any system to suggest as a substitute, to which he replied 'No. Leave that to posterity. Why should posterity have nothing to do? It will certainly know what it requires better than we do?'" Two [Arthur George Liddon Rogers], *Home Life with Herbert Spencer* (Bristol: J. W. Arrowsmith, 1910), 195.
105. Spencer, *Various Fragments*, 170.
106. Herbert Spencer, *The Man versus the State* (New York: D. Appleton and Company, 1885).
107. Spencer, *Various Fragments*, 151.
108. Spencer, *Various Fragments*, 165.
109. Spencer, *Various Fragments*, 234, 148, 165.
110. Spencer, *Various Fragments*, 237, 227.
111. Spencer, *Various Fragments*, 167–68, 228.
112. Spencer, *Various Fragments*, 234, 236–37.
113. "Herbert Spencer's Will," *New York Times*, January 14, 1904.
114. "Doudecimalisms," *Tucson Daily Citizen*, April, 28, 1904; "Spencer and the Metric," *San Jose Mercury News*, January 23, 1904. One exception was William Benjamin Smith (a professor of mathematics and amateur historian of early Christianity), who brought back Spencer's idea in an article in *Science* in 1919—without acknowledging Spencer, though. See "Not Ten but Twelve," *Science* 50 (1919): 239–42.
115. For a couple of examples see: A. V. Draper, "Herbert Spencer's Opposition to the Metric System," *Manufacturers Record*, August 26 (1920): 109–12; *The American Machinist* 52 (1920): 1222. On the opposite direction, see "A Word for the Metric System," *Springfield Republican*, January 7, 1906.

116. Spencer's whole phrase was "[the] survival of the fittest, implies multiplication of the fittest." He used it to express, "in mechanical terms," what Darwin called "natural selection." Herbert Spencer, *Principles of Biology* (London: Williams & Norgate, 1864), 444–45.
117. "Spencer's Unusual Will," *Baltimore Sun,* July 11, 1920.
118. Samuel Dale to John W. Gaines, January 11, 1904. CU, Dale Papers, box 4. And a few years later, in a newspaper article, Dale came back to this topic and vocabulary: "Our English system is not the arbitrary scheme of a few men who, meeting behind closed doors in Paris, devised the metric system on the theory that they knew better what the world needed than did the world itself. Our system is the product of natural law working through the ages to meet the daily needs of unnumbered generations of men. It is [...] an adjunct of a league, which finds its roots in the hearts and minds of the people." See Samuel S. Dale, "Yards, Gallons, and Grains: The Economic and Moral Aspects of English Weights and Measures," *The Christian Endeavor World,* November 6, 1919, 106. This kind rhetoric was so prevalent in America that even pro-metric people used it. Industrialist Albert Herbert (a sponsor of pro-metric organizations), in a discussion on the savings in railway costs by switching to the metric system closed up an argument by saying that "the survival of the fittest means the survival of the most economical." Albert Herbert to Melvil Dewey, February 27, 1897. Melvil Dewey Papers, box 66.
119. Richard Hofstadter, *Social Darwinism in American Thought* (Boston: Bacon Press, 1955), 31–50.
120. T. C. Mendenhall, "The Metric System," *Popular Science Monthly* 49 (1896): 721–34.
121. This was before the creation of the National Bureau of Standards in 1901.
122. Mendenhall, "The Metric System," 721–22.
123. See John Herschel, "The Yard, the Pendulum, and the Metre," in *Familiar Lectures on Scientific Subjects* (London: Alexander Strahan, 1867), 419–51.
124. Mendenhall, "The Metric System," 733.
125. E. E. Slosson, "Decimal Numeration in the United States," *Science* 4 (1896): 59–62; Oscar Oldberg, "Herbert Spencer vs. the Metric System," *The Bulletin of Pharmacy* 10 (1896): 292–98. Oldberg, a professor of pharmacy at Northwestern University, was a veteran in the promotion, teaching, and defense of metrication; he was the author of *The Metric System in Medicine* (Philadelphia: Presley Blackiston, 1881), and *A Manual of Weights, Measures and Specific Gravity* (Chicago: n.p., 1885). For other articles discussing Spencer, see Florence Yaple, "Herbert Spencer and the Metric System," *American Journal of Pharmacy* 76 (1904): 125–28.
126. Hofstadter, *Social Darwinism*, 33.
127. Edward Livingston Youmans, ed., *Herbert Spencer on the Americans and the Americans on Herbert Spencer* (New York: Appleton, 1883), 87.
128. William Thompson, "Electrical Units of Measurement," *Popular Lectures and Addresses* (London: Macmillan, 1891), 80.
129. Quoted in Silvanus P. Thompson, *The Life of Lord Kelvin* (New York: Chelsea Publishing Company, 1976), 436.

130. "The Metric System," *The Times*, April 30, 1896; Kelvin also sent a brief addition to the newspaper, *The Times*, May 1, 1896. The first of these contributions was republished in the United States a month later: "Lord Kelvin on the Metric System," *Science* (May 22, 1896): 765–66.
131. Duncan, *The Life and Letters of Herbert Spencer*, 379.
132. Quoted in Thompson, *The Life of Lord Kelvin*, 1121–22.
133. "Lord Kelvin a Witness: Appears Before a House Committee and Indorses the Metric System Bill," *New York Times* April 25 (1902): 1; see also Harold Issadore Sharlin, *Lord Kelvin, the Dynamic Victorian* (University Park: Pennsylvania State University Press, 1979), 234. On that occasion George Westinghouse, the American inventor and entrepreneur, also testified as well in that session in favor of metrication. For some negative reactions to Kelvin and metrication, see *Springfield Republican*, April 26, 1902.
134. For some negative reactions to Kelvin and metrication, see *Springfield Republican*, April 26, 1902.
135. Quoted in Paul Tunbridge, *Lord Kelvin, His Influence on Electrical Measurements and Units* (London: Institution of Electrical Engineers, 1992), 13 (see also in this book 8–16, 69–70, 88–91). For an example of the echo of Kelvin in the metric debate in the America almost two decades later: *San Jose Mercury News*, August 20, 1919.
136. Mihm, "Inching toward Modernity," 58–59.
137. Pierre Thuillier, *El saber ventrílocuo*, Mexico: Fondo de Cultura Económica, 1995.
138. Jon Bosak, *The Old Measure: An Inquiry into the Origins of the U.S. Customary System of Weights and Measures* (Ithaca: Pinax Publishing, 2010), 105.

Chapter 4

1. "Congressional Consideration," *Measurement* 1, no. 3 (1926): 1.
2. "World Quart," *Time*, February 22 (1926), 11.
3. "Plea for Metric System: English Weights and Measures, Says advocate of Decimal Method, Were Devised and Scrapped by Germans," *New York Times*, October 26, 1919.
4. For a summary of the arguments against the bill, see "World Quart," *Time*, March 29 (1926), 8; Treat, *A History of the Metric System Controversy*, 217–26.
5. On "social knowledge" and economic production, see Karl Marx, *Grundrisse: Foundations of the Critique of Political Economy* (London: Penguin, 1993), 706.
6. Douglass C. North, Review of *Measures and Men*, by Witold Kula. *The Journal of Economic History* 47 (1987): 593–94.
7. Sharon Zukin and Paul DiMaggio, eds., *Structures of Capital: The Social Organization of the Economy* (New York: Cambridge University Press, 1990), 15–16.
8. Bruce G. Carruthers, "The Sociology of Money and Credit," in *The Handbook of Economic Sociology*, ed. N. Smelser and R. Swedberg (Princeton, NJ: Princeton University Press, 2005), 358.
9. Jack Goody, *The Logic of Writing and the Organization of Society* (New York: Cambridge University Press, 1986), 46.

10. Thomas Crump, *The Anthropology of Numbers* (Cambridge: Cambridge University Press, 1992), 72.
11. Douglass C. North, "Transaction Costs in History," *Journal of European Economic History* 14 (1985): 558; Douglass C. North, Review of *Measures and Men*, by Witold Kula. *Journal of Economic History* 47 (1987): 594.
12. North, "Transaction Costs in History," 560–66. For a particular historical example of this process see Masaru Iwahashi, "The Institutional Framework of the Tokugawa Economy," in *Emergence of Economic Society in Japan, 1600–1859*, ed. A. Hayami, O. Saito, and R. P. Toby (New York: Oxford University Press, 2004), 96–98.
13. Douglass C. North, "Institutions," *The Journal of Economic Perspectives* 5 (1991): 100.
14. Peter Swann, "The Economics of Metrology and Measurement," Report for National Measurement Office, Department for Business, Innovation and Skills (2009): iv (see also 60–64).
15. Bruce Carruthers and Sarah Babb, *Economy/Society: Markets, Meanings, and Social Structure* (Thousand Oaks: Pine Forge Press, 2000), 7.
16. Hector Vera, "Economic Rationalization, Money and Measures: A Weberian Perspective," in *Max Weber Matters: Interweaving Past and Present*, eds. David Chalcraft, Fanon Howell, Marisol Lopez M., and Hector Vera (London: Ashgate, 2008), 135–36.
17. Max Weber, *Economy and Society* (Berkeley: University of California Press, 1978), 85.
18. Weber, *Economy and Society*, 81, 107.
19. Weber, *Economy and Society*, 101.
20. Weber, *Economy and Society*, 81, 83.
21. Weber, *Economy and Society*, 100–107.
22. Witold Kula, *Measures and Men* (Princeton, NJ: Princeton University Press, 1986).
23. On some of these problems see Witold Kula, "Money and the Serfs in Eighteenth Century Poland," in *Peasants in History*, ed. E. J. Hobsbawm *et al.* (Calcutta: Sameeksha Trust, 1980), 30–41.
24. Vera, "Economic Rationalization, Money and Measures," 137.
25. See the works of Virginia García Acosta, "Weights and Prices of Bread in Eighteenth-Century Mexico." *Cahiers de Métrologie* 11–12 (1993–1994): 45–57.
26. To see a more detailed description of the function of measures in pre-capitalist economy see Kula, *Measures and Men*, 102–110.
27. Vera, "Economic Rationalization, Money and Measures," 137.
28. John Gourville and Jonathan J. Koehler, "Downsizing Price Increases: A Greater Sensitivity to Price than Quantity in Consumer Markets," Harvard Business School Working Paper, No. 04–042, 2004.
29. "Downsized! More and More Products Lose Weight," *Consumer Reports*, February (2011): 18–19.
30. Kula, *Measures and Men*, 31.
31. Kula, *Measures and Men*, 30.
32. See Andro Linklater, *Measuring America* (New York: Walker and Company, 2002), 21–28. As we can see, it is not an accident that the changes in how land was

measured coincided exactly in time and space with the "primitive accumulation of capital" described by Marx in *Capital*.

33. Clifford Geertz, "The Bazaar Economy: Information and Search in Peasant Marketing," *Supplement to the American Economic Review* 68 (1978): 29.
34. Sidney W. Mintz, "Standards of Value and Units of Measure in the Fond-des-Nègres Market Place, Haiti," *The Journal of the Royal Anthropological Institute of Great Britain and Ireland* 91 (1961): 25.
35. Mintz, "Standards of Value and Units of Measure," 23.
36. Theodore Porter, *Trust in Numbers* (Princeton, NJ: Princeton University Press, 1995), 25. See also Kula, *Measures and Men*, 114–19.
37. Quoted in Debdas Banerjee, *Colonialism in Action* (New Delhi: Orient Longman Limited, 1999), 49–50.
38. Quoted in Virginia D. Harrington, *The New York Merchant in the Eve of the Revolution* (New York: Columbia, 1935), 79. For a more detailed analysis on this problem, see John J. McCusker, "Weight and Measures in the Colonial Sugar Trade: The Gallon and the Pound and Their Equivalents," in *Essays in the Economic History of the Atlantic* (London: Routledge, 1997), 76–101.
39. Douglass C. North, "Institutions," *The Journal of Economic Perspectives* 5 (1991), 106.
40. Tomás Antonio de Marien y Arróspide, *Tratado general de monedas, pesas, medidas y cambios de todas las naciones reducidas a las que se usan en España* (Madrid: Imprenta de D. Benito Cano, 1789).
41. Ken Alder, "A Revolution to Measure: The Political Economy of the Metric System in France," in *The Values of Precision*, ed. N. Wise (Princeton, NJ: Princeton University Press, 1994), 39–71; Bruce Curtis, "From the Moral Thermometer to Money: Metrological Reform in Pre-Confederation Canada," *Social Studies of Science* 28 (1998): 547–49.
42. Randall Collins, "Weber's Last Theory of Capitalism: A Systematization," in *The Sociology of Economic Life*, ed. M. Granovetter and R. Swedberg (Cambridge: Westview Press, 2001), 384.
43. Robert L. Heilbroner, *The Worldly Philosophers* (New York: Simon & Schuster, 1967), 20.
44. Karl Marx and Friedrich Engels, *The Communist Manifesto* (London: Penguin, 2002), 223–24.
45. Jürgen Osterhammel and Neils P. Petersson, *Globalization: A Short History* (Princeton, NJ: Princeton University Press, 2009), 77.
46. Bernd Hausberger, *La globalización temprana* (Mexico: Colegio de México, 2018), 24.
47. Eric Hobsbawm, *The Age of Capital, 1848–1875* (New York: Charles Scribner's Sons, 1975), 15.
48. Hobsbawm, *The Age of Capital*, 48–68; Akira Iriye, *Cultural Internationalism and World Order* (Baltimore: The Johns Hopkins University Press, 1997), 28.
49. Sidney Pollard, "Capitalism and Rationality: A Study if Measurement in British Coal Mining, ca. 1750–1850," *Explorations in Economic History* 20 (1983): 110–29.

50. Eric Helleiner, "National Currencies and National Identities," *American Behavioral Scientist* 41 (1998): 1414.
51. Helleiner, "National Currencies," 1414–15.
52. Helleiner, "National Currencies," 1418.
53. See for example James Vincent, "The Battle of the Standards: Why the US and UK Can't stop Fighting the Metric System," *The Verge*, January 16, 2023, https://www.theverge.com/2023/1/16/23507199/us-uk-anti-metric-sentiment-beyond-measure-james-vincent-excerpt; Ian Traynor, "EU Provides Reprieve for Mile and Pint," *The Guardian*, September 12, 2007. On the cultural intricacies of money, see Zelizer, *The Social Meaning of Money*, 199–216; Hector Vera, "Money: Instrument of Quantification, Agent of Rationalization, Cultural Object," *Digithum* 24 (2019): 6–9.
54. John Quincy Adams, *Report of the Secretary of State, upon Weights and Measures* (Washington, DC: Gales & Seaton, 1821), 55.
55. Thomas Frederick Wilson, *The Power "to Coin" Money: The Exercise of Monetary Powers by the Congress* (Armonk, NY: M. E. Sharp, 1992); Stephen Mihm, *A Nation of Counterfeiters* (Cambridge: Harvard University Press, 2007), 340–59.
56. Viviana A. Zelizer, "The Creation of Domestic Currencies," *The American Economic Review* 84 (1994): 138; *The Social Meaning of Money* (New York: Basic Book, 1994), 13–18; Tschoegl, "The International Diffusion of an Innovation," 104–105.
57. Edward Cox, "The Metric System: A Quarter-Century of Acceptance (1851–1876)," *Isis* 13 (1958): 362–71.
58. I follow here the accounts made by Walter T. K. Nugent, *Money and American Society, 1865–1880* (New York: Free Press, 1968), 67–90; Martin H. Greyer, "One Language for the World: The Metric System, International Coinage, Gold Standard, and the Rise of Internationalism, 1850–1900," in *The Mechanics of Internationalism*, ed. M. Geyer and J. Paulmann (New York: Oxford University Press, 2001), 55–92; Steven P. Reti, *Silver and Gold: The Political Economy of International Monetary Conferences, 1867–1892* (Westport: Greenwood Press, 1998), 33–59; Charles P. Kindleberger, "International Monetary Reform in the Nineteenth Century," in *Keynesianism vs. Monetarism and Other Essays in Financial History* (Boston: G. Allen & Unwin, 1985), 213–25; and Luca Einaudi, *Money and Politics: European Monetary Unification and the International Gold Standard, 1865–1873* (New York: Oxford University Press, 2001), 147–50.
59. Greyer, "One Language for the World," 81.
60. T. A. Tefft, *Universal Currency: A Plan for Obtaining a Common Currency in France, England, and America, Based on the Decimal System; with Suggestions for Rendering the French Decimal System of Weight and Measure More Simple and Popular* (London: Effingham Wilson, 1858), iii.
61. John Sherman to Samuel Ruggles, May 18, 1867, in *John Sherman's Recollections of Forty Years in the House, Senate and Cabinet: An Autobiography* (Chicago: The Werner Company, 1896), 348.
62. United States. Congress. Senate. *Committee on Finance, International Coinage: I. Report of Senator Sherman. II. Report of Senator Morgan. III. Bill to establish a uniform coinage. IV. Report of Mr. S. B. Ruggles* (Washington, DC: Government

Printing Office, 1868), 4. On later work in Congress on these topics see House of Representatives, *Metric Coinage Report* (Washington, DC: Government Printing Office, 1880).

63. Nugent, *Money and American Society*, 70.
64. Nugent, *Money and American Society*, 70.
65. Still in 1879 there were conversations in the United States Congress about metric coinage, see *Report of the Committee on Coinage, Weights, and Measures: Part 1, on the Adoption of the Metric System of Weights and Measures, Together with Documents and Statistics Relating to the Subject; Part 2, on Metric Coinage* (Washington, DC: Government Printing Office, 1879), 203–10.
66. A. E. Haynes "Metric Department; Progress of the Metric System of Weights and Measures," *Hillsdale Standard*, December 4, 1877; see also *Indianapolis News*, April 28, 1921.
67. *The Metric Bulletin* 7 (1877): 102. On some of the work made by Elliott as part of his work in the Treasury Department, see E. B. Elliot, "On the relative value of Gold and Silver for Series of Years," in *Proceedings of the American Association for the Advancement of Science* 17 (1868): 122–23; "Report of Mr. E. B. Elliott on the Metric System of Coinage," *New York Times*, February 18, 1869.
68. Frederick A. P. Barnard, *International Coinage* (London: William Clowes, 1874); *Mono-Metallism, Bi-Metallism, and International Coinage* (New York: The S. W. Green Type-setting Machines, 1879); *The Possibility of an Invariable Standard of Value* (New York, 1879); *The Regulation of Time; International Coinage; the Unification of Weights and Measures; Sea-Signals* (London: W. Clowes and Sons, 1881); "On the Relation to the Public Welfare of Changes in the Volume of Money and on Money Standards," *Proceedings of the American Metrological Society* 2 (1882): 202–30.
69. *Constitution of the American Metrological Society,* as revised in 1888. For other examples of the work of the AMS: American Metrological Society, *The Metric System: Detailed Information as to Laws, Practice, Etc.* (New York: American Metrological Society, 1891); George Eastburn, *The Metric System* (New York: American Metrological Society, 1892).
70. Charles W. Stone, "A Common Coinage for All Nations," *The North American Review* 476 (1896): 55.
71. James F. Vivian, "The Pan American Conference Act of May 10, 1888: President Cleveland and the Historians," *The Americas* 27 (1970): 185–86.
72. Edward Franklin Cox, "A History of the Metric System of Weights and Measures: With Emphasis on Campaigns for its Adoption in Great Britain and in the United States prior to 1914" (PhD diss., Indiana University, 1956), 557; David Healy, *James G. Blaine and Latin America* (Columbia: University of Missouri Press, 2002), 145. On Pan-Americanism within the longer history of regionalism in Latin America, see Mark J. Petersen, *The Southern Cone and the Origins of Pan America, 1888–1933* (Notre Dame: University of Notre Dame Press, 2022).
73. James Brown Scott, *The International Conferences of American States, 1889–1928* (New York: Oxford University Press, 1931), 6–47.
74. "An Important Conference," *Evening Transcript*, July 6, 1888.

75. Healy, *James G. Blaine and Latin America*, 146.
76. Healy, *James G. Blaine and Latin America*, 147.
77. International American Conference, "Report of the Committee on Weights and Measures," in *Reports of Committees and Discussion Thereon* (Washington, DC: Government Printing Office, 1890), 77–80.
78. On the participation of Romero in the Conference see Chester C. Kaiser, "México en la primera Conferencia Panamericana," *Historia Mexicana* 11 (1961): 56–80.
79. International American Conference, "Report of the Committee on Weights and Measures," 80.
80. Cox, "A History of the Metric System of Weights and Measures," 559.
81. See "Proposed Legislation in Regard to the Metric System," *Science* 65 (1896): 457–63.
82. H.R. Report No. 795, quoted in Cox, "A History of the Metric System of Weights and Measures," 572.
83. Treat, *A History of the Metric System Controversy*, 102–12; Cox, "A History of the Metric System of Weights and Measures," 577.
84. The members of the Committee on Language, Weights and Measures were Albert Herbert (chairman of the committee), president of Hub Gore Makers; Charles A. Schieren, of C. A. Schieren & Co. and ex-mayor of Brooklyn; Charles H. Harding, of Erben Harding & Co.; Henry Fairbanks, vice-president of E. and T. Fairbanks Scale Co.; Theodore C. Search, Manufacturer, president and founder of American National Association of Manufacturers; and Andrew Carnegie.
85. A copy of the report can be seen in Melvil Dewey Papers, box 67. A somewhat modified version of the text was reproduced in *World Metric Standardization: An Urgent Issue*, edited by Aubrey Drury (San Francisco: World Metric Standardization Council, 1922), 38–44.
86. Melvil Dewey received generous donations from Andrew Carnegie to continue with his metric efforts (and also for his campaign in favor of spelling simplification). See Hector Vera, "Melvil Dewey, Metric Apostle," *Metric Today* 45 (2010): 1, 4–6.
87. Melvil Dewey Papers, box 67.
88. "Manufacturers Meeting: Opening of the Third Annual Convention of the Association in Masonic Temple," *New York Times*, January 26, 1898.
89. Cox, "A History of the Metric System of Weights and Measures," 584–87. For a profile of Stratton life see A. E. Kennelly, *Biographical Memoir of Samuel Wesley Stratton, 1861–1931* (National Academy of Sciences, 1935).
90. National Board of Trade, *Proceedings of the Thirty-First Annual Meeting of the National Board of Trade* (Philadelphia: John R. McFetridge & Sons, 1901), 55, 73–74, 299–300. That same resolution was supported again in 1902, see National Board of Trade, *Proceedings of the Thirty-Second Annual Meeting of the National Board of Trade* (Philadelphia: John R. McFetridge & Sons, 1902), 137. In a related issue that implicates the National Board of Trade and plans for metrological reform, the Board was involved in the 1860s and 1870s on the diffusion of the so called "cental system." The cental was a unit of one hundred pounds and was devised to replace the different existing kinds of bushel, especially for the commerce of grain. The

cental was used for some time in mayor American cities (and there are reports of its adoption in Liverpool and other English cities). See T. A. Bryce, *The American Commercial Arithmetic* (n.p.: C. G. Swensberg, 1873), 366; John Groesbeck, *The Crittendon Commercial Arithmetic and Business Manual: Designed for the Use of Merchants, Business Men, Academies, and Commercial Colleges* (Philadelphia, PA: Eldredge & Brother, 1871), 224; "The Metric System," *The Shareholder: A Railway, Banking and Investors' Gazette*, December 26, 1879, 3; *Daily Evening Traveler*, July 23, 1879.

91. *Hearings before the Committee, Feb. 6-March 3, 1902, on Bill H. R. 2054* (Washington, DC: Government Printing Office, 1902), 1–5.
92. *Supplemental Hearing on the Subject of the Metric System of Weights and Measures: Hearings before the United States House Committee on Coinage, Weights, and Measures, Fifty-Seventh Congress, First Session, on Apr. 24, 1902* (Washington, DC: Government Printing Office, 1902), 1–11.
93. "Report of the Committee Appointed to Discuss the Arguments in Favor and Against the Metric System." *Transactions of the American Society of Mechanical Engineers* 24 (1903): 630–712.
94. As illustrations of their positions on metrication: Samuel S. Dale, *The Foreign Attack on Our Weights and Measures* (Boston, 1926); Frederick A. Halsey, "The Metric System," *Transactions of the American Society of Mechanical Engineers* 24 (1903): 397–629.
95. F. A. Halsey, "The Premium Plan of Paying for Labor," *Transactions of the American Society of Mechanical Engineers* 12 (1890): 755–80. On Halsey's career, see Robert R. Jenks, "Frederick Arthur Halsey," *American National Biography Online*. Feb. 2000.
96. For a couple of examples of Halsey's non-specialized anti-metric articles, see "Disputes Metric Success," *New York Times*, August 23, 1925; and "Continuing the Metric War," *New York Times*, June 5, 1927.
97. Dale also amassed a large collection of books on historical metrology that he later donated to Columbia University, alongside his enormous correspondence.
98. Samuel S. Dale to Charles H. Harding, March 26, 1919. Samuel Dale Papers, vol. 7.
99. To see an explicit use of Spencer's ideas in their writings, see Samuel Dale, *The World Trade Club of San Francisco and its Metric Propaganda* (Boston: Textiles, 1920).
100. The first edition of Halsey's *The Metric Fallacy* was published in a single volume with Dale's *The Metric Failure in the Textile Industry*. The second edition of the book, this time without Dale's text, was published as *The Metric Fallacy: An Investigation of the Claims Made for the Metric System and Especially of the Claim that Its Adoption Is Necessary in the Interest of Export Trade* (New York: The American Institute of Weights and Measures, 1920).
101. Frederick A. Halsey and Samuel S. Dale, *The Metric Fallacy* and *The Metric Failure in the Textile Industry* (New York: D. Van Nostrand, 1904), 16–17; see also Cox, "A History of the Metric System of Weights and Measures," 606–17.
102. Halsey, *The Metric Fallacy*, 127. On these issues, see "Over-Standardization," *Bulletin of the American Institute of Weights and Measures*, October 1, 1923, 3; and in that same publication "Elusiveness of World Uniformity," July 1, 1924, 14.

103. Charles S. Peirce, "The Metric Fallacy," *The Nation* 78, March 17 (1904), 215–16; reprinted in *Charles Sanders Peirce: Contributions to* The Nation. *Part Three: 1901–1908*, eds. K. L. Ketner and J. E. Cook. (Lubbock: Texas Tech University Press, 1979), 156–61. Despite its present fame as a philosopher, at that time Peirce was a marginal figure. In those years he earned money doing some engineering consulting and writing for magazines like *The Nation*. He had worked for decades in the United States Coast Survey and held for some years a nontenured position at Johns Hopkins University. Curiously, he lost those two jobs as a result of the actions of a couple of important pro-metric figures. In 1884 he was dismissed from Johns Hopkins due the maneuvers of eminent astronomer Simon Newcomb, who defended the metric cause in Congress, during some hearings on the metric system before the Coinage, Weights and Measures Committee. And in 1891 Peirce had to resign from the Coast and Geodetic Survey at the request of Thomas Mendenhall, superintendent of the Survey (and author, in 1893, of the *Mendenhall Order*, which directed the change of the fundamental standards of length and mass of the United States from customary to metric standards). See Joseph Brent, *Charles Sanders Peirce: A Life* (Bloomington: Indiana University Press, 1993), 150–54, 198–202; Congress. House. Committee on Coinage, Weights, and Measures, "Statement of Prof. Simon Newcomb" *The Metric System of Weights and Measures* (Washington, DC: Government Printing Office, 1902), 71–74; Lewis V. Judson and Louis E. Barbrow, *Weights and Measures Standards of the United States: A Brief History* (Washington, DC: National Bureau of Standards, 1976), 28–29; Thomas C. Mendenhall, "The United States Fundamental Standards of Length and Mass," *Science* 56 (1922): 377–80.
104. Peirce, "The Metric Fallacy," 216. Peirce was not right about the development of interchangeable parts in Europe; see Ken Alder, *Engineering the Revolution: Arms and Enlightenment in France, 1763–1815* (Princeton, NJ: Princeton University Press, 1999), 221–49.
105. Resolution and Protest of the National Machine Tool Builders Association, 1902.
106. Quoted in "Manufacturers and the Metric System," *New York Times*, April 25, 1902. See also Albert K. Steigerwalt, *The National Association of Manufacturers, 1895–1914* (Ann Arbor: University of Michigan, 1964), 93–94.
107. *The Metric System: Hearings before the Committee on Coinage, Weights, and Measures on H. R. 93 (58th Congress, 1st Session); H. R. 2054 (58th Congress, 2d Sessions), and H. R. 8988 (59th Congress, 1st Sessions)* (Washington, DC: Government Printing Office, 1906), 1–19.
108. For an analysis on the distribution of costs of going metric (for a different national case, though) see Roger Faith, Robert McCormick, and Robert Tollison, "Economics and Metrology: Give'em an Inch and They'll Take a Kilometre," *International Review of Law and Economics* 1 (1981): 207–21.
109. For some conjectures to this question see Coleman Sellers, *The Metric System: Is It Wise to Introduce It into Our Machine Shops?* (Philadelphia: American Society of Mechanical Engineers, 1880); Monte A. Calvert, *The Mechanical Engineer in America, 1830–1910: Professional Cultures in Conflict* (Baltimore: Johns Hopkins Press, 1967), 179–86; Bruce Sinclair, *A Centennial History of the American Society*

of Mechanical Engineers, 1880–1980 (Toronto: University of Toronto Press, 1980), 46–60; Robert R. Jenks, "Governing in the Absence of Government: The Birth and Development of the United States Industrial Standards System" (PhD diss., University of California, Santa Barbara, 1999), 128–59. Also of interest in this context: Leo Marx, *The Machine in the Garden: Technology and the Pastoral Ideal in America,* (New York: Oxford University Press; 1964), 202 ff.

110. On Southard's activity in metric legislation, see James Southard, "The Metric System of Weights and Measures," *Science* 15 (1902): 829–36.

111. Samuel S. Dale to Marshall Cushing October 6, 1906; Marshall Cushing to Samuel S. Dale, October 8, 1906 (Samuel Dale Papers, box 4).

112. For access to a copy of the minutes of that first meeting of the American Metric Association (today U.S. Metric Association): https://usma.org/wp-content/uploads/2016/01/American_Metric_Association_first-meeting.pdf?x30984.

113. As he detailed in a letter: "I have written more than 100,000 individual letters on the metric topic, besides sending out several million form letters and pamphlets, and have written also several million words in metric articles and publications." Aubrey Drury to Karl E. Ettinger, not dated, Drury Papers, box 3.

114. For a taste of this debates in the popular press, see Aubrey Drury "Modern Trade—Antiquated Tools: A Plea for International Commodity Quantity Standards in Trade and Industry," *The Rotarian* 26 (1925): 2: 29, 66–69, and the response by C. C. Stutz, "The Tools of Our Industry: Shall We Scrap Them?," *The Rotarian* 26 (1925): 6: 19, 87–91; Stutz was secretary of the American Institute of Weights and Measures and member of the American Society of Mechanical Engineers. Also in that magazine, see the exchange between pro-metric Hilton Ira Jones (a chemist) and anti-metric Henry D. Sharpe (president of Brown & Sharpe Co.), in the editorial section A Debate-of-the-Month on "Adopt the Metric System?" *The Rotarian* 70 (April, 1927): 28–30.

115. Around those years inventiveness in the history of weights and measures started to get out of hand. In daily papers, journalists displayed a peculiar historical exuberance. New and astonishing hypotheses about the genesis of the metric system were presented to the public. The *Buffalo News* reported a "discovery" made by archeologists in Mexico, suggesting that "the early Aztecs probably used a metric system. Study of their buildings and inscriptions has led to the belief that they divided all measurements into units of ten." The *Christian Science Monitor* and the Kansas City's *The Star* stated that the meter was "in reality much older than [the French Revolution], for the Egyptian priests are said to have known it, only they kept their knowledge a secret." Some metric enthusiasts, it seemed, wanted to steal a page from the old Pyramid metrologists' book! See "Fighting for the Metric System," *Buffalo News,* June 13 (1927); "Metric System Spreads. Countries Are Adopting Method of Measure," *The Star* (Kansas City MO), September 24 (1928).

116. News clipping and typed manuscript by Aubrey Drury titled "James Watt and Metric," in Aubrey Drury Papers, File Metric - History - Char - Advantages, Box 3. See a modern reinvention of history—suggesting that "metric

is a British invention" because it was devised by seventeenth-century philosopher John Wilkins—in "By any Yardstick, Metric Units Beat Imperial," *The Guardian*, January 4, 2024, https://www.theguardian.com/politics/2024/jan/04/by-any-yardstick-metric-units-beat-imperial.
117. W. E. Hague, "Plea for Metric System: English Weights and Measures, Says advocate of Decimal Method, Were Devised and Scrapped by Germans," *New York Times*, October 26 (1919), X8; see also W. E. Hague, *The Universal Language of Quantity, Meter-Liter-Gram* (San Francisco: World Trade Club, 1919).
118. See, for example, "Pros and Cons of Metric System," *Pittsburgh Post*, June 26, 1927.
119. Frederick A. Halsey, *The Metric Fallacy: An Investigation of the Claims Made for the Metric System and Especially of the Claim That its Adoption Is Necessary in the Interest of Export Trade* (New York: The American Institute of Weights and Measures, 1920), 6.
120. Halsey, *The Metric Fallacy*, second edition, 151. See also, "Non-Metric Reform of Weights and Measures," *Literary Digest*, April 10 (1920), 128–30.
121. Halsey, *The Metric Fallacy*, second edition, 151.
122. See for example the papers presented by Drury, the Metric Association, Dale, and W. R. Ingalss in the second Pan American standardization meeting: Inter American High Commission, *Report of the Second Pan American Standardization Conference* (Washington, DC: Government Printing Office, 1927), 75–87.
123. John Lind, *The Mexican People* (Minneapolis: The Bellman, n.d.), 29–30. A similar case, but in Bolivia, is described in Hilton Ira Jones, "Why Not Now?," *School Science and Mathematics* 19 (1919): 512.
124. William Wells, "The Metric System from the Pan-American Standpoint," *The Scientific Monthly* 4 (1917): 196–202.
125. "To Start New Fight on Metric System: Manufacturers See Insidious Move to Make It a Feature in Pan-Americanism," *New York Times*, March 30, 1916.
126. Frederick A. Halsey, *The Weights and Measures of Latin America* (New York: American Society of Mechanical Engineers, 1918).
127. "The Weights and Measures of Latin America," *Decimal Educator* 2 (1919–1920): 178–86, 218–19, 259.
128. George Kunz, "The Metric System as a Factor in Pan American Unity," in *Report of the Second Pan American Commercial Conference: Pan American Commerce; Past, Present, Future from the Pan American Viewpoint*, ed. John Barret (Washington, DC: Pan American Union, 1919): 270. See also George Kunz, "The International Language of Weights and Measures," *The Scientific Monthly* 4 (1917): 215–19.
129. George Kunz to Melvil Dewey, December 18, 1924. Melvil Dewey Papers, box 66.
130. Frederick A. Halsey, "Pan Americanism in Weights and Measures," in *Report of the Second Pan American Commercial Conference: Pan American Commerce*, ed. John Barret (Washington, DC: Pan American Union, 1919), 274.
131. Drury even published some articles in Spanish. On his plan for continental metrological unification, see Aubrey Drury, *Un mismo sistema de pesas y medidas para toda la América* (Washington, DC: La Unión Panamericana, 1927); Aubrey Drury,

One Standard for All America ([San Francisco]: American Standards Council, [1927]); also of interest is Aubrey Drury, "Making this World Metric," *Pan Pacific: A Magazine of International Commerce,* June 20 (1920): 84–85.

132. Samuel S. Dale, "Uniformity or Confusion in Pan America?" and a letter by A. F. Del Solar to the First Pan American Standardization Conference, December 19, 1924. Melvil Dewey Papers, box 67.
133. From Ministry of Industry, Colombia to Aubrey Drury, November 5, 1927 (Aubrey Drury Papers, box 3).
134. Quoted in *Hearings Before the Committee, Feb. 6-March 3, 1902, on Bill H. R. 2054* (Washington, DC: Government Printing Office, 1902), 163.
135. Confederación de Cámaras de Comercio de los Estados Unidos Mexicanos, "Urgente necesidad de hacer universal el uso del sistema métrico decimal. Estudio para la 9na. Asamblea General de Cámaras de Comercio de la República. Septiembre de 1926." CU, Aubrey Drury Papers, box 3.
136. Lyman J. Briggs to J. T. Johnson, April 9, 1943 (a copy of the letter was sent to the Vice President of the United States, Henry A. Wallace). CU, Aubrey Drury Paper, box 3.
137. In 1988 the National Bureau of Standards changed its name to National Institute of Standards and Technology (NIST).
138. *Carta Metrológica* 2 (1980): 7–21.
139. Financial and technical support for the creation of the Centro Nacional de Metrología also came from the Organization of American States and the Physikalisch-Technische Bundesanstalt (the German institute for science and technology for the field of metrology, which also helps some other Latin American countries to fund their respective metrological centers).
140. Louis Jourdan, *La grande métrication* (Nice: France Europe éditions, 2002), 166.
141. George Sarton, *The Study of the History of Mathematics* (Cambridge: Harvard University Press, 1936), 15.

Conclusion

1. Richard Deming, *Metric Power: Why and How We Are Going Metric* (Nashville, TN: Thomas Nelson Publishers, 1974).
2. U. K. Metric Association, "Two Kinds of Country?," *Metric Views*, October 16, 2018, https://metricviews.uk/2018/10/16/two-kinds-of-country/.
3. Benedict Anderson, *A Life Beyond Boundaries* (New York: Verso, 2016), 127.
4. Werner Sombart, *Why Is there No Socialism in the United States?* (White Plains, NY: M. E. Sharpe, 1976), 3–15.
5. Andro Linklater, *Measuring America*, 248–49.
6. Andrei S. Markovits and Steven L. Hellerman, *Offside: Soccer and American Exceptionalism* (Princeton, NJ: Princeton University Press, 2001).
7. George H. Gallup, *The Gallup Poll: Public Opinion, 1972–1977* (Wilmington, DE: Scholarly Resources, 1978), 967–68; Malcolm W. Browne, "Growing Opposition to Conversion to Metric System Is Found in Poll," *New York Times*, November 24 (1977): 38.

8. Grace Ellen Watkins and Joel Best, "Successful and Unsuccessful Diffusion of Social Policy: The United States, Canada, and the Metric System," in *How Claims Spread: Cross-National Diffusion of Social Problems,* Joel Best ed. (New York: Aldine de Gruyter, 2001), 280.
9. It is interesting to contras Gallup poll numbers with surveys in England, see Aashish Velkar, "'Imperial Folly': Metrication, Euroskepticism, and Popular Politics in Britain, 1965–1980," *The Journal of Modern History* 92 (2020): 573–79.
10. On the difference between prediction and social forecast, and their perils, see Daniel Bell, *The Coming of Post-Industrial Society: A Venture in Social Forecasting* (New York, Basic Books, 1973), 3–9.
11. Christopher J. Phillips, "The New Math and Midcentury American Politics," *The Journal of American History* 101 (2014): 458.
12. Fernand Braudel, "History and the Social Sciences: The Longue Durée," in *The Longue Durée and World-Systems Analysis,* Richard E. Lee ed., (State University of New York Press, 2012), 249–51.
13. Linklater, *Measuring America,* 133–142; Keith Martin, "Pirates of the Caribbean (Metric Edition)," *Taking Measure: Just a Standard Blog,* September 19, 2017, https://www.nist.gov/blogs/taking-measure/pirates-caribbean-metric-edition; Larry Getlen, "How Pirates Blocked the US from Adopting the Metric System," *New York Post,* December 7, 2019, https://nypost.com/2019/12/07/how-pirates-blocked-the-us-from-adopting-the-metric-system-and-other-weird-facts/; Blake Stilwell, "Why Pirates Might Be the Reason the United States Doesn't Use the Metric System," *Military.com,* April 29, 2021, https://www.military.com/history/why-pirates-might-be-reason-united-states-doesnt-use-metric-system.html.
14. John Marciano, *Whatever Happened to the Metric System?* (New York: Bloomsbury, 2014), 266.
15. NASA Safety Center, "Lost in Translation," *System Failure Case Studies* 3 (2009): 1–4.
16. Penn State Milton S. Hershey Medical Center, "Metric Units Make for More Accurate Medication Doses," April 15, 2015, https://www.psu.edu/news/hershey/story/medical-minute-metric-units-make-more-accurate-medication-doses/.
17. Norbert Elias, "Towards a Theory of Social Processes," in *Essays III: On Sociology and the Humanities* (Dublin: University College Dublin Press, 2009), 11–17.

Appendix

1. The main bibliographical sources are: Frederick Barnard, *The Metric System of Weights and Measures* (New York: Columbia College, 1872); Burdun, "Worldwide Dissemination of the Metric System," 1147–51; Aubrey Drury, "The Metric Advance," *Measurement* 1, no. 3 (1926): 6–7; E. Lewis Frasier, "Improving an Imperfect Metric System," *Bulletin of the Atomic Scientists* 30, no. 2 (1974): 9–12; Letter of the Secretary of the Treasury, in Response to A Resolution of the House of Representatives, Transmitting certain Reports in Reference to the Adoption of the Metric System (Washington, DC: Government Printing Office, 1878); Ch.-Ed. Guillaume,

Les récents progrès du système métrique (Paris: Gauthier-Villars, 1913); William Hallock and Herbert Treadwell Wade, *Outlines of the Evolution of Weights and Measures and the Metric System* (New York: Macmillan, 1906); Arthur E. Kennelly, *Vestiges of Pre-Metric Weights and Measures Persisting in Metric-System Europe* (New York: The Macmillan Company, 1928); Bruno Kisch, *Scales and Weights: A Historical Outline* (New Haven, CT: Yale University Press, 1965); Henri Moreau, *Le système métrique* (Paris: Chiron, 1975); Musée National des Techniques, *L'aventure du mètre* (Paris: Le Musée, 1989); National Industrial Conference Board, *The Metric versus the English System of Weights and Measures* (New York: The Century Co., 1912); Albert Gustave Léon Pérard, *Les récents progrès du système métrique, 1948* (Paris: Gauthier-Villars, 1949); United Nations, *World Weights and Measures: Handbook for Statisticians* (New York: United Nations, 1966); U.S. Metric Association. "Metrication in Other Countries," https://usma.org/metrication-in-other-countries#locale-notification; Albino Zertuche, *Estudio sobre pesas y medidas en los países centroamericanos* (United Nations, 1958); Ronald E. Zupko, "Worldwide Dissemination of the Metric System during the 19th and 20th Centuries," *Metric System Guide Bulletin* 2, no. 2 (1974): 14–25; Ronald E. Zupko, *Revolution in Measurement: Western European Weights and Measures since the Age of Science* (Philadelphia: American Philosophical Society, 1990). *Encyclopedia of the Nations*, 2020, http://www.nationsencyclopedia.com.

ARCHIVES

AGN – Archivo General de la Nación (Mexico)
 CU – Rare Book & Manuscript Library, Columbia University
   Melvil Dewey Papers
   Samuel S. Dale Papers
   Aubrey Drury Papers
 LOC – Library of Congress
NYHS – New York Historical Society
 SASB – Stephen A. Schwarzman Building; New York Public Library
 SIBL – Science, Industry, and Business Library; New York Public Library

BIBLIOGRAPHY

Acharya, Anil Kumar. *History of Decimalisation Movement in India*. Calcutta: Indian Decimal Society, 1958.

"Act to Authorize the Use of the Metric System of Weights and Measures." In *The Statutes at Large, Treaties, and Proclamations of the United States of America from December 1865, to March 1867*, vol. XIV, edited by George P. Sanger, 339–40. Boston: Little, Brown, and Company, 1886.

Adam, George. "Alternatives to the Metric System, Based on British Unit." *Decimal Educator* (1935), 17:43–46, 18:11–12, 20–21, 28–29.

Adams, John Quincy. *Report of the Secretary of State, upon Weights and Measures, Prepared in Obedience to a Resolution of the House of Representatives of the fourteenth of December 1819*. Washington: Gales & Seaton, 1821.

"Address by the Secretary of Commerce, Hon. Herbert Hoover." In *Weights and Measures: Sixteenth Annual Conference of Representatives from Various States held at the Bureau of Standards*. Washington: Government Printing Office, 1923.

Agnew, P. G. "The Work of the Bureau of Standards." *Annals of the American Academy of Political and Social Science* 82 (1919): 278–88.

Alder, Ken. "A Revolution to Measure: The Political Economy of the Metric System in France." In *The Values of Precision*, edited by Norton Wise, 39–71. Princeton, NJ: Princeton University Press, 1994.

Alder, Ken. *Engineering the Revolution: Arms and Enlightenment in France, 1763–1815*. Princeton, NJ: Princeton University Press, 1999.

Alder, Ken. *The Measure of all Things: The Seven Years Odyssey and Hidden Error that Transformed the World*. New York: The Free Press, 2002.

American Enterprise Institute for Public Policy Research. *Metric Conversion Bills*. Washington: American Enterprise Institute, 1974.

American Metric Bureau. *Medical Group Circular of the Metric Bureau*. Boston: American Metric Bureau, 1878.

American Metric Bureau. "International Measures." *Bulletin of the American Metric Bureau* 20 (1878): 316.

American Metrological Society. *The Metric System: Detailed Information as to Laws, Practice, Etc.* New York: American Metrological Society, 1891.

American State Papers: Documents, Legislative and Executive of the Congress of the United States. Class X, Miscellaneous: Volume II. Washington, Gales and Seaton, 1834.

Anderson, Benedict. *Imagined Communities: Reflections on the Origin and Spread of Nationalism.* London: Verso, 1991.

Anderson, Benedict. *A Life Beyond Boundaries.* New York: Verso, 2016.

Andrews, George Gordon. "Making the Revolutionary Calendar." *American Historical Review* 36 (1931): 515–32.

Annual Report of the Governor of Porto Rico to the Secretary of War. Washington: Government Printing Office, 1919.

"Apostle of Simplification." *Journal of Calendar Reform* 2 (1932): 1–18.

Ariouet, Céline Fellag. "Marie Curie, the International Radium Standard and the BIPM." *Applied Radiation and Isotopes* 168 (2021): 109528.

Arnold, Peri E. "The 'Great Engineer' as Administrator: Herbert Hoover and Modern Bureaucracy." *The Review of Politics* 42 (1980): 329–48.

Arnold-Foster, H. O. *The Coming of the Kilogram or the Battle of the Standards: Plea for the Adoption of the Metric System of Weights and Measures.* London: Cassell, 1898.

Ashworth, William J. "Metrology and the State: Science, Revenue, and Commerce." *Science* 19 (2004): 1314–17.

Asimov, Isaac. "Forget It!" In *Asimov on Numbers*, 131–46. New York: Pocket Books, 1978.

Asimov, Isaac. "How Many Inches in a Mile?" In *Today and Tomorrow and . . .*, 147–54. London: Abelard-Schuman, 1974.

Aslit, J. S. H. *Decimal Coinage: With a Proposal for Decimalizing our Weights, and Measures of Length and Capacity.* London: Silverlock, 1854.

Assmann, Jan. *The Mind of Egypt: History and Meaning in the Time of the Pharaohs.* New York: Metropolitan Books, 1996.

Baczko, Bronislaw. "Le calendrier républicain." In *Les lieux de mémoire*, edited by Pierre Nora, 67–106. Paris: Gallimard, 1997.

Baczko, Bronislaw. "Rationaliser revolutionnairement." In *Les mesures et l'histoire*, edited by Institut D'Historie Moderne et Contemporaine Centre National de la Recherche Scientifique, 55–70. Paris: Éditions du Centre National de la Recherche Scientifique, 1984.

Baker, Keith Michael. "Science and Politics at the end of the Old Regime." In *Inventing the French Revolution: Essays on French Political Culture in the Eighteenth Century*. Cambridge: Cambridge University Press, 1990.

Balinski, Michel L., and H. Peyton Young. *Fair Representation: Meeting the Ideal of One Man, One Vote*. New Haven, CT: Yale University Press, 1982.

Bancroft, Randy. "Brand-ing in 30 Seconds?" *Metric Maven*, October 10 (2017). https://themetricmaven.com/brand-ing-in-30-seconds.

Banerjee, Debdas. *Colonialism in Action: Trade, Development, and Dependence in Late Colonial India*. New Delhi: Orient Longman Limited, 1999.

Barman, Roderick J. "The Brazilian Peasantry Reexamined: The Implications of the Quebra-Quilo Revolt, 1874–1875." *Hispanic American Historical Review* 53 (1977): 401–24.

Barnard, Frederick A. P. *The Imaginary Metrological System of the Great Pyramid of Gizeh*. New York: John Wiley & Sons, 1884.

Barnard, Frederick A. P. "On the Relation to the Public Welfare of Changes in the Volume of Money and on Money Standards." *Proceedings of the American Metrological Society* 2 (1882): 202–30.

Barnard, Frederick A. P. *The Regulation of Time; International Coinage; the Unification of Weights and Measures; Sea-Signals*. London: W. Clowes and Sons, 1881.

Barnard, Frederick A. P. *Mono-Metallism, Bi-Metallism, and International Coinage*. New York: The S. W. Green Type-setting Machines, 1879.

Barnard, Frederick A. P. *The Possibility of an Invariable Standard of Value*. New York, 1879.

Barnard, Frederick A. P. *International Coinage*. London: William Clowes, 1874.

Barnard, Frederick A. P. *The Metric System of Weights and Measures*. New York: Columbia College, 1872.

Bartky, Ian. "The Adoption of Standard Time." *Technology and Culture* 30 (1989): 25–56.

Baxter, John. "Measuring New Mexico's Irrigation Water: How Big Is a Surco?" *New Mexico Historical Review* 75, no. 3 (2000): 397–413.

Beigbeder Atienza, Federico. *Manual de pesos, medidas y monedas del mundo con equivalencias al sistema métrico decimal*. Madrid: Castilla, 1959.

Bell, Daniel. *The Coming of Post-Industrial Society: A Venture in Social Forecasting*. New York, Basic Books, 1973.

Bensel, Richard Franklin. *Yankee Leviathan: The Origins of Central State Authority in America, 1859–1877*. New York: Cambridge University Press, 1990.

Berger, Peter. "The Cultural Dynamics of Globalization." In *Many Globalizations: Cultural Diversity in the Contemporary World*, edited by Peter Berger, 1–16. Oxford: Oxford University Press, 2002.

Berger, Peter, and Thomas Luckmann. *The Social Construction of Reality.* New York: Doubleday, 1966.

Berriman, A. E. *Historical Metrology: A New Analysis of the Archaeological and the Historical Evidence Relating to Weights and Measures.* New York: Greenwood Press, 1953.

Bloch, Marc. "Le témoignage des mesures agraires." *Annales d'historie économique et sociale* 6 (1934): 280–82.

Bodin, Jean. *On Sovereignty.* New York: Cambridge University Press, 1992.

Bonaparte, Napoleon. *Mémoires pour servir a l'histoire de France, sous Napoleon écrits a Saint-Hélène, par les generaux qui ont partagé sa captivité et publiés sur les manuscrits entiérement corrigés de la main de Napoléon*, vol. IV. Paris: Firmin Didot Pere et Fils Libraries, 1823.

Boorstin, Daniel J. "Afterlives of the Great Pyramid." *Wilson Quarterly* 16 (1992): 130–38.

Bosak, Jon. *The Old Measure: An Inquiry into the Origins of the U.S. Customary System of Weights and Measures.* Ithaca: Pinax Publishing, 2010.

Bouk, Dan. *Democracy's Data: The Hidden Stories in the U.S. Census and How to Read Them.* New York: Farrar, Straus and Giroux, 2022.

Bourdieu, Pierre. "Rethinking the State: Genesis and Structure of the Bureaucratic Field." In *State/Culture: State-Formation after the Cultural Turn*, edited by George Steinmetz, 53–75. Ithaca: Cornell University Press, 1999.

Bowman, J. N. "Weights and Measures of Provincial California." *California Historical Society Quarterly* 30 (1951): 315–38.

Brand, Stewart. "Stopping Metric Madness!" *New Scientist* 88 (October 30, 1980), 315.

Branscomb, Lewis M. "The Metric System in the United States." *Proceedings of the American Philosophical Society* 116 (1972): 294–300.

Braudel, Fernand. "History and the Social Sciences: The Longue Durée." In *The Longue Durée and World-Systems Analysis*, edited by Richard E. Lee, 241–76. Albany: State University of New York Press, 2012.

Brent, Joseph. *Charles Sanders Peirce: A Life.* Bloomington: Indiana University Press, 1993.

Browne, Malcolm W. "Growing Opposition to Conversion to Metric System Is Found in Poll." *New York Times*, November 24, 1977: 38.

Browne, Malcolm W. "Kinder, Gentler Push for Metric Inches Along." *New York Times*, June 4, 1996.

Brück, H. A. and M. T. Brück. *The Peripatetic Astronomer: The Life of Charles Piazzi Smyth.* Philadelphia, PA: Adam Hilger, 1988.

Bryce, T. A. *The American Commercial Arithmetic*. N. p.: C. G. Swensberg, 1873.

Burdun, G. D. "Worldwide Dissemination of the Metric System." *Measurement Techniques* 9 (September 1968): 1147–51.

Bureau of Standards. *Metric Manual for Soldiers*. Washington, DC: Government Printing Office, 1918.

Bureau of Standards. *War Work of the Bureau of Standards*. Washington, DC: Government Printing Office, 1921.

Busch, Lawrence. "Herbert Hoover and the Construction of Modernity." *Journal of Innovation Economics & Management* 22 (2017): 29–55.

Busch, Lawrence. "The Moral Economy of Grades and Standards." *Journal of Rural Studies* 16 (2000): 273–83.

Calvert, Monte A. *The Mechanical Engineer in America, 1830–1910: Professional Cultures in Conflict*. Baltimore, MD: Johns Hopkins Press, 1967.

Carpentier, Alejo. *Explosion in a Cathedral*. Boston: Little, Brown and Company, 1963.

Carroll, Patrick. *Science, Culture, and Modern State Formation*. Berkeley: University of California Press, 2006.

Carruthers, Bruce G. "The Sociology of Money and Credit." In *The Handbook of Economic Sociology*, edited by Neil J. Smelser and Richard Swedberg, 355–77. Princeton, NJ: Princeton University Press, 2005.

Carruthers, Bruce G. and Sarah L. Babb. *Economy/Society: Markets, Meanings, and Social Structure*. Thousand Oaks, CA: Pine Forge Press, 2000.

Castells, Manuel. *The Information Age, vol. III: End of Millennium*. Oxford: Blackwell, 2010.

Champagne, Ruth Inez. "The Role of Five Eighteenth Century French Mathematicians in the Development of the Metric System." PhD diss., Columbia University, 1979.

"Charles Latimer," *Virtual American Biographies*, copyright 2001. http://famousamericans.net/charleslatimer.

Chartier, Roger. *The Cultural Uses of Print in Early Modern France*. Princeton, NJ: Princeton University Press, 1988.

Chassen López, Francie R. *From Liberal to Revolutionary Oaxaca: The View from the South, Mexico 1867–1911*. University Park: Pennsylvania State University Press, 2004.

Childe, Gordon. *Man Makes Himself*. Bradford-on-Avon: Moonraker, 1981.

Childe, Gordon. *What Happened in History*. Baltimore, MD: Penguin, 1961.

Ciscar, Gabriel. *Memoria elemental sobre los nuevos pesos y medidas decimales fundados en la naturaleza*. Madrid: Imprenta Real, 1800.

Cline, Sheldon S. "Metric System Urged for U.S. and England, Only Two Civilized Nations Retaining Ancient and Cumbersome Weights and Measures." *Evening Star*, June 15, 1919.

Cochrane, Rexmond C. *Measures for Progress: A History of the National Bureau of Standards*. Washington, DC: US Dept. of Commerce, 1966.

Cock-Starkey, Claire. *The Curious History of Weights & Measures*. Chicago: University of Chicago Press, 2023.

Cohen, I. B. *Science and the Founding Fathers*. New York: Norton, 1995.

Cohen, I. B. *The Triumph of Numbers: How Counting Shaped Modern Life*. New York: W. W. Norton, 2005.

Coles, Jessie V. *The Consumer-Buyer and the Market*. New York: John Wiley & Sons, 1938.

Collins, Randall. "Weber's Last Theory of Capitalism: A Systematization." In *The Sociology of Economic Life*, edited by Mark Granovetter and Richard Swedberg, 379–99. Cambridge, MA: Westview Press, 2001.

Colonial Laws of New York, from the Year 1664 to the Revolution. Albany, NY: James B. Lyon, 1894.

Comte, Auguste. *Cours de philosophie positive 6: Le complément de la philosophie sociale et les conclusions generals*. Paris: Baillière, 1864.

Comte, Auguste. *The Positive Philosophy of Auguste Comte: Freely Translated and Condensed by Harriet Martineau*, vol. III. London: Geroge Bell & Sons, 1896.

Condorcet. "Observations on the Twenty-Ninth Book of *The Spirit of Laws*" annexed in Antoine Louis Claude Destutt de Tracy, *A Commentary and Review of Montesquieu's Spirit of Laws*, 261–82. Philadelphia, PA: William Duane, 1811.

Confederación de Cámaras de Comercio de los Estados Unidos Mexicanos. "Urgente necesidad de hacer universal el uso del sistema métrico decimal. Estudio para la 9na. Asamblea General de Cámaras de Comercio de la República." 1926.

Congress. House. Committee on Coinage, Weights, and Measures. "Statement of Prof. Simon Newcomb." *The Metric System of Weights and Measures* (Washington, DC: Government Printing Office, 1902), 71–74.

Constant, Benjamin. *Political Writings*. New York: Cambridge University Press, 1988.

Cousins, James J. "Weights and Measures in England versus the Decimal and Metric Systems." *Science* 25 (1892): 298.

Cox, Edward Franklin. "A History of the Metric System of Weights and Measures: With Emphasis on Campaigns for its Adoption in Great Britain and in the United States prior to 1914." PhD diss., Indiana University, 1956.

Cox, Edward Franklin. "The International Institute: First Organized Opposition to the Metric System." *Ohio Historical Quarterly* 58 (1959): 54–83.

Cox, Edward Franklin. "The Metric System: A Quarter-Century of Acceptance (1851–1876)." *Isis* 13 (1958): 358–79.

Crease, Robert P. *World in the Balance: The Historic Quest for an Absolute System of Measurement*. New York: W. W. Norton, 2011.

Crosby, Alfred W. *The Measure of Reality: Quantification and Western Society, 1250–1600*. New York: Cambridge University Press, 1997.

Crosland, Maurice. "The Congress on Definitive Metric Standards, 1798–1799: The First International Scientific Conference?" *Isis* 60 (1969): 226–31.

Crump, Thomas. *The Anthropology of Numbers*. Cambridge: Cambridge University Press, 1992.

Curtis, Bruce. "From the Moral Thermometer to Money: Metrological Reform in Pre-Confederation Canada." *Social Studies of Science* 28 (1998): 547–70.

Dale, Samuel S. "Yards, Gallons, and Grains: The Economic and Moral Aspects of English Weights and Measures." *The Christian Endeavor World*, November 6, 1919, 106.

Dale, Samuel S. *The World Trade Club of San Francisco and its Metric Propaganda; with Additions to April 2, 1920*. Textiles: Boston, 1920.

Dale, Samuel S. *The Foreign Attack on Our Weights and Measures*. Boston: self-published, 1926.

David, Paul. "Clio and the Economics of QWERTY." *American Economic Review* 75 (1995): 332–37.

Dawe, Grosvenor. *Melvil Dewey, Seer: Inspirer: Doer, 1851–1931*. Lake Placid Club, NY: Melvil Dewey Biografy, 1932.

Deming, Richard. *Metric Power: Why and How We Are Going Metric*. Nashville, TN: Thomas Nelson Publishers, 1974.

Department of Consumer Affairs, Mayor's Bureau of Weights and Measures, City of New York. *What the Purchasing Public Should Know*. New York: J. W. Pratt Company, 1911.

Department of Weights and Measures, City of Newark. *What Every Housewife Should Know*. Newark, 1911.

Dewey, Melvil. "Efficiency Society." In *The Encyclopedia Americana: A Library of Universal Knowledge*, 9: 719–20. New York: Encyclopedia Americana Corporation, 1918.

Dewey, Melvil. *A Classification and Subject Index, for Cataloguing and Arranging the Books and Pamphlets of a Library*. Amherst: n. p., 1876.

Diderot, Denis et al. *Encyclopédie, ou Dictionnaire raisonné des sciences, des arts et des métiers, par une société des gens de letters*. Paris: De l'imprimerie de Le Breton, imprimeur ordinaire du Roy, 1765.

Donovan, Frank. *Prepare Now for a Metric Future*. New York: Weybright and Talley, 1970.

"Doudecimalisms." *Tucson Daily Citizen*, April 28, 1904.

Draper, A. V. "Herbert Spencer's Opposition to the Metric System." *Manufacturers Record*, August 26, 1920: 109–12.

Draper, A. V. "Herbert Spencer's Opposition to the Metric System." *Manufacturers Record*, August 26, 1920: 109–12.

Droz, Yves, and Joseph Flores, eds. *Les heures révolutionnaires*. Besançon: Association Française des Amateurs d'Horlogerie Ancienne, 1989.

Drury, Aubrey, ed. *World Metric Standardization: An Urgent Issue*. San Francisco, CA: World Metric Standardization Council, 1922.

Drury, Aubrey. "Making This World Metric." *Pan Pacific: A Magazine of International Commerce*, June 20, 1920: 84–85.

Drury, Aubrey. "Modern Trade—Antiquated Tools: A Plea for International Commodity Quantity Standards in Trade and Industry." *The Rotarian* 26 (1925): 2: 29, 66–69.

Drury, Aubrey. "The Metric Advance." *Measurement* 1 (1926): 6–7.

Drury, Aubrey. *One Standard for All America*. [San Francisco]: American Standards Council, [1927].

Drury, Aubrey. *Un mismo sistema de pesas y medidas para toda la América*. Washington, DC: La Unión Panamericana, 1927.

Dudley Duncan, Otis. *Notes on Social Measurement*. New York: Russell Sage Foundation, 1984.

Dudzik, Michael. "The Decimalisation of Republican Time. The French Revolution Which Failed (1793–1795)." *Acta Polytechnica* 64 (2024): 182–93.

Duncan, David. *The Life and Letters of Herbert Spencer*. London: Methuen & Co., 1908.

Dupree, A. Hunter. "Metrication as Cultural Adaptation." *Science* 185 (1974): 208.

Dupree, A. Hunter. "The Measuring Behavior of Americans." In *Nineteenth-Century American Science: A Reappraisal*, edited by George H. Daniels, 22–37. Evanston, IL: Northwestern University Press, 1972.

Dupree, A. Hunter. *Science in the Federal Government: A History of Policies and Activities*. Baltimore: Johns Hopkins University Press, 1986.

Dupree, A. Hunter. "Measurement." In *The History of Science in the United States: An Encyclopedia*, edited by Marc Rothenberg, 338–40. New York: Garland Publishing, 2001.

Easley, David and Jon Kleinberg. *Networks, Crowds, and Markets: Reasoning about a Highly Connected World.* New York: Cambridge University Press, 2010.

Eastburn, George. *The Metric System.* New York: American Metrological Society, 1892.

Edwin, Williams, ed. *Addresses and Messages of the Presidents of the United States, Inaugural, Annual, and Special, from 1789 to 1846.* New York: Edward Walker, 1846.

Einaudi, Luca. *Money and Politics: European Monetary Unification and the International Gold Standard, 1865-1873.* New York: Oxford University Press, 2001.

Elias, Norbert. "Towards a Theory of Social Processes." In *Essays III: On Sociology and the Humanities*, 9-39. Dublin: University College Dublin Press, 2009.

Elias, Norbert. *The Civilizing Process.* Oxfrod: Blackwell, 2000.

Elias, Norbert. *Time: An Essay.* Oxford: Blackwell, 1992.

Elias, Norbert. "Knowledge and Power." In *Knowledge and Society*, edited by N. Stehr and V. Meja, 251-91. New Brunswick, NJ: Transaction Books, 1984.

Elliot, E. B. "On the Relative Value of Gold and Silver for a Series of Years." *Proceedings of the American Association for the Advancement of Science* 17 (1868): 122-23.

Ellis, Keit. *Man and Money.* London: Priory Press, 1973.

Emigh, Rebecca Jean. "The Power of Negative Thinking: The Use of Negative Case Methodology in the Development of Sociological Theory." *Theory and Society* 26 (1997): 649-84.

Engels, Frederick. *The Peasant War in Germany.* New York: Routledge, 2015.

Ervin, Michael A. "Statistics, Maps, and Legibility: Negotiating Nationalism in Post-Revolutionary Mexico." *The Americas* 66 (2009): 155-79.

Ervin, Michael A. "The 1930 Agrarian Census in Mexico: Agronomists, Middle Politics, and the Negotiation of Data Collection." *Hispanic American Historical Review* 87 (2007): 537-70.

Everett, Caleb. *Numbers and the Making of Us: Counting and the Course of Human Cultures.* Cambridge, MA: Harvard University Press, 2017.

Faith, Roger, Robert McCormick, and Robert Tollison. "Economics and Metrology: Give 'em an Inch and They'll Take a Kilometre." *International Review of Law and Economics* 1 (1981): 207-21.

Fauchois, Yann. "Centralization." In *A Critical Dictionary of the French Revolution*, edited by F. Furet and M. Ozouf, 629-39. Cambridge, MA: Harvard University Press, 1989.

Felton, John. *The Decimal System: An Argument for American Consistency.* New York: G. P. Putnam & Co., 1857.

Fena, Donald. *Elsevier's Encyclopedic Dictionary of Measures.* New York: Elsevier, 1998.

Fischer, Louis Albert. "History of the Standard Weights and Measures of the United States." *Bulletin of the Bureau of Standards* 1 (1905): 365–81.

Fischer, Louis Albert. "Recent Developments in Weights and Measures in the United States." *Popular Science Monthly* 84 (1914): 345–69.

Foner, Eric. *Reconstruction: America's Unfinished Revolution, 1863–1877.* New York: Harper, 2014.

"Foreign Literature: France." *American Monthly Review*, February 1795: 195–98.

Fourcade, Marion and Kieran Healy. *The Ordinal Society.* Cambridge: Harvard University Press, 2024.

Frasier, E. Lewis. "Improving an Imperfect Metric System." *Bulletin of the Atomic Scientists* 30 (1974): 9–12.

Frazier, Arthur H. *United States Standards of Weights and Measures, their Creation and Creators.* Washington: Smithsonian Institution Press, 1978.

Friguglietti, James. "Gilbert Romme and the Making of the French Republican Calendar." In *The French Revolution in Culture and Society,* edited by N. Andrews, 13–22. New York: Greenwood Press, 1991.

Frost, Douglas V. "Logical Steps to Metric Conversion." *Poultry Science* 44 (1965), 1227–36.

Furet, F., and M. Ozouf, eds. *A Critical Dictionary of the French Revolution.* Cambridge, MA: Harvard University Press, 1989.

Galison, Peter. *Einstein's Clocks, Poincaré's Maps: Empires of Time.* New York: Norton, 2003.

Gallup, George H. *The Gallup Poll: Public Opinion, 1972–1977.* Wilmington, DE: Scholarly Resources, 1978.

García Acosta, Virginia. "Weights and Prices of Bread in Eighteenth-Century Mexico." *Cahiers de Métrologie* 11–12 (1993–1994): 45–57.

Geertz, Clifford. "The Bazaar Economy: Information and Search in Peasant Marketing." *American Economic Review* 68 (1978): 28–32.

Geiger, Fred A. "Why the Metric System Must Not be Made Compulsory." *Machinery*, May 1920.

Genschel, Philipp. "Path-Dependence." In *International Encyclopedia of Economic Sociology*, edited by J. Beckert and M. Zafirocski, 507–8. London: Routledge, 2005.

Geyer, Martin H. "One Language for the World: The Metric System, International Coinage, Gold Standard, and the Rise of Internationalism, 1850–1900." In *The*

Mechanics of Internationalism: Culture, Society, and Politics from the 1840s to the First World War, edited by Martin H. Geyer and Johannes Paulmann, 55–92. New York: Oxford University Press, 2001.

Giddens, Anthony. *The Nation-State and Violence.* Berkeley: University of California Press, 1987.

Gillispie, Charles Coulston. *Science and Polity in France: The Revolutionary and Napoleonic Years.* Princeton, NJ: Princeton University Press, 2004.

Gillispie, Charles Coulston. *Pierre-Simon Laplace, 1749–1827: A Life in Exact Science.* Princeton, NJ: Princeton University Press, 1997.

Giunta, Carmen J. *A Brief History of the Metric System: From Revolutionary France to the Constant-Based SI.* Cham: Springer Nature, 2023.

Gooday, Graeme J. N. *The Morals of Measurement: Accuracy, Irony, and Trust in Late Victorian Electrical Practice.* New York: Cambridge University Press, 2004.

Goodridge, R. E. W. *On the Proposed Change of Time Marking to a Decimal System: A Plea that the Duodecimal System be Retained.* Winnipeg: Manitoba Daily Free Press, 1886.

Goody, Jack. *The Logic of Writing and the Organization of Society.* New York: Cambridge University Press, 1986.

Gómez Martínez, José Ramón, María Teresa Sánchez Trujillano, and José Antonio Tirado Martínez, *¿Y esto en onzas cuánto es?: 1853–2003.* Logroño: La Rioja, 2003.

Gordin, Michael D. "Measure of All the Russians: Metrology and Governance in the Russian Empire." *Kritika: Explorations in Russian and Eurasian History* 4 (2003): 783–815.

Gourville, John, and Jonathan J. Koehler. "Downsizing Price Increases: A Greater Sensitivity to Price than Quantity in Consumer Markets." Harvard Business School Working Paper, No. 04–042 (2004).

Groesbeck, John. *The Crittendon Commercial Arithmetic and Business Manual: Designed for the Use of Merchants, Business Men, Academies, and Commercial Colleges.* Philadelphia: Eldredge & Brother, 1871.

Grossman, Jonathan H. "Standardization (Standardisation)." *Critical Inquiry* 44 (2018): 447–478.

Guedj, Denis. *Le mètre du monde: Historie politique, scientifique et philosophique de l'invention du système métrique decimal.* Paris: Editions du Seuil, 2000.

Guedj, Denis. *The Measure of the World.* Chicago: University of Chicago Press, 2001.

Guillaume, C. E. *Les récents progrès du système métrique.* Paris: Gauthier-Villars, 1913.

Günergun, Feza. "Introduction of the Metric System to the Ottoman State." In *Transfer of Modern Science and Technology to the Muslim World,* edited by Ekmeleddin Ihsanoglu, 297–316. Istanbul: Research Centre for Islamic History, Art, and Culture, 1992.

Gurvitch, Georges. *The Social Frameworks of Knowledge.* New York: Harper, 1972.

Gutiérrez Cuadrado, Juan. *Metro y kilo: El sistema métrico decimal en España.* Madrid: Akal, 1997.

Hague, W. E. "Plea for Metric System: English Weights and Measures, Says Advocate of Decimal Method, Were Devised and Scrapped by Germans." *New York Times,* October 26, 1919: 8.

Hague, W. E. *The Universal Language of Quantity, Meter-Liter-Gram.* San Francisco, CA: World Trade Club, 1919.

Hallock, William, and Herbert Treadwell Wade. *Outlines of the Evolution of Weights and Measures and the Metric System.* New York: Macmillan, 1906.

Halsey, Frederick. "Continuing the Metric War." *New York Times,* June 5, 1927.

Halsey, Frederick. "Disputes Metric Success." *The New York Times,* August 23, 1925.

Halsey, Frederick A. *The Metric Fallacy: An Investigation of the Claims Made for the Metric System and Especially of the Claim That its Adoption Is Necessary in the Interest of Export Trade,* 2nd ed. New York: American Institute of Weights and Measures, 1920.

Halsey, Frederick A. "Pan Americanism in Weights and Measures." In *Report of the Second Pan American Commercial Conference: Pan American Commerce; Past, Present, Future from the Pan American Viewpoint,* edited by John Barret, 270–74. Washington, DC: Pan American Union, 1919.

Halsey, Frederick A. *The Weights and Measures of Latin America.* New York: American Society of Mechanical Engineers, 1918.

Halsey, Frederick A., and Samuel S. Dale. *The Metric Fallacy* and *The Metric Failure in the Textile Industry.* New York: D. Van Nostrand, 1904.

Halsey, Frederick A. "The Metric System." *Transactions of the American Society of Mechanical Engineers* 24 (1903): 397–629.

Halsey, Frederick A. "The Premium Plan of Paying for Labor." *Transactions of the American Society of Mechanical Engineers* 12 (1890): 755–80.

Hanley, Anne. "Men of Science and Standards: Introducing the Metric System in Nineteenth-Century Brazil." *Business History Review* 96 (2022): 17–45.

Harrington, Virginia D. *The New York Merchant in the Eve of the Revolution.* New York: Columbia, 1935.

Hart, David. "Herbert Hoover's Last Laugh: The Enduring Significance of the 'Associative State.'" *Journal of Policy History* 10 (1998): 419–44.

Hausberger, Bernd. *La globalización temprana*. Mexico: Colegio de México, 2018.

Haynes, A. E. "Metric Department; Progress of the Metric System of Weights and Measures." *Hillsdale Standard*, December 4, 1877.

Healy, David. *James G. Blaine and Latin America*. Columbia: University of Missouri Press, 2002.

Hearings before the Committee on Coinage, Weights, and Measures on H. R. 93 (58th Congress, 1st Session); H. R. 2054 (58th Congress, 2d Sessions), and H. R. 8988 (59th Congress, 1st Sessions). Washington, DC: Government Printing Office, 1906.

Hearings before the Committee, Feb. 6–March 3, 1902, on Bill H. R. 2054. Washington, DC: Government Printing Office, 1902.

Heilbron, John L. "The Politics of the Meter Stick." *American Journal of Physics* 57 (1989): 988–92.

Heilbron, John L. *Weighing Imponderables and Other Quantitative Science Around 1800*. Berkeley: University of California Press, 1993.

Heilbroner, Robert L. *The Worldly Philosophers*. New York: Simon & Schuster, 1967.

Helleiner, Eric. "National Currencies and National Identities." *American Behavioral Scientist* 41 (1998): 1409–36.

Hellman, C. Doris. "Jefferson's Efforts towards Decimalization of United States Weights and Measures." *Isis* 16 (1931): 266–314.

"Herbert Spencer Opposes Metric System." *New York Times*, June 22, 1896.

Herschel, John. *Familiar Lectures on Scientific Subjects*. London: Alexander Strahan, 1867.

Higby, Gregory J., and Glenn Sonnedecker. "Adoption of the Metric System by the U.S. Pharmacopoeia." *Journal of the History of Medicine and Allied Sciences* 40 (1985): 207–13.

Hill, Tom M. "Imperial Nomads: Settling Paupers, Proletariats, and Pastoralists in Colonial France and Algeria, 1830–1863." PhD diss., University of Chicago, 2006.

Hobsbawm, Eric. *The Age of Capital, 1848–1875*. New York: Charles Scribner's Sons, 1975.

Hobsbawm, Eric. *The Age of Extremes: A History of the World, 1914–1991*. New York: Pantheon, 1994.

Hocquet, Jean-Claude, and B. Garnier, eds. *Genèse et diffusion du système métrique*. Caen: Editions-Diffusion du Lys, 1990.

Hocquet, Jean-Claude. "Weights and Measures in Mexico." In *Encyclopaedia of the History of Science, Technology, and Medicine in Non-Western Cultures*, edited by Helaine Selin, 1023–25. Dordrecht: Kluwer Academic Publishers, 1997.

Hofstadter, Richard. *Social Darwinism in American Thought*. Boston: Beacon Press, 1955.

Hogeback, Jonathan. "Why Doesn't the U.S. Use the Metric System?" *Encyclopedia Britannica*, November 17, 2016. https://www.britannica.com/story/why-doesnt-the-us-use-the-metric-system.

Holden, Constance. "Metrication: Craft Unions Seek to Block Conversion Bill." *Science* 184 (1974): 48–50, 94.

Holland, Simon J. "The Decimal System: The Ten Essentials of a Complete Decimal System of Measures, Weights and Money." *Mechanics' Magazine* 65 (1856): 461–63.

Hong, Heong, and Tan Miau Ing. "Contested Colonial Metrological Sovereignty: The *Daching* Riot and the Regulation of Weights and Measures in British Malaya." *Modern Asian Studies* 56, no. 1 (2022): 407–26.

Hopkins, Richard L. *Origin of the American Point System for Printers' Type Measurement*. Terra Alta, WV: Hill & Dale Press, 1976.

House of Commons. *Papers by Command*, vol. 77. London: Darlyn & Son, 1906.

House of Representatives. *Metric Coinage Report*. Washington, DC: Government Printing Office, 1880.

Hubbard, Henry D. "Measurements of Tomorrow." Talk at the Summer Meeting of the Metric Association, Friday, June 24, 1927.

Hugo, Victor. *Les Misérables*. New York: Kelmscott Society, 1887.

Immerwahr, Daniel. *How to Hide an Empire: A History of the Greater United States*. New York: Farrar, Straus and Giroux, 2020.

Inter American High Commission. *Report of the Second Pan American Standardization Conference*. Washington, DC: Government Printing Office, 1927.

International American Conference. *Reports of Committees and Discussion Thereon*. Washington, DC: Government Printing Office, 1890.

International Conference Held at Washington for the Purpose of Fixing a Prime Meridian and a Universal Day, October, 1884: Protocols of the Proceedings. Washington, DC: Gibson Bros., 1884.

Iriye, Akira. *Cultural Internationalism and World Order*. Baltimore, MD: Johns Hopkins University Press, 1997.

Iwahashi, Masaru. "The Institutional Framework of the Tokugawa Economy." In *Emergence of Economic Society in Japan, 1600–1859*, edited by A. Hayami, O. Saito, and R. P. Toby, 85–104. New York: Oxford University Press, 2004.

Iwata, Shigeo. "Weights and Measures in Japan." In *Encyclopaedia of the History of Science, Technology, and Medicine in Non-Western Cultures*, edited by Helaine Selin, 1019–23. Boston: Kluwer Academic, 1997.

"James Watt: Pioneer of Decimal Systems." *Decimal Educator* 19 (1936): 9–10.

"Japan Does Not Favor Adoption of the Metric System in That Country." *American Machinist*, July 14, 1921.

Jefferson, Thomas. *The Papers of Thomas Jefferson*, edited by Julian P. Boyd. Princeton, NJ: Princeton University Press, 1953–71.

Jefferson, Thomas. *The Writings of Thomas Jefferson Being His Autobiography, Correspondence, Reports, Messages, Addresses, and Other Writings, Official and Private*, edited by H. A. Washington. Washington, DC: Taylor & Maury, 1854.

Jenkins-Smith, Hank, Carol Silva, and Geoboo Song. *Health Policy Survey 2010: A National Survey on Public Perceptions of Vaccination Risks and Policy Preferences*. Norman, OK: University of Oklahoma, 2010.

Jenks, Robert R. "Frederick Arthur Halsey." *American National Biography Online*, February 2000. https://doi.org/10.1093/anb/9780198606697.article.1300688.

Jenks, Robert R. "Governing in the Absence of Government: The Birth and Development of the United States Industrial Standards System." PhD diss., University of California, Santa Barbara, 1999.

Jessop, William Henry. *A Complete Decimal System of Money and Measures*. Cambridge: Deighton, 1855.

Johnson, J. T. "Three Studies on the Effect of Compulsory Metric Usage." *Journal of Educational Research* 37 (1944): 587–92.

Jones, Hilton Ira. "Why Not Now?" *School Science and Mathematics* 19 (1919): 512–19.

Jones, Sarah Ann. "Weights and Measures in Congress: A Historical Summary of Events Culminating in the Weights and Measures Act of 1836." MA diss., George Washington University, 1935.

Josephus, Flavius. *Josephus: The Essential Writings*. New York: Kregel, 1990.

Jourdan, Louis. *La grande métrication*. Nice: France Europe Éditions, 2002.

Judson, Lewis V. and Louis E. Barbrow. *Weights and Measures Standards of the United States: A Brief History*. Washington, DC: National Bureau of Standards, 1976.

Kaiser, Chester C. "México en la primera Conferencia Panamericana." *Historia Mexicana* 11 (1961): 56–80.

Kennedy, Emmet. *A Cultural History of the French Revolution*. New Haven, CT: Yale University Press, 1989

Kennelly, A. E. *Biographical Memoir of Samuel Wesley Stratton, 1861–1931.* Washington, DC: National Academy of Sciences, 1935.

Kennelly, Arthur E. *Vestiges of Pre-Metric Weights and Measures Persisting in Metric-System Europe.* New York: Macmillan Company, 1928.

Kindleberger, Charles P. *Keynesianism vs. Monetarism and Other Essays in Financial History.* Boston: G. Allen & Unwin, 1985.

Kisch, Bruno. *Scales and Weights: A Historical Outline.* New Haven, CT: Yale University Press, 1965.

Knight, Charles. *The Popular History of England*, vol. II. London: Bradbury and Evans, 1856.

Knight, Isabel F. *The Geometric Spirit: The Abbé de Condillac and the French Enlightenment.* New Haven, CT: Yale University Press, 1968.

Koenig, Joan. "Uniformity in Weights and Measures Laws and Regulations." In *A Century of Excellence in Measurements, Standards, and Technology*, edited by David Lide, 368–70. New York: CRC Press, 2002.

Krajewski, Markus. *Paper Machines: About Cards & Catalogs, 1548–1929.* Cambridge, MA: MIT Press, 2011.

Kula, Witold. "Money and the Serfs in Eighteenth Century Poland." In *Peasants in History: Essays in Honor of Daniel Thorner*, edited by E. J. Hobsbawm et al., 30–41. Calcutta: Sameeksha Trust, 1980.

Kula, Witold. *Measures and Men.* Princeton, NJ: Princeton University Press, 1986.

Kunz, George F. "New International Metric Diamond Carat of 200 Milligrams." *Science* 38 (1913): 523–24.

Kunz, George F. "The New International Metric Diamond Carat of 200 Milligrams (Adopted July 1, 1913, in the United States)." *Bulletin of the American Institute of Mining and Metallurgical Engineers* 79 (1913): 1225–45.

Kunz, George F. "The International Language of Weights and Measures." *Scientific Monthly* 4 (1917): 215–19.

Kunz, George F. "The Metric System as a Factor in Pan American Unity." In *Report of the Second Pan American Commercial Conference: Pan American Commerce; Past, Present, Future from the Pan American Viewpoint*, edited by John Barret, 266–270. Washington, DC: Pan American Union, 1919.

Langevin, Luce. "The Introduction of the Metric System: The First Example of Scientific Rationalization by Society." *Impact of Science on Society* 11 (1961): 77–95.

Latimer, Charles. *The French Metric System, or, The Battle of the Standards: A Discussion of the Comparative Merits of the Metric System and the Standards of the Great Pyramid*. Chicago: Thomas Wilson, 1880.

Latimer, Charles. *The Divining Rod: Virgula Divina-Baculus Divinatorius (Water-Witching)*. Cleveland: Fairbanks, Benedict & Co., 1876.

Latour, Bruno. *Reassembling the Social: An Introduction to Actor-Network-Theory*. Oxford: Oxford University Press, 2005.

Laws of the State of New York in Relation to Weights and Measures and Uniform Rules and Regulations. Albany, NY: J. B. Lyon Company, 1916.

Lenzen, Victor F. "The Contributions of Charles S. Peirce to Metrology." *Proceedings of the American Philosophical Society* 109 (1965): 29–46.

Lester, Toby. "New-Alphabet Disease?" *Atlantic Monthly* 280, no. 1 (1997): 20–27.

Letter of the Secretary of the Treasury, in Response to A Resolution of the House of Representatives, Transmitting certain Reports in Reference to the Adoption of the Metric System. Washington, DC: Government Printing Office, 1878.

Lewis Frasier, E. "Improving an Imperfect Metric System." *Bulletin of the Atomic Scientists* 30, no. 2 (1974): 9–12.

Lide, David R., ed. *A Century of Excellence in Measurements, Standards, and Technology*. New York: CRC Press, 2002.

Lind, John. *The Mexican People*. Minneapolis: The Bellman, n.d.

Linklater, Andro. *Measuring America: How an Untamed Wilderness Shaped the United States and Fulfilled the Promise of Democracy*. New York: Walker and Company, 2002.

"Lord Kelvin a Witness: Appears Before a House Committee and Indorses the Metric System Bill." *New York Times*, April 25 1902: 1.

"Lord Kelvin on the Metric System." *Science*, May 22, 1896: 765–66.

Loveman, Mara. "The Modern State and the Primitive Accumulation of Symbolic Power." *American Journal of Sociology* 110 (2005): 1651–83.

Luckmann, Thomas. "Common Sense, Science and the Specialization of Knowledge." *Phenomenology + Pedagogy* 1 (1983): 59–73.

Lugli, Emanuele. *The Making of Measure and the Promise of Sameness*. Chicago: University of Chicago Press, 2022.

MacFarquhar, Neil. "Libya Under Qaddafi: Disarray Is the Norm." *New York Times*, February 14, 2001.

Machabey, Armand. "Techniques of Measurement." In *A History of Technology & Invention: II. The First Stages of Mechanization*, edited by Maurice Daumas, 306–43. New York: Crown, 1969.

Maestro, Marcello. "Going Metric: How It All Started." *Journal of the History of Ideas* 3 (1980): 479–86.

Maher, John C. *Multilingualism: A Very Short Introduction*. Oxford: Oxford University Press, 2017.

Malinowski, Bronislaw, and Julio de la Fuente. *Malinowski in Mexico: The Economics of a Mexican Market System*. Boston: Routledge, 1982.

Mandavilli, Sujay Rao. "Metrication in India." July 22, 2009. https://usma.org/metrication-in-other-countries#india.

Mann, Michael. "The Autonomous Power of the State: Its Origins, Mechanisms and Results." *European Journal of Sociology* 25 (1984): 185–213.

Mann, W. Wilberforce. *A Decimal Metric New System Founded on the Earth's Polar Diameter, and Designed for the Adoption of all Civilized Nations as the One Common System*. New York: University Publishing Company, 1872.

Marciano, John. *Whatever Happened to the Metric System? How America Kept Its Feet*. New York: Bloomsbury, 2014.

Marec, Yannick. "Autor des résistances au sysème métrique." In *Genèse et diffusion du système métrique*, edited by J.-C. Hocquet and B. Garnier, 135–44. Caen: Editions-Diffusion du Lys, 1990.

Marec, Yannick. "L'ambition revolutionnaire: Mesurer toutes choses rationnellement." In *La révolution Française et les processus de socialisation de l'homme moderne*, 691–700. Paris: Éditions Messidor, 1989.

Markoff, John. *The Abolition of Feudalism: Peasants, Lords, and Legislators in the French Revolution*. University Park: Pennsylvania State University Press, 1996.

Markovits, Andrei S., and Steven L. Hellerman. *Offside: Soccer and American Exceptionalism*. Princeton, NJ: Princeton University Press, 2001.

Macqueen. C. E. *The Advantages of a Complete Decimal System of Money, Weights and Measures*. Liverpool: Financial Reform Association, 1855.

Marquet, Louis. "24 heures ou 10 heures? Un essai de division décimale du jour (1793–1795)." *L'Astronomie* 103 (1989): 285–90.

Marquet, Louis. "Condorcet et la creation du système métrique décimal." In *Condorcet, mathématicien, économiste, philosophe, homme politique*, edited by Pierre Crépel and Christian Gilain, 52–62. Paris: Minerve, 1989.

Marx, Karl. *Capital: A Critique of Political Economy*. New York: Penguin, 1990.

Marx, Karl. *Grundrisse: Foundations of the Critique of Political Economy*. London: Penguin, 1993.

Marx, Karl, and Friedrich Engels. *The Communist Manifesto*. London: Penguin, 2002.

Marx, Leo. *The Machine in the Garden: Technology and the Pastoral Ideal in America*. New York: Oxford University Press; 1964.

Mask, Deirdre. *The Address Book: What Street Addresses Reveal about Identity, Race, Wealth, and Power*. New York: St. Martin Press, 2020.

Mau, Steffen. *The Metric Society: On the Quantification of the Social*. London: Polity, 2019.

Mayer, Joseph. "Parliament and the Metric System. Comments." *Isis* 57 (1966): 117–19.

McAdam, Doug, Sidney Tarrow, and Charles Tilly. *Dynamics of Contention*. New York: Cambridge University Press, 2001.

McCourbrey, Arthur O. "Measures and Measuring Systems." In *The Encyclopedia Americana, International Edition*, 18: 584–97. Danbury: Grolier, 1993.

McCusker, John J. "Weight and Measures in the Colonial Sugar Trade: The Gallon and the Pound and Their Equivalents." In *Essays in the Economic History of the Atlantic*, 76–101. London: Routledge, 1997.

Mendenhall, T. C. "The Metric System." *Popular Science Monthly* 49 (1896): 721–34.

Mendenhall, T. C. "The United States Fundamental Standards of Length and Mass." *Science* 56 (1922): 377–80.

Mennell, Stephen. *The American Civilizing Process*. Cambridge, MA: Polity, 2007.

Menninger, Karl. *Number Words and Number Symbols: A Cultural History of Numbers*. Cambridge, MA: MIT Press, 1969.

Merton, Robert K. "Self-Fulfilling Prophecy." In *Social Theory and Social Structure*, 475–90. New York: The Free Press, 1968.

"Mesure." In *Encyclopédie, ou Dictionnaire raisonné des sciences, des arts et des métiers, par une société des gens de letters*, edited by Denis Diderot and Jean le Rond d'Alembert, 10: 408–25. Paris: De l'imprimerie de Le Breton, imprimeur ordinaire du Roy, 1765.

"Metric Advantages: Melvil Dewey Scores Opponents of Decimal Measures." *New York Herald Tribune*, February 16, 1925.

"Metric System of Weights and Measures." *Chicago Tribune*, May 25, 1866.

Michell, John. "A Defense of Sacred Measures." *The CoEvolution Quarterly* 17 (Spring 1978): 2–6.

Mihm, Stephen. "Inching toward Modernity: Industrial Standards and the Fate of the Metric System in the United States." *Business History Review* 96 (2022): 47–76.

Mihm, Stephen. "Whole Foods and the Old Thumb on the Scale." *Bloomberg*, July 2, 2015. https://www.bloomberg.com/view/articles/2015-07-02/whole-foods-and-the-old-thumb-on-the-scale.

Mihm, Stephen. *A Nation of Counterfeiters: Capitalists, Con Men, and the Making of the United States*. Cambridge, MA: Harvard University Press, 2007.

Mintz, Sydney W. "Standards of Value and Units of Measure in the Fond-des-Nègres Market Place, Haiti." *Journal of the Royal Anthropological Institute of Great Britain and Ireland* 91 (1961): 23–38.

Moles, Frank J. "An Auxiliary System for the Measurement of Time." In *The Metric System of Weights and Measures: The National Council of Teachers of Mathematics*, 225–33. New York: Teachers College, 1948.

Montesquieu, *The Spirit of Laws*. Cambridge: Cambridge University Press, 1989

Montessori, Maria. "The House of Children. Lecture, Kodaikanal, 1944." *North American Montessori Teachers Association Journal* 38 (2013): 11–19.

Moreau, Henri. *Le système métrique: Des anciennes mesures au Système International d'Unités*. Paris: Chiron, 1975.

Moreno Fraginals, Manuel. *The Sugarmill: The Socioeconomic Complex of Sugar in Cuba 1760–1860*. New York: Monthly Review Press, 1976.

Morrison, Tessa. *Isaac Newton's Temple of Solomon and his Reconstruction of Sacred Architecture*. Cham: Springer Science & Business Media, 2010.

Muller, Jerry Z. *The Tyranny of Metrics*. Princeton, NJ: Princeton University Press, 2019.

Musée National des Techniques. *L'aventure du mètre*. Paris: Le Musée, 1989.

"Myanmar to Adopt Metric System." *Eleven Myanmar*, October 10, 2013.

NASA Safety Center. "Lost in Translation." *System Failure Case Studies* 3 (2009): 1–4.

National Board of Trade. *Proceedings of the Thirty-First Annual Meeting of the National Board of Trade*. Philadelphia: John R. McFetridge & Sons, 1901.

National Board of Trade. *Proceedings of the Thirty-Second Annual Meeting of the National Board of Trade*. Philadelphia: John R. McFetridge & Sons, 1902.

National Bureau of Standards. *A Metric America: A Decision whose Time has Come*. Washington, DC: National Bureau of Standards, 1971.

National Bureau of Standards. *Federal and State Laws Relating to Weights and Measures*. Washington, DC: Bureau of Standards, 1926.

National Bureau of Standards. *Laws Concerning the Weights and Measures of the United States*. Washington, DC: Government Printing Office, 1904.

National Bureau of Standards. *U.S. Metric Study Report*. Washington, DC: Government Printing Office, 1971.

National Council of Teachers of Mathematics. *The Metric System of Weights and Measures*. New York: Teachers College, 1948.

National Geographic Society. "A Wonderland of Science." *National Geographic Magazine* 27 (1915): 153–96.

National Industrial Conference Board. *The Metric versus the English System of Weights and Measures*. New York: The Century Co., 1912.

National Machine Tool Builders Association. *Resolution and Protest of the National Machine Tool Builders Association*, n.p., 1902.

Newton, Isaac. "A Dissertation upon the Sacred Cubit of the Jews and the Cubits of the Several Nations." In *Miscellaneous Works of Mr. John Greaves* vol. 2: 405–33. London: J. Hughs, 1737.

Noel, Edward. *Science of Metrology or Natural Weights and Measures: A Challenge to the Metric System*. London: Edward Stanford, 1889.

North, Douglass C. "Institutions." *Journal of Economic Perspectives* 5 (1991): 97–112.

North, Douglass C. "Review of *Measures and Men*, by Witold Kula." *Journal of Economic History* 47 (1987): 593–95.

North, Douglass C. "Transaction Costs in History." *Journal of European Economic History* 14 (1985): 557–76.

Nugent, Walter T. K. *Money and American Society, 1865–1880*. New York: Free Press, 1968.

Nystrom, John William. *Project of a New System of Arithmetic, Weight, Measure and Coins: Proponed to Be Called the Tonal System, with Sixteen to the Base*. Philadelphia: J. B. Lippincott & Co., 1862.

Oldberg, Oscar. "Herbert Spencer vs. the Metric System." *Bulletin of Pharmacy* 10 (1896): 292–98.

Oldberg, Oscar. *A Manual of Weights, Measures and Specific Gravity*. Chicago: n.p., 1885.

Oldberg, Oscar. *The Metric System in Medicine*. Philadelphia, PA: Presley Blackiston, 1881.

Ortega y Gasset, José. *The Modern Theme*. New York: Harper, 1961.

Osterhammel, Jürgen, and Neils P. Petersson. *Globalization: A Short History*. Princeton, NJ: Princeton University Press, 2009.

Ozouf, Mona. "Revolutionary Calendar." In *A Critical Dictionary of the French Revolution,* edited by F. Furet and M. Ozouf, 538–46. Cambridge, MA: Harvard University Press, 1989.

Pace, Lisha. "Nystrom." History-Computer, July 25, 2023. https://history-computer.com/nystrom.

Page, Chester Hall, and Paul Vigoureux, eds. *The International Bureau of Weights and Measures 1875–1975*. Washington, DC: National Bureau of Standards, 1975.

Paine, Thomas. *Common Sense: Addressed to the Inhabitants of America*. London: D. Jordan, 1792.

Pan American Standardization Conference. *Primera Conferencia Panamericana para la Uniformidad de Especificaciones, reunida en Lima, Perú, del 23 de diciembre de 1924 al 6 de enero de 1925. Actas y demás documentación*. Lima: Libraría e Imprenta E. Moreno, 1927.

Paz, Octavio. *The Labyrinth of Solitude*. New York: Grove Press, 1961.

Peirce, Charles S. "Review of Noel's *Science of Metrology*." In *Writings of Charles S. Peirce: A Chronological Edition*, edited by Charles Hartshorne and Paul Weiss, 6: 377–79. Bloomington: Indiana University Press, 1982.

Peirce, Charles S. "Testimony on the Organization of the Coast Survey." In *Writings of Charles S. Peirce: A Chronological Edition*, edited by Charles Hartshorne and Paul Weiss, 3: 149–61. Bloomington: Indiana University Press, 1982.

Peirce, Charles S. "The Metric Fallacy." *The Nation* 78 (March 17, 1904): 215–16.

Peirce, Charles S. *Charles Sanders Peirce: Contributions to* The Nation. *Part Three: 1901–1908*, edited by Kenneth Laine Ketner and James Edward Cook. Lubbock: Texas Tech University Press, 1979.

Pérard, Albert Gustave Léon. *Les récents progrès du système métrique, 1948*. Paris: Gauthier-Villars, 1949.

Perovic, Sanja. *The Calendar in Revolutionary France: Perceptions of Time in Literature, Culture, Politics*. Cambridge: Cambridge University Press, 2015.

Perramond, Eric. "How Metrics Shape Water Politics in New Mexico: From Quantifying Governance to Active Monitoring." *Journal Water Alternatives* 17 (2024): 455–68.

Perry, John. *The Story of Standards*. New York: Funk & Wagnalls, 1955.

Petersen, Mark J. *The Southern Cone and the Origins of Pan America, 1888–1933*. Notre Dame, IN: University of Notre Dame Press, 2022.

Phillips, Christopher J. "The New Math and Midcentury American Politics." *Journal of American History* 101 (2014): 454–79.

Piazzi Smyth, Charles. "Weights and Measures." In Edward Hine, *Flashes of Light. Part 2 of Forty-Seven Identifications of the British Nation with the Lost Ten Tribes of Israel*. London: W. H. Guest, 1876.

Piazzi Smyth, Charles. *Life and Work at The Great Pyramid*, vol. 3. Edinburgh: Edmonston and Douglas, 1867.

Piazzi Smyth, Charles. *Our Inheritance in the Great Pyramid*. London: Alexander Strahan & Co., 1864.

Poincaré, Henri. "Rapport sur les résolutions de la commission chargeé de l'étude des projects de décimalisation du temps et de la circonférence." In *Oeuvres de Henri Poincaré*, 8: 648–64. Paris: Gauthier-Villars et Cie, 1952.

Poirier, Jean-Pierre. *Lavoisier: Chemist, Biologist, Economist*. Philadelphia, PA: University of Pennsylvania Press, 1998.

Pollard, Sidney. "Capitalism and Rationality: A Study of Measurement in British Coal Mining, ca. 1750–1850." *Explorations in Economic History* 20 (1983): 110–29.

Poole, Robert. "'Give Us Our Eleven Days!': Calendar Reform in Eighteenth-Century England." *Past and Present* 149 (1995): 95–139.

Porter, Theodore M. *Trust in Numbers: The Pursuit of Objectivity in Science and Public Life*. Princeton, NJ: Princeton University Press, 1995.

"Proposed Legislation in Regard to the Metric System." *Science*, March 27 1896: 457–63.

Puente Feliz, Gustavo. "El Sistema métrico decimal: Su importancia e implantación en España." *Cuadernos de Historia Moderna y Contemporánea* 3 (1982): 95–125.

Pujals de la Bastida, Vicente. *Sistema métrico perfecto ó docial: Y demostración de sus inmensas ventajas sobre el decimal y sobre todo otro sistema de medidas, pesos y monedas*. Madrid: Imprenta de la Esperanza, 1862.

Quinn, Terry. *From Artifacts to Atoms: The BIPM and the Search for Ultimate Measurement Standards*. New York: Oxford University Press, 2012.

Register of Debates in Congress: Comprising the Leading Debates and Incidents of the Second Session of the Eighteenth Congress. Washington, DC: Gales and Seaton, 1826.

Reisenauer, Eric Michael. "'The Battle of the Standards': Great Pyramid Metrology and British Identity, 1859–1890." *The Historian* 65 (2003): 931–78.

"Report of the Committee Appointed to Discuss the Arguments in Favor and Against the Metric System." *Transactions of the American Society of Mechanical Engineers* 24 (1903): 630–712.

Report of the Committee on Coinage, Weights, and Measures: Part 1, on the Adoption of the Metric System of Weights and Measures, Together with Documents and Statistics Relating to the Subject; Part 2, on Metric Coinage. Washington, DC: Government Printing Office, 1879.

Report of the Joint Special Committees of the Chamber of Commerce and American Geographical and Statistical Society on the Extension of the Decimal System to Weights and Measures of the United States. New York: W. C. Bryant & Co. Printers, 1857.

Report of the Secretary of the Treasury of the Construction and Distribution of Weights and Measures. Washington, DC: A. O. P. Nicholson, 1857.

Report of the Superintendent of the United States Coast Survey, 1890. Washington, DC: Government Printing Office, 1891.

Reports of the Committees of the House of Representatives, Made during the First Session Thirty-Ninth Congress, 1865–'66. Washington, DC: Government Printing Office, 1866.

Reti, Steven P. *Silver and Gold: The Political Economy of International Monetary Conferences, 1867–1892.* Westport, CT: Greenwood Press, 1998.

Richardson, Kim. *Quebra-Quilos and Peasant Resistance: Peasants, Religion, and Politics in Nineteenth-Century Brazil.* Lanham: University Press of America, 2012.

Robson, Eleanor. *Mathematics in Ancient Iraq: A Social History.* Princeton, NJ: Princeton University Press, 2009.

Rogers, Everett M. *Diffusion of Innovations.* New York: The Free Press, 1995.

Rooney, David. "Public Standards of Length." In: *Extinct: A Compendium of Obsolete Objects,* edited by Barbara Penner, Adrian Forty, Olivia Horsfall Turner and Miranda Critchley, 260–63. London: Reaktion Books, 2021.

Rosenthal, Caitlin. *Accounting for Slavery: Masters and Management.* Cambridge, MA: Harvard University Press, 2018.

Russell, Andrew L. *Open Standards and the Digital Age: History, Ideology, and Networks.* Cambridge: Cambridge University Press, 2014.

Russell, Andrew. "The American System: A Schumpeterian History of Standardization." Progress on Point Paper 12. Washington, DC: Progress & Freedom Foundation, 2005.

Sarton, George. *The Life of Science: Essays in the History of Civilization.* Bloomington: Indiana University Press, 1960.

Sarton, George. *The Study of the History of Mathematics.* Cambridge, MA: Harvard University Press, 1936.

Schaffer, Simon. "Metrology, Metrication, and Victorian Values." In *Victorian Science in Context,* edited by Bernard Lightman, 438–74. Chicago: University of Chicago Press, 1997.

Schlieben-Lange, Brigitte. *Ideologie, revolution et uniformité de la langue.* Hayen: Mardaga, 1996.

Schmandt-Besserat, Denise. *How Writing Came About*. Austin: University of Texas Press, 1996.

Scott, James Brown, ed. *The International Conferences of American States, 1889–1928: A Collection of the Conventions, Recommendations, Resolutions, Reports, and Motions Adopted by the First Six International Conferences of the American States*. New York: Oxford University Press, 1931.

Scott, James C. *Weapons of the Weak: Everyday Forms of Peasant Resistance*. New Haven, CT: Yale University Press, 1985.

Scott, James C. *Seeing Like a State: How Certain Schemes to Improve the Human Condition Have Failed*. New Haven, CT: Yale University Press, 1998.

Sellers, Coleman. *The Metric System: Is It Wise to Introduce It into Our Machine Shops?* Philadelphia, PA: American Society of Mechanical Engineers, 1880.

Semmel, Bernard. "Parliament and the Metric System." *Isis* 54 (1963): 125–33.

Sharlin, Harold Issadore. *Lord Kelvin, the Dynamic Victorian*. University Park: Pennsylvania State University Press, 1979.

Shaw, Matthew John. *Time and the French Revolution: The Republican Calendar, 1789–Year XIV*. Woodbridge: Royal Historical Society, 2011.

Sheldon, Richard, Adrian Randall, Andrew Charlesworth, and David Walsh. "Popular Protest and the Persistence of Customary Corn Measures: Resistance to the Winchester Bushel in the English West." In *Markets, Market Culture and Popular Protest in Eighteenth-Century Britain and Ireland*, edited by A. Randall and A. Charlesworth, 25–45. Liverpool: Liverpool University Press, 1996.

Sherman, John. *John Sherman's Recollections of Forty Years in the House, Senate and Cabinet: An Autobiography*. Chicago: The Werner Company, 1896.

Shirokov, K. P. "Fifty Years of the Metric System in the USSR." *Measurement Techniques* 11 (1968): 1141–46.

Shost'in, N. A. "History of Russian Metrology: D. I. Mendeleev and the Metric System of Measures." *Measurement Techniques* 11 (1968): 429–31.

Shusterman, Noah. *Religion and the Politics of Time: Holidays in France from Louis XIV through Napoleon*. Washington, DC: Catholic University of America Press, 2010.

Sinclair, Bruce. *A Centennial History of the American Society of Mechanical Engineers, 1880–1980*. Toronto: University of Toronto Press, 1980.

Skocpol, Theda. *Social Policy in the United States: Future Possibilities in Historical Perspective*. Princeton, NJ: Princeton University Press, 1994.

Skocpol, Theda. *States and Social Revolutions*. New York: Cambridge University Press, 1979.

Skowronek, Stephen. *Building a New American State: The Expansion of National Administrative Capacities, 1877–1920.* New York: Cambridge University Press, 1982.

Slosson, E. E. "Decimal Numeration in the United States." *Science* 4 (1896): 59–62.

Smith, Jeanette C. "Take Me to Your Liter: A History of Metrication in the United States." *Journal of Government Information* 25 (1998): 419–38.

Smith, Paul. "La division décimale du jour: L'Heure qu'il n'est pas." In *Genèse et diffusion du système métrique*, edited by J.-C. Hocquet and Bernard Garnier, 123–34. Caen: Editions-diffusion du Lys, 1990.

Smith, Ralph W. *The Federal Basis for Weights and Measures: A Historical Review of Federal Legislative Effort, Statutes, and Administrative Action in the Field of Weights and Measures in the United States. National Bureau of Standards Circular 593.* Washington, DC: National Bureau of Standards, 1958.

Smith, Ralph Weir. *Weights and Measures Administration.* Washington, DC: National Bureau of Standards, 1962.

Smith, Robert Frederick. "Cognitive Effects of the Introduction of a New Measurement Language into American Culture." PhD diss., New York University, 1978.

Smith, William Benjamin. "Not Ten but Twelve." *Science* 50 (1919): 239–42.

Sombart, Werner. *Why Is There No Socialism in the United States?* White Plains, NY: M. E. Sharpe, 1976.

Southard, James. "The Metric System of Weights and Measures." *Science* 15 (1902): 829–36.

Spang, Rebecca L. *Stuff and Money in the Time of the French Revolution.* Cambridge, MA: Harvard University Press, 2015.

"Spencer and the Metric." *San Jose Mercury News*, January 23, 1904.

Spencer, Herbert. *Principles of Biology.* London: Williams & Norgate, 1864.

Spencer, Herbert. *The Man versus the State.* New York: D. Appleton and Company, 1885.

Spencer, Herbert. "The Metric System." *The Times*, April 4, 7, 9 and 25, 1896.

Spencer, Herbert. "The Metric System." *Popular Science Monthly* 49 (1896): 186–99.

Spencer, Herbert. "The Metric System." *The Times*, March 28, April 4, 8, and 13, 1899.

Spencer, Herbert. *An Autobiography.* New York: D. Appleton, 1904.

Spencer, Herbert. *Various Fragments.* New York: D. Appleton and Company, 1914.

"Spencer's Unusual Will." *Baltimore Sun*, July 11, 1920.

Starobinski, Jean. *1789, The Emblems of Reason.* Charlottesville: University Press of Virginia, 1982.

Statistical Office of the United Nations. *World Weights and Measures: Handbook for Statisticians.* New York: United Nations, 1966.

Steigerwalt, Albert K. *The National Association of Manufacturers, 1895–1914.* Ann Arbor: University of Michigan, 1964.

Stevens, W. Le Conte. "The Metric System: Shall It Be Compulsory?" *Popular Science Monthly* 64 (1904): 394–405.

Stewart, John Hall. *A Documentary Survey of the French Revolution.* New York: Macmillan, 1951.

Stone, Charles W. "A Common Coinage for All Nations." *North American Review* 476 (1896): 47–55.

Stratton, S. W. "Address to the Conference." In *Weights and Measures: Thirteenth Annual Conference of Representatives from Various States held at the Bureau of Standards,* 13–18. Washington, DC: Government Printing Office, 1921.

Struik, Dirk Jan. "The Prohibition of the Use of Arabic Numerals in Florence." *Archives Internationales d'Histoire des Sciences* 21 (1968): 291–94.

Struik, Dirk Jan. *A Concise History of Mathematics.* New York: Dover Publications, 1987.

Stubbs, A. J. "The Necessity of Compulsion." *Decimal Educator* 2 (1919): 155.

Stutz, C. C. "The Tools of Our Industry: Shall We Scrap Them?" *The Rotarian* 26 (1925): 6: 19, 87–91.

Sumner, Charles. "The Metric System of Weights and Measures." In *The Works of Charles Sumner,* X: 524–539. Boston: Lee and Shepard, 1876.

Sumner, William Graham. *Folkways: A Study of the Sociological Importance of Usages, Manners, Customs, Mores, and Morals.* New York: Dover, 1906.

Supplemental Hearing on the Subject of the Metric System of Weights and Measures: Hearings before the United States House Committee on Coinage, Weights, and Measures, Fifty-Seventh Congress, First Session, on Apr. 24, 1902. Washington, DC: Government Printing Office, 1902.

Swann, Peter. "The Economics of Metrology and Measurement." Report for National Measurement Office, Department for Business, Innovation and Skills, 2009.

Tallent, Gustave. *Histoire du Système Métrique.* Paris: Librairie H. Le Soudier, 1910.

Tavernor, Robert. *Smtoot's Ear: The Measure of Humanity.* New Haven, CT: Yale University Press, 2007.

Taylor, Alfred B. "Octonary Numeration, and its Application to a System of Weights and Measures." *Proceedings of the American Philosophical Society* 24 (1887): 296–366.

Taylor, John. *The Battle of the Standards: The Ancient, of Four Thousand Years, against the Modern, of the Last Fifty Years—The Less Perfect of the Two*. London: Longman, Green, Longman, Roberts & Green, 1864.

Tefft, T. A. *Universal Currency: A Plan for Obtaining a Common Currency in France, England, and America, Based on the Decimal System; with Suggestions for Rendering the French Decimal System of Weight and Measure More Simple and Popular*, 2nd ed. London: Effingham Wilson, 1858.

Ten, Antonio E. "El sistema métrico decimal y España." *Arbor: Ciencia, Pensamiento y Cultura* 527–528 (1989): 101–22.

"The Colonial Conference and the Metric System" *London Times*, August 23, 28, 29, September 3, 4, 11, and October 13, 1902.

The Europa World Year 2004. London: Europa Publications, 2004.

"The Metric System." *Hudson Star*, December 5, 1927.

The Metric System: A Compilation, Consisting of Extracts from the Report of the Committee of the House of Representatives, and Law of Congress Adopting the System; and a Translation of a Portion of a Work Entitled "The Legal System Of Weights And Measures," By M. Lamotte. Philadelphia, PA: J. B. Lippincott & Co., 1867.

"The Proposed Metric System." *Scientific American* 15 (July 21, 1866), 50.

Thompson, E. P. *Customs in Common: Studies in Traditional Popular Culture*. New York: The New Press, 1993.

Thompson, Silvanus P. *The Life of Lord Kelvin*. New York: Chelsea Publishing, 1976.

Thompson, William. *Popular Lectures and Addresses*. London: Macmillan, 1891.

Thuillier, Pierre. *El saber ventrílocuo*. Mexico: Fondo de Cultura Económica, 1995.

Tilly, Charles. *European Revolutions, 1492–1992*. Cambridge, MA: Blackwell, 1993.

Tocqueville, Alexis de. *The Ancien Régime and the French Revolution*. Cambridge: Cambridge University Press, 2011.

Todd, Jan. *For Good Measure: The Making of Australia's Measurement System*. Crows Nest, NSW, Aus.: Allen & Unwin, 2004.

Torpey, John C. *The Invention of the Passport: Surveillance, Citizenship, and the State*. New York: Cambridge University Press, 2000.

Totten, Charles A. L. "The Metrology of the Great Pyramid." *Van Nostrand's Engineering Magazine* 31 (1884): 226–34.

Totten, Charles A. L. *An Important Question in Metrology Based upon Recent and Original Discoveries: Challenge to "The Metric System," and an Earnest Word with the English-Speaking Peoples on their Ancient Weights and Measures*. New York: John Wiley & Sons, 1884.

Toynbee, Arnold J. *A Study of History: Abridgement of Volumes VII–X*. New York: Oxford University Press, 1987.

Traynor, Ian. "EU Provides Reprieve for Mile and Pint." *The Guardian*, September 12, 2007.

Treat, Charles F. *A History of the Metric System Controversy in the United States*. Washington, DC: National Bureau of Standards, 1971.

Trillin, Calvin. "Anti-Metrics." *The New Yorker*, August 31, 1981: 82–85.

Tschoegl, Adrian. "The International Diffusion of an Innovation: The Spread of Decimal Currency." *Journal of Socio-Economics* 39 (2010): 100–109.

Tunbridge, Paul. *Lord Kelvin, His Influence on Electrical Measurements and Units*. London: Institution of Electrical Engineers, 1992.

Two [Arthur George Liddon Rogers]. *Home Life with Herbert Spencer*. Bristol: J. W. Arrowsmith, 1910.

Tyler-McGraw, Marie. *An African Republic Black and White Virginians in the Making of Liberia*. Chapel Hill: University of North Carolina Press, 2009.

U.K. Metric Association. "Two Kinds of Country?" *Metric Views*, October 16, 2018. https://metricviews.uk/2018/10/16/two-kinds-of-country.

U.S. Metric Association. "Metrication in Other Countries." August 3, 2020. https://usma.org/metrication-in-other-countries#locale-notification.

United Nations. *World Weights and Measures: Handbook for Statisticians*. New York: United Nations, 1966.

United States. Congress. Senate. *Committee on Finance. International Coinage: I. Report of Senator Sherman. II. Report of Senator Morgan. III. Bill to establish a uniform coinage. IV. Report of Mr. S. B. Ruggles*. Washington, DC: Government Printing Office, 1868.

United States. General Accounting Office. *Metric Conversion: Future Progress Depends Upon Private Sector and Public Support*. Washington, DC: Government Printing Office, 1994.

Velkar, Aashish. "'Imperial Folly': Metrication, Euroskepticism, and Popular Politics in Britain, 1965–1980." *Journal of Modern History* 92, no. 3 (2020): 561–601.

Velkar, Aashish. *Markets and Measurements in Nineteenth-Century Britain*. Cambridge: Cambridge University Press, 2012.

Vera, Hector. "Quantitative Measurement and the Production of Meaning." In *The Oxford Handbook of Symbolic Interactionism*, edited by Wayne H. Brekhus, Thomas DeGloma, and William Ryan Force, 104–21. New York: Oxford University Press, 2023.

Vera, Hector. "Money: Instrument of Quantification, Agent of Rationalization, Cultural Object." *Digithum* 24 (2019): 1–12.

Vera, Hector. "Counting Measures: The Decimal Metric System, Metrological Census, and State Formation in Revolutionary Mexico, 1895–1940." *Histoire & Mesure* 32 (2017): 121–40.

Vera, Hector. "Weights and Measures." In *Blackwell Companion to the History of Science*, edited by Bernard Lightman, 459–71. Oxford: Blackwell Publishers, 2016.

Vera, Hector. "The Social Construction of Units of Measurement: Institutionalization, Legitimation and Maintenance in Metrology." In *Standardization in Measurement: Philosophical, Historical and Sociological Issues*, edited by L. Huber and O. Schlaudt, 173–87. London: Pickering and Chatto, 2015.

Vera, Hector. "The Social Life of Measures: Metrication in the United States and Mexico, 1789–2004." PhD diss., New School for Social Research, 2012.

Vera, Héctor. "Medidas de resistencia: Grupos y movimientos sociales en contra del sistema métrico." In *Metros, leguas y mecates: Historia de los sistemas de medición en México*, edited by H. Vera and V. García Acosta, 181–99. Mexico: CIESAS, 2011.

Vera, Hector. "Melvil Dewey, Metric Apostle." *Metric Today* 45 (2010): 1, 4–6.

Vera, Hector. "Decimal Time: Misadventures of a Revolutionary Idea, 1793–2008." *KronoScope: Journal for the Study of Time* 9 (2009): 29–48.

Vera, Hector. "Economic Rationalization, Money and Measures: A Weberian Perspective." In *Max Weber Matters: Interweaving Past and Present*, edited by David Chalcraft, Fanon Howell, Marisol Lopez Menendez, and Hector Vera, 135–47. London: Ashgate, 2008.

Vera, Héctor. *A peso el kilo: Historia del sistema métrico decimal en México*. Mexico: Libros del Escarabajo, 2007.

Vera, Héctor, and Virginia García Acosta, eds. *Metros, leguas y mecates: Historia de los sistemas de medición en México*. Mexico: CIESAS, 2011.

Verman, Lal C., and Jainath Kaul, eds. *Metric Change in India*. New Delhi: Indian Standards Institution, 1970.

Vincent, James. *Beyond Measure: The Hidden History of Measurement*. London: Faber & Faber, 2022.

Vivian, James F. "The Pan American Conference Act of May 10, 1888: President Cleveland and the Historians." *The Americas* 27 (1970): 185–92.

Wade, Herbert T. "Weights and Measures." In *Encyclopaedia of the Social Sciences*, edited by Edwin R. A. Seligman and Alvin Johnson, 15: 389–392, New York: MacMillan Company, 1930.

Walthall, D. Alex. *Sicily and the Hellenistic Mediterranean World: Economy and Administration during the Reign of Hieron II*. Cambridge: Cambridge University Press, 2024.

Watkins, Grace Ellen, and Joel Best. "Successful and Unsuccessful Diffusion of Social Policy: The United States, Canada, and the Metric System." In *How Claims Spread: Cross-National Diffusion of Social Problems*, edited by Joel Best, 267–81. New York: Aldine de Gruyter, 2001.

Watkins, Grace Ellen Grove. "Measures of Change: A Constructionist Analysis of Metrication in the United States." PhD diss., Southern Illinois University, 1998.

Weber, Eugen. *Peasants into Frenchmen: The Modernization of Rural France 1870–1914*. Stanford, CA: Stanford University Press, 1976.

Weber, Max. *Economy and Society: An Outline of Interpretive Sociology*. Berkeley: University of California Press, 1978.

Weber, Max. *For Max Weber: Essays in Sociology*. New York: Oxford University Press, 1958.

"Weights and Measures (Metric System) Bill." In *The Parliamentary Debates: Fourth Series, Second Session of the Twenty-Eighth Parliament of the United Kingdom of Great Britain and Ireland*, volume 171, 1311–64. London: Wyman and Sons, 1907.

"Weights and Measures of Latin America." *Decimal Educator* 2 (1919–1920): 178–86, 218–219, 259.

Weights and Measures: . . . Annual Conference of Representatives from Various States held at the Bureau of Standards. Washington. DC: Government Printing Office, 1910–1926.

Wells, William C. "The Metric System from the Pan-American Standpoint." *Scientific Monthly* 4 (1917): 196–202.

Wexler, Jay. *The Odd Clauses: Understanding the Constitution through Ten of Its Most Curious Provisions*. Boston: Beacon Press, 2011.

Whitelaw, Ian. *A Measure of All Things: The Story of Man and Measurement*. New York: St Martin's Press, 2007.

Whorf, Benjamin Lee. *Language, Thought, and Reality: Selected Writings*. Cambridge, MA: Technology Press of the MIT, 1956.

Wiegand, Wayne A. *Irrepressible Reformer: A Biography of Melvil Dewey*. Chicago: American Library Association, 1996.

Wigglesworth, Edward. "The Metric System in a Nutshell." *Boston Medical and Surgical Journal* 98 (1878): 766–69.

Wigglesworth, Edward. "Why Should We Use the Metric System?" *Western Lancet: A Journal of Medicine, Surgery, and Reviews of Medical Literature* 7 (1878): 424–27.

Wilkinson, Endymion. *Chinese History: A Manual*. Cambridge, MA: Harvard University Asia Center, 2000.

Wilson, Guy M. "Review of *The Metric System of Weights and Measures*, by National Council of Teachers of Mathematics." *Journal of Educational Research* 43 (1949): 74–75.

Wilson, Thomas Frederick. *The Power "to Coin" Money: The Exercise of Monetary Powers by the Congress*. Armonk, NY: M. E. Sharp, 1992.

Wimmer, Andreas, and Brian Min. "From Empire to Nation-State: Explaining Wars in the Modern World, 1816–2001." *American Sociological Review* 71 (2006): 867–97.

Yaple, Florence. "Herbert Spencer and the Metric System." *American Journal of Pharmacy* 76 (1904): 125–28.

Yorke, Gertrude Cushing. "Three Studies on the Effect of Compulsory Metric Usage." *Journal of Educational Research* 37 (1944): 343–51.

Youmans, Edward Livingston, ed. *Herbert Spencer on the Americans and the Americans on Herbert Spencer*. New York: Appleton, 1883.

Young, H. Peyton. "The Economics of Convention." *Journal of Economic Perspectives* 10 (1996): 105–22.

Zelizer, Viviana A. *The Social Meaning of Money*. New York: Basic Book, 1994.

Zelizer, Viviana A. "The Creation of Domestic Currencies." *American Economic Review* 84 (1994): 138–42.

Zengerle, Jason. "Waits and Measures." *Mother Jones* 24 (1999): 68–71.

Zertuche, Albino. *Estudio sobre pesas y medidas en los países centroamericanos*. n.p.: United Nations, 1958.

Zerubavel, Eviatar. *Ancestors and Relatives: Genealogy, Identity, and Community*. New York: Oxford University Press, 2013.

Zerubavel, Eviatar. *Time Maps: Collective Memory and the Social Shape of the Past*. Chicago: Chicago University Press, 2003.

Zerubavel, Eviatar. *The Seven Day Circle: History and Meaning of the Week*. New York: The Free Press, 1985.

Zerubavel, Eviatar. "The Standardization of Time: A Sociological Perspective." *American Journal of Sociology* 88 (1982): 1–23.

Zerubavel, Eviatar. "The French Republican Calendar: A Case Study in the Sociology of Time." *American Sociological Review* 42 (1977): 868–77.

Zolberg, Aristide R. *A Nation by Design: Immigration Policy in the Fashioning of America.* New York: Russell Sage Foundation, 2006.

Zukin, Sharon, and Paul DiMaggio, eds. *Structures of Capital: The Social Organization of the Economy.* New York: Cambridge University Press, 1990.

Zupko, Ronald E. "Worldwide Dissemination of the Metric System during the 19th and 20th Centuries." *Metric System Guide Bulletin* 2 (1974): 14–25.

Zupko, Ronald E. *French Weights and Measures before the Revolution: A Dictionary of Provincial and Local Units.* Bloomington: Indiana University Press, 1968.

Zupko, Ronald E. *Revolution in Measurement: Western European Weights and Measures since the Age of Science.* Philadelphia, PA: American Philosophical Society, 1990.

INDEX

Page numbers in *italic* refer to illustrations

acre, 22, 140, 155, 167, 260
Afghanistan, *41*, 225
Albania, *41*, 225
Algeria, *40*, 53, 54, 56, 223, 246n52
Al-Khwarizmi, 34
All-American Standards Council, 192
Amateur Athletic Union, 65, 233
American Academy of Pediatrics, 217
American Association for the Advancement of Sciences, 129, 152, 192, 233
American Engineering Standards Committee, 102
American Geographical and Statistical Society, 127, 230
American Institute of Weights and Measures (AIWM), 192, 194, 197, 233, 276n114
American Library Association, 132, 133
American Metric Association, 1, 23, 115, 121, 192, 233, 263n52, 276n112
American Metric Bureau (AMB), 110, 129–36, 180, 231
American Metrological Society (AMS), 35, 93, 97, 132, 180, 231, 232, 263n52
American Philosophical Society, 104, 216
American Protective Tariff League, 159
American Society of Mechanical Engineers, 97, 129, 134, 142, 187, 232, 276n114
Americans for Customary Weight and Measures, 115

Anderson, Benedict, 80, 208
Andorra, *41*, 45, 225
Angola, *41*, 224, 246n52
Antigua and Barbuda, *41*, 50, 227
anti-metric stance, 108, 110, 142, 151, 154, 161, 187, 204
   engineers and scientists, 129, 142
   groups, 18, 99, 115, 151, 188, 212
   movements, 16–17, 98, 110, 116, 151, 187, 190, 191
   organizations, 27, 136, 156, 191, 197. 205
   personalities, 17, 116, 141, 144, 151, 157, 187, 191, 199
   revolts, 84
Argentina, *40*, 57, 223
Armenia, *41*, 225
Articles of Confederation (1777), 85, 86, 218
Asimov, Isaac, 118, 135, 259n7
Atatürk, 49
Australia, *41*, 193, 194, 226, 234
Austria, *40*, 58, 59, 223, 245
Automobile Manufacturers Association, 106
aversion to compulsion, 21, 28, 109, 110, 209, 213
Azerbaijan, *41*, 55, 225

backwardism, American, 207
Bahamas, *41*, 50, 226
Bahrain, *41*, 226
Bangladesh, *41*, 227
Barbados, *41*, 50, 227

317

Barnard, Frederick, 118, 132, 142, 153, 154, 180, 231
Beccaria, Cesare, 126
Belarus, *41*, 225
Belgium, *40*, 52–56, 223, 230
Belize, *41*, 50, *57*, 224
Benin, *40*, 224, 246n52
Bismarck, Otto von, 48
Blaine, James, 181, 182, 184
Bloch, Mark, 6
Bodin, Jean, 75
Bolívar, Simón, 208
Bolivia, *40*, 57, *57*, 223
Bonaparte, Napoleon, 31, 90
Borda, Jean-Charles de, 91
Bosak, Jon, 157
Bosnia and Herzegovina, *40*, 224
Botswana, *41*, 226, 248n85
Bourdieu, Pierre, 76, 79, 80
Brand, Stewart, 116
Branscomb, Lewis M., 104–5, 235
Brazil, *40*, 44, 56, *57*, 84, 121, 193, 223
Briggs, Lyman J., 201
British Empire, 2, 128, 139, 204
  adoption of metric system, 59–61, 65, 201
  colonies, 60–62, 65, 86
  unification of measures with North and Latin America, 196, 199, 218
British Israelism, 137, 139, 141, 156
Britten, Fred A., 159, *160*, 161, 233
Brown & Sharpe Manufacturing Co., 192, 233, 276n114
Brunei, *41*, 62, 227
Bulgaria, *40*, 48, 224
Burke, Edmund, 12
Burkina Faso, *41*, 44, 226, 248n85
Burroughs, William, 116, *117*
Burundi, *41*, 224, 246n52
Bush, George H. W., 107
bushel, 69, *71*, 83, 130, 167, 274n90
  decimal, 88
  Winchester, 83, 252n51
Bhutan, *41*, 225, 244n30

calendar
  Gregorian, 12, 17, 38–39, 46, 49, 59, 130, 135, 246n58
  Julian, 46, 49, 59
  republican, 11–12, 28, 30, 37–39, 229, 230, 243n16
Cambodia, *41*, 54, 225, 246n52
Cameroon, *41*, 42, 226, 248n85
Canada, 18, *41*, 67, 103, 214, 227, 244n30
  commercial agreements, 23, 51, 212
  and English-speaking countries, 177, 210
  in metric maps, 193–94
  made metric optional, 245n45
Cape Verde, *41*, 224, 246n52
carat, 88, 121–23
Caribbean Community (CARICOM), 50
Carlson, Tucker, 16
Carnegie, Andrew, 118, 185, 192
Carpentier, Alejo, 12
Castellanos, Jacinto, 183
Central African Republic, *41*, 226, 248n85
Central American Convention (1910), 60
centralization, 9, 84–85, 99, 109, 173
  loose, 19, 29, 109, 209, 211
  and metric system, 15, 26, 52
  political, 48, 74, 75, 214
  of states, 26, 80, 170
  and uniformity, 9, 245n47
  in US, 13, 14, 26, 70, 175
Chad, *40*, 224, 246n52
Charlemagne, 59, 74, 245n47
Chevalier, Michael, 178
Childe, Gordon, 73
Chile, *40*, 56, *57*, 58, 199, 223
China, 25, 61, 66, 171, 224, 244n30, 246n60
  and metric system, *41*, 47, *57*, 67, 233
Cisalpine Republic, 52, 240n51, 242n3
Císcar, Gabriel, 118, 247n71
Civil War, 25, 63, 65, 93, 96, 153, 214

Coast and Geodetic Survey, 22, 97, 98, 127, 152, 275n103
Office of Weights and Measures, 97
Cochrane, Rexmond, 98
cognition and measurement, 27, 36, 79–80, 123, 124, 162
  cognitive barriers, 35, 36, 124
coinage, 48, 75–76, 91, 165, 173, 175, 177, 196
  monopoly of state, 185, 173, 175
  as right of US Congress, 85, 86, 229
Colombia, 40, 49, 56, 57, 58, 141, 200, 223, 244n30
colonialism, 39, 42–43, 46, 52–56, 60–65, 80, 166, 204
  in Africa, 25, 53, 248n85
  in America, 86
  British, 42, 60, 62–63, 65, 86, 148, 150, 214, 219
  Spanish, 56, 58, 167, 199
Commonwealth of Nations, 25, 42, 45, 62, 75, 99, 103
Comoros, 41, 225, 246n52
Comte, Auguste, 118, 144, 145
Condorcet, Nicolas de, 12, 31, 32, 88, 118, 254n76
Conference of the Caribbean Free Trade Association, 50
Congress on Definite Metric Standards (1799), 31, 36, 43, 56, 91, 229
Constant, Benjamin, 51
Constitution, US, 63, 84–87, 89, 254n83
Convention of the Metre, 172, 231
cosmopolitanism, 28, 188
Costa Rica, 40, 49, 57, 224, 244n30, 245n44
Côte d'Ivoire, 40, 224, 246n52
Croatia, 40, 223
Cuba, 40, 56, 57, 223, 246n52
cubit, 73, 137, 138, 156
  ancient, 138
  sacred, 137, 138, 156
Curie, Marie, 118

currency, 175, 179–80, 189, 198
  British, 235
  Carolingian, 59
  decimal, 13, 17, 39, 59–60, 62, 131, 175, 240n53
  decimal currency in US, 87–89, 131, 134, 153, 177–79, 229
  duodecimal, 39
  international metric gold coinage, 177, 180–82, 184, 231
  metric, 176–79, 180–81, 184, 194
  monometallism & bimetallism, 179, 182
  national, 173–76, 229, 231
  unification in US, 175, 186
  universal, 176–81, 186
  vigesimal, 39
Cushing, Marshall, 191
Cyprus, 41, 227
Czech Republic, 40, 223

Dale, Samuel S., 16, 110, 157, 187–94, 196–200, 233, 267n118, 274n97
Darwin, Charles, 16, 267n116
De la Fuente, Julio, 123
decagramists, 178, 179
decimal, 13, 37, 83, 89, 135, 143, 229, 261n29
  arithmetic, 72, 87, 145, 265n88
  decimal classification, 131, 231
  despotism, 111
  fractions, 13, 14, 32
  notation, 32, 122, 146
  point, 32, 123, 218, 265n88
  principle, 52, 127, 131, 143, 266n97
  progression, 89, 124
  time, 30, 38, 39, 243n20, 266n97
  weights and measures, 13–14, 33, 88, 109, 146
Decimal Day, 59, 235
Decimal Indian Society, 62
Decimal Society (UK), 60
decimalization, 13, 38, 60, 62, 115, 126, 153, 240n49, 243n17, 262n29

decolonization, 61–62
Del Río, Andrés Manuel, 118
Delambre, Jean-Baptiste Joseph, 10, 91
Deming, Richard, 207
Democratic Republic of the Congo, *41*, 224, 246n52
Denmark, *41*, 224, 242n3
Department for Business and Trade, 55
Department of Commerce, 235, 237
Dewey, Melvil, 118, 119, 129–36, 141–42, 150, 218
    Dewey's Decimal Classification, 13, 131, 231
Díaz Covarrubias, Francisco, 118
Diderot, Denis, 30
diffusion networks, 43, 45
Djibouti, *40*, 224, 246n52
dollar, 87–88, 131, 161, 173, 175, 179–81, 229
    decimal, 87, 186, 229
    Liberian, 63
Dombey, Joseph, 216
Dominica, *41*, 50, 226
Dominican Republic, *40*, 50, 226
Dresser, Solomon, 191
Drury, Aubrey, 192, 194–97, 200, 233, 277n
Dudley Duncan, Otis, 7

Eastman, George, 118
Ecuador, *40*, 56, *57*, 223
Edison, Thomas A., 118
Education Amendments, 105, 235
Edward I, 128
El Salvador, *40*, 49, *57*, 183, 224, 244n30
Elias, Norbert, 76, 250n23
Elliott, Edward Bishop, 180, 272
Engels, Friedrich, 171
England, 16, 25, 75, 174, 204, 218, 229, 231
    customary measures in, 83, 95, 103, 167, 194, 249n
    and European Union, 25, 50, 55, 59, 212
    metric system adoption in, 57, 60, 140, 181, 218, 232, 234, 244n30
    metric system transition in, 23, 59, 65, 96, 117, 126, 155, 212, 214
    reform of weights and measures in, 59, 103, 126
    reluctance to collaborate with France and US, 31
Equatorial Guinea, *40*, 224, 246n52
Eritrea, *41*, 226
Estonia, *41*, 225
Ethiopia, *41*, 226, 248n85
European Economic Community, 95
European Union, 50, 59, 212
exceptionalism, American, 17, 207, 209

Fibonacci, 34
Fiji, *41*, 227
Finland, *40*, 224
First International Conference of American States (1899–1890), 50, 181, 182, 203, 232
First Pan American Standardization Conference, 200
foot, 5, 66, 86, 145, 153, 199, 214
    Bushwick foot, 5
    decimal, 88
    Foot Contest, 115
    Foot of 26th ward, 5
    international foot, 234, 236
    United States foot, 5
    US survey foot, 103, 234, 236
    Williamsburg foot, 5
Ford, Gerald, 116
Ford, Henry, 118
France, 8, *40*, 97, 195, 223, 229, 230, 244n30, 245n44
    cognitive penetration of state in, 80, 108
    colonialism, 42, 52–54, 56, 92, 230
    compulsory metric units, 83, 111, 230, 252n54
    currency in, 177, 179–80
    customary measures in, 8, 19, 86, 167, 213, 218–19
    international currency in, 178
    metric system reintroduction in, 75, 144

France (*continued*)
    mixture of measurement systems in, 90, 213, 265n93
    uniform standard of weights and measures in, 254–55
French Revolution, 37, 46, 88, 195, 211, 276
    creation of metric system and, 8–13, 31, 33, 47, 59, 144–45, 170
    metrological reform, 90, 229
French West Africa, 40

Gabon, *41*, 226, 248n85
Gagarin, Yuri, 215
gallon, 2, 69, 83, 111, 129, 162, 165, 197
Gallup, 106, 213, 279n9
Gambia, *41*, 227, 248n85
Garfield, James, 181
genealogical imagination, 142, 194
General Conference on Weights and Measures, 97, 202
geometric spirit, 9, 11–13, 136
Georgia, *41*, 225
Germany, *40*, 48, 86, 171, 192, 196, 197, 223
    metric system in, 48, 67, 98, 196, 197, 231
Ghana, *41*, 227, 248n85
Giza pyramid, 137, 138
Goodlatte, Robert, 85
Graham Bell, Alexander, 118
Graham Sumner, William, 118, 144
gram, 1, 14, 122, 161, 199, 230, 256n103
    in currency, 178–80
Great Britain, 31, 95, 193, 194, 196, 254n83
Greaves, John, 138
Greece, *41*, 226
Greenwich, 35, 36
    mean time, 17, 140, 232
Grenada, *41*, 50, 226
Guatemala, *41*, 45, 49–51, 57, 224
Guinea, *41*, 45, 224, 246n52
Guinea-Bissau, *41*, 224, 246n52
Guyana, *41*, 45, 50, 54, 57, 227

Hague, W. E., 196
Haiti, *41*, 50, 54
Halsey, Frederick A., 117, 187–93, 196–99, 204, 218, 233, 258n145
Hannibal, 55
Harrison, Benjamin, 97, 184, 232
Hassler, Ferdinand Rudolph, 91 92, 118, 216, 230
Hellerman, Steven L., 210
Henry I, 75
Henry VIII, 167
Herbert, Albert, 185, 192, 193
Herschel, John, 118, 138, 152
Hindu-Arabic numerals, 26, 32–36, 49, 94, 134–35, 185–86
Hine, Edward, 139
Hobsbawm, Eric, 46, 265n83
Honduras, *40*, 49, 57
Hoover, Herbert, 101, 102, 233
House of Representatives, 159, 230, 235
    Committee on Coinage, Weights and Measures, 93, 95, 99, 134, 155, 178, 184–86, 191
Hubbard, Henry, 69
Hugo, Victor, 11
Hungary, *40*, 223
Hutton, Frederick R., 108

Iannelli, Gerry, 107
Iceland, *40*, 224
Immerwahr, Daniel, 23
imperial measures, 42, 50, 55, 60–61, 103, 140–41, 199, 218
    as better for industrial processes, 189
    complexity of, 155, 197
    divine origin of, 138, 153
inch, 2, 15, 22–23, 66, 112–17, 128, 140, 145
    Anglo-Saxon, 141
    decimal, 88, 153
    English, 22, 103, 138 139
    pyramid, 138
    US, 103, 212
India, 34, *41*, 73, 169, 171
    metric system in, 62, 67, 194, 226, 234

Indonesia, *41*, 225, 246n52
Inter-American Metrology System (SIM), 202
International Brotherhood of Electrical Workers, 106, 116
International Bureau of Weights and Measures (BIMP), 97, 231
International Institute for Preserving and Perfecting Weights and Measures, 129, 136, 140, 156, 231
International Meridian Conference, 17, 232
International Meteorological Organization, 172
International Monetary Conference (1867), 176, 231
International System of Units (SI), 8, 120, 234
International Telegraph Union, 172
international time zones, 26, 33–36
internationalism, 35, 129, 161, 177, 179
Iran, *41*, 225
Iraq, *41*, 225
Ireland, 25, 59, 86, 193
    metric system in, *41*, 50, 67, 103, 194, 212, 226, 234–35
Israel, *41*, 139, 140, 225
Italy, 25, 40, 48, 52, 53, 56, 126, 223

Jackson, Michael, 29
Jamaica, *41*, 50, 227
Japan, 5, 21, *41*, 61, 63–64, 225, 234, 247n64
Jefferson, Thomas, 96, 126, 208, 229, 260n9
    and decimal currency, 87–89, 131, 175, 218, 219
    interest in metric system, 31, 108, 118, 213, 216, 254n76
Jenks, Robert, 20
Johnson, Andrew, 93
Johnson, Boris, 55
Johnson, John T., 1, 23, 201
Jomard, Edme-François, 137

Jordan, *41*, 225
Joule, James, 118

Kasson, John A., 93, 95, 130, 180, 262n37
Kazakhstan, *41*, 225
Kelley, William D., 178
Kelvin, lord (Thomson, William), 27, 118, 119, 143, 154–56, 186, 232
Kennelly, Arthur, 52, 53
Kenya, *41*, 50, 193, 226, 248n85
Khan, Oghuz, 55
kilogram, 1, 17, 67, 71, 124–25, 127, 144, 161, 174
    as basic unit of metric system, 118, 120, 233–34, 236
    equivalency, 12, 103
    as exclusive unit, 194, 198
    as gauging tool, 32, 84
    kilograms number 4 & 20, 97, 232
    as legal unit, 46, 93
    as metrological innovation, 122
    in mixed systems, 58–59
    as replacement of customary measures, 5, 8, 125, 159
    in science, 217
    as standard, 37, 58, 61, 97, 152, 216, 232
Kiribati, *41*, 227
Krakel, Dean, 16
Kula, Witold, 7, 77
Kunz, George, 121, 192, 197
Kuwait, *41*, 62, 226
Kyrgyzstan, *41*, 225

laissez-faire, 108, 161
    policies, 21, 101, 105, 205
    principle, 13, 104, 177, 190, 213
    standards, 15, 98, 211
Lalande, Joseph, 91
language, 16, 29, 34, 73, 80, 139, 188, 202
    American and Anglo-Saxon, 28
    of commerce, 159, 170, 172, 181, 185
    customary measures as part of, 151
    Dewey's ideas on, 135, 150
language (*continued*)

extinction, 60
  metric system as, 8, 19, 32, 36, 46, 72, 112
  metric systems as universal, of science, 8, 26, 30, 32, 88, 161
  metrological dialects, 81, 128
  metrological language, 3, 57, 67
  systems of measurement as, 80, 122, 128, 134
  uniform, 9, 12, 51
  universal, 81, 94, 144, 150, 161, 170–72, 176, 186
Laos, 41, 226
Laplace, Pierre-Simon, 118
Latimer, Charles, 16, 119, 129, 136, 140–43, 156–57, 218, 231
Latin America, 25, 56–58, 94, 96, 181, 196
  adoption of English units in, 199, 218
  alternative system of units in, 199, 200, 218
  customary measures in, 200–201
  immigrants from, 67
  metric system in, 27, 39, 42, 50, 200, 231
  metrication of, 64, 199, 232
  national measurement systems in, 202, 203, 278
  trade and, 159, 182–84, 192, 193, 197, 198, 201
  voluntary adopter of metric system, 56–58
Latvia, 41, 225
Lavoisier, Antoine, 88, 91, 118
Lebanon, 41, 225
Leslie, Seaver, 115
Lesotho, 41, 227, 248n85
Liberia, 6, 41, 43, 62, 63, 227, 230, 236, 244n26
Libya, 41, 55, 225, 246n52
Liechtenstein, 40, 223
Lincoln, Abraham, 93
liter, 71, 93, 111–12, 122–25, 144, 160, 235
  as basic unit, 5, 8, 12, 17, 29, 161, 199
  as exclusive legal unit, 46, 194, 198
  as gauging tool, 32, 37
  as metrological innovation, 122
Lithuania, 41, 225
Littauer, Lucius, 191
Louis XI, 75
Louis XVIII, 75
Luxembourg, 40, 56, 223, 230

MacArthur, Douglas, 64, 234
Macau, 41, 53, 226, 246n52
Macedonia, 40, 224
Macqueen, C. E., 127
Madagascar, 41, 226, 248n85
Madison, James, 89, 230
Malawi, 41, 227, 248n85
Malaysia, 41, 227
Maldives, 41, 226
Mali, 41, 44, 226, 248n85
Malinowski, Bronislaw, 123
Malta, 41, 224, 246n52
Marien y Arróspides, Tomás Antonio de, 170
Markovits, Andrei, 210
Mars Climate Orbiter, 217, 235
Marshall Islands, 6, 41, 43, 53, 62–63
Marx, Karl, 8, 171, 270n32
Mauritania, 40, 224, 246n52
Mauritius, 40, 223, 246n52
Maximilian I of Mexico, 58, 59
Maxwell, James, 118
Mayo, Charles H., 118
means of measurement, 26
  legitimate, 26, 70, 72, 76, 77
  monopoly of, 76, 78, 173
Mecca Time, 35
Méchain, Pierre, 10
Mendeleev, Dmitri, 118
Mendenhall, Thomas Corwin, 97, 115, 118, 152–53, 156, 218, 232
Mendenhall order, 85, 97, 152, 232, 275n103
mesures usuelles, 90, 144, 213–14, 230, 265n93

meter, 14, 45, 59, 66, 103, 112, 123, 146
    as attack on identity, 16
    as basic unit, 5, 8, 12, 17, 29, 31, 194
    definition, 10
    final magnitude, 51, 91, 229, 234, 235
    as gaugin tool, 32
    global character of, 174, 177
    as legal unit, 46, 93
    meter prototype, 216
    meter prototypes 21 & 27, 92, 232
    as metrological innovation, 122
    political and economic power of, 127
    as replacement of customary measures, 113, 211
    and schooling, 124
    and science, 118, 120
    as standard of measurement, 51, 58, 61, 92, 97, 152
Metre Convention (1875), 6, 21, 64, 97, 231–32
Metric Act (1866), 85, 93–95, 180, 231
Metric Conversion Act (1975), 16, 85, 105–6, 107, 116, 235
Metric Study Act, 103–4, 234
metric system
    anti-compulsory argument for, 153
    compulsory adoption of, 36–37, 72, 149–50, 187–88, 190–91, 245
    compulsory legislation for, 18, 51, 85, 94–95, 109–11, 143, 156, 218
    compulsory use of, 20, 42, 52, 108, 149, 212, 230–31
    as general and special knowledge, 122
    history of, 53, 104, 123, 125, 126, 135, 192
    imposition of as anti-democratic, 109–11, 149, 156
    legislation of, 6, 27, 49, 50, 59, 84, 99, 122, 134, 198
    literacy, 19, 34, 36, 66, 78, 101, 162, 110
    non-compulsory, 42, 56, 58, 93, 94, 107, 111, 221
    opposition to, 28, 85, 111, 140, 151, 187, 191, 212, 221

    rational opposition to, 147, 151, 187–88, 190
    reform, 35, 46, 49, 90, 99, 103, 123, 132, 133, 135
    spontaneous adoption, 213
metrication, 16, 20–22, 25, 49, 113, 149, 208, 211
    advocates of, 142, 154, 155, 218
    anti-democratic, 111
    by colonization, 53, 54, 58, 60–62
    "from below," 47–48, 213
    full, 18, 19, 26, 71, 72, 109, 194, 211
    and modernization, 25, 96, 214
    in science and engineering, 27, 39, 126, 128
    total, 18
    US resistance to, 219, 213, 217
    voluntary, 105, 106
metrology, 120, 122, 156, 202
    authority, 15, 47, 78, 79, 209
    autonomy, 75, 211
    education, 77
    history, 128, 140, 151, 170
    and religion, 157
    scientific, 120, 152, 156
    as special knowledge, 120
    in US, 218
Mexico, 18, 37, 40, 57, 197, 203, 223, 244n30
    compulsory metric laws in, 51, 108
    Constitution (1857), 49
    customary measures in, 5, 123, 193, 199, 200, 201
    metric system in, 45, 49, 56, 223, 231, 244n30, 245n44
    trade, 50, 51, 200, 201, 212
    transition to metric system, 58, 84, 100, 101, 123
Michell, John, 116
Micronesia, 6, 41, 43, 53, 62–63, 227, 234, 236
Mihm, Stephen, 156
mile, 2, 15, 50, 88, 111, 155, 235

Mintz, Sidney, 168
Moldova, *41*
Monaco, *40*, 56, 223
monetary system, 16, 76, 179
    Carolingian, 59
    decimal, 179
    duodecimal, 39
    hexadecimal, 127
    homogenization, 176
    monopoly of, 175
    universal, 16
    vigesimal, 39
Monge, Gaspard, 118
Mongolia, *41*, 57
Montenegro, *40*, 223
Montessori, Maria, 118, 192
Moreno Franginals, Manuel, 125
Morocco, *41*, 225, 246n52
Mozambique, *41*, 224, 246n52
Myanmar, *41*, 42, 49, 61, 227, 244n26

Namibia, *41*, 226, 248n85
Napoleon III, 58, 179
NASA, 215, 217, 235
nation-states, 26, 39, 47, 56, 61, 71–72, 79, 168
    metric system adoption of, 44, 221, 222
National Academy of Sciences, 93, 129
National Association of Manufacturers (NAM), 185–86, 190–92, 233, 273n84
National Board of Trade, 186
National Bureau of Standards (NBS), 20, 64, 67, 98, 186, 202, 232, 235
    destruction of defective instruments by, 68, 69
    Division of Simplified Practice, 101, 102, 233
National Conference on Weights and Measures, 99, 100, 233
National Efficiency Society, 135
National Institute of Standards and Technology (NIST), 15, 99, 107, 235

National Machine Tool Builders Association, 190
National Spatial Reference System, 103, 236
Nauru, *41*, 227, 244n30
Nepal, *41*, 226
Netherlands, *40*, 52, 54, 56, 75, 223
New York, 1–2, 69–70, 115, 136, 153, 169, 187, 207
New Zealand, *41*, 44, 57, 103, 117, 210, 226, 234
    in Colonial Conference, 60
    and independence of Samoa, 6
    in metric maps, 193–94
Newton, Isaac, 138, 153
Nicaragua, *40*, 49, 57, 224
Niger, *40*, 224, 246n52
Nigeria, *41*, 60, 62, 226, 248n85
Niyazov, Saparmurat, 55, 246n58
Noel, Edward, 127, 128
non-metric countries, 6, 23, *41*, 43, 62, 222, 244n26
North, Douglas, 162
North America, 51, 86, 182, 196, 202, 203
    as soft metric zone, 245n45
North American Free Trade Agreement (NAFTA), 50, 202, 203
    United States-Mexico-Canada Agreement (USMCA), 50, 212
North Korea, *41*, 66
Norway, *40*, 224
number systems, 122, 146, 148
    binary number system, 120, 175
    decimal, 32, 60, 62, 87, 89, 123–24, 135, 145–47, 151–52
    duodecimal 143, 145–48, 151, 218
    Hindu-Arabic, 26, 32–36, 49, 94, 134, 135, 186
    non-decimal, 34
    octonary, 127
    Roman, 34, 135
Nystrom, John William, 127

Office of Standard Weights and Measures, 92, 97–99
Oldberg, Oscar, 153
Oman, *41*, 227
O'Mealia, Leo, 2, 3
ounce, 55, 88, 145, 166, 197, 199
    fluid, 66, 168, 257n121

Paine, Thomas, 14
Pakistan, *41*, 226
Palau, 6, *41*, 43, 53, 62–63, 227, 234, 236
Panama, *41*, *57*, 225
Pan-American Conferences (1890s), 21, 50, 64, 182, 184, 197, 198, 202
Panero, James, 16
Papua New Guinea, *41*, 227
Paraguay, *40*, *57*, *57*, 224
path-dependence, 21–23, 148, 216
Peirce, Charles Sanders, 22, 29, 118, 127, 189, 275n103
Peru, *40*, 44, 56, *57*, 200, 223
Peter the Great, 34
Philip the Fair, 75
Philip the Tall, 75
Philippines, 21, *41*, 64, 67, 224, 233, 246n52
Piazzi, Giuseppe, 137
Piazzi-Smyth, Charles, 118, 136–42, 232
pint, 2, 36, 69, 103, 140–41, 197, 234
    decimal pint, 88
    demi-pint, 88
    US pint, 257n121
Poincaré, Henri, 115
Poland, *41*, 225, 245n50
Porter, Theodore, 52
Portugal, *40*, 54, 121, 141, 233
pound, 2, 20, 55, 86, 118, 199, 140–41, 152
    avoirdupois, 153, 256n103
    as customary measure, 67, 103, 129, 171
    decimal, 88
    derived from metric standards, 96, 103, 217, 232
    international, 103, 234
    as measure of currency, 59, 175, 186, 197, 235
    as measure of weight, 95, 113, 117, 130, 258n145, 262n38, 264n79
    sterling, 179
    troy, 91, 145
    world, 160, 233
pro-metric, 109, 181, 186, 204, 205, 212, 275n103
    associations, 132, 136, 180, 213, 233
    campaigns, 117, 196
    groups, 102, 112, 160, 184–86, 188, 194
    legislation, 151, 186
    organizations, 35, 93, 98, 185, 186
    personalities, 52, 110, 191, 267n118, 276n114
    scientists, 116, 129, 153, 154, 188
protectionism, 161, 188
Public Land Survey System, 21, 87, 253n70
Puerto Rico, *40*, 56, 224, 246n52
pyramid metrology, 137–40, 142, 143, 156, 194, 232
pyramidology, 136

Qaddafi, Muammar, 55
Qatar, *41*, 62, 227
quantification, 6, 7, 123, 154, 170
quantitative literacy, 72, 162, 166
quart, 69, 111, 130, 161
    world quart, 160, 233
Quebra-Quilo movement, 84
Queen Isabel II, 58
Quincy Adams, John, 4, 91, 96, 108, 118, 175, 218
    report on weights and measures (1820), 89, 213, 230

Reagan, Ronald, 85, 106, 218, 235
Reconstruction, 85, 93, 96, 212, 214, 214
regeneration, 9, 10, 13, 14
Republic of the Congo, *40*, 224, 246n52
Romania, *40*, 223, 244n30

Romero, Matías, 183, 273n78
Ruggles, Samuel, 177, 179, 180
Russia, 5, 23, 25, 34, 46, 55, 121, 194
  metric system in, *41*, 47, 49, 67, 181, 225, 233
Rwanda, *41*, 224, 246n52
Ryff, Andreas, 171

Saint Kitts and Nevis, *41*, 50, 227, 244n30
Saint Lucia, *41*, 50, 227
Saint Vincent and the Grenadines, *41*, 227
Samoa, 6, 25, *41*, 42, 227
San Marino, *41*, 224
Sao Tome and Principe, *40*, 224, 246n52
Sarton, George, 204
Saudi Arabia, 35, *41*, 226
Schaffer, Simon, 142
Sellers, Coleman, 142, 232
Senate, 21, 93, 94, 97, 177, 230, 231
Senegal, *40*, 53, 56, 2230, 230, 246n52
Serbia, *40*, 223
Seychelles, *40*, 224, 246n52
Sherman, John, 177
Sierra Leone, *41*, 45, 60, 227, 248n85
Singapore, *41*, 226
Sistema Interamericano de Metrología y Calibración (SIMYC), 202
Slosson, Edwin Emery, 153
Slovakia, *40*, 223
Slovenia, *40*, 223
Smoot, Oliver R., 119, 260n12
Society for the Promotion of Engineering Education, 108
sociology of measurement, 7
Solomon Islands, *41*, 227
Somalia, *41*, 61, 226, 248n85
Sombart, Werner 209–10
South Africa, *41*, 103, 117, 193, 194, 210, 226, 248n85
South Korea, *41*, 222
South Sudan, *41*, 227
Southard, James H., 134, 191

sovereignty, 7, 25, 47, 75, 77–78, 204, 215
Spain, *40*, 45, 58, 61, 92, 141, 167, 223, 247n70, 261n28
  colonialism, 54
  in Congress on Definite Metric Standards, 56, 242n3
spelling reform, 130, 132, 133, 262n38, 273n86
Spencer, Herbert, 27, 116, 118–19, 143–57, 188, 218, 232, 265n92, 266n114
Sri Lanka, *41*, 227
standard time, 35, 134, 172, 180
standardization, 101–2, 129, 202
  conferences for, 197, 200
  of customary measures, 92, 124, 163, 199
  as federal process, 20, 186, 191–92, 211
  in industry, 102, 121–22, 129, 156, 205
  international, 51, 65, 172, 195
  of national languages, 173
  of products, 15, 101, 167, 169, 198
  as tool of legibility, 80
  voluntary, 20, 102
  weak or lacking, 86, 91, 100, 169, 170
  of weights and measures, 18, 92, 132, 161, 164, 169–71, 230
state
  administrative functions of, 214
  activist, 213
  American, 14, 19, 176, 211
  associative, 101, 102
  authority of, 93
  capacity, 19, 96
  centralized, 9, 13, 26, 28, 74, 80, 170, 211
  cognitive penetration of, 80, 124
  and compulsion, 71, 72, 113, 149
  concentration of capital by, 79
  formation, 168
  homogenization and, 90, 164, 205
  intervention, 9, 18, 107. 188
  legitimacy of, 21, 214
  and markets, 28
  measurement, and, 73, 84–85,
  and metrication, 19, 38

state (*continued*)
    metrological capacities of, 24, 26, 69, 71
    metrological standardization and, 71, 72, 81, 190
    and monopolies, 76
    monopoly of means of measurement of, 70, 72, 76, 77, 78, 82, 173, 190, 211
    monopoly to coin money, 173
    parochial, 74
    power of, 11, 18, 19, 26, 71, 72, 76, 143
    as social frameworks of knowledge, 37, 79, 80
    sovereignty of, 25, 53
    state-sanctioned knowledge, 78
    symbolic power, 19–21
    time and space quantification by, 79
    universal, 74
Stevin, Simon, 32, 33
Stone, Charles W., 181, 184
Stratton, Samuel, 98, 186, 192, 232
Studebaker, Clement, 183
Sudan, *41*, 61, 226, 248n85
Sumner, Charles, 93, 94, 180
Sumner, William Graham, 118, 144, 145
Suriname, *40*, 45, 56, *57*, 223, 246n52
Survey of the Coast, 92, 230
Swaziland, *41*, 62, 226, 248n85
Sweden, *40*, 223
Switzerland, *40*, 53, 91, 223
symbolic capital, 72, 79, 211, 214
Syria, *41*, 54, 225, 246n52
systems of measurement, 3, 5, 7–8, 20, 102, 109, 150–51, 197–98
    and capitalism, 167–71
    and cognition, 122–23
    customary, 244
    as economic institutions, 162, 163
    Egypt, *41*, *57*, 73, 116, 137–38, 225, 247n65, 248n85, 249n9
    Egyptian, 139
    French, 139
    Hebrew, 130
    legal, 30, 76, 94, 203, 217, 218
    modern, 121, 163, 166, 167, 170
    rational, 31, 145, 170, 172
    scientific, 118, 125
    unified, 48, 71–72, 75, 80, 170, 173–74, 218, 222
    universal, 90, 96

Taiwan, *41*, 225
Tajikistan, *41*, 225
Tanzania, *41*, 50, 226, 248n85
tariffs, 28, 187, 190
taxation, 4, 48, 72, 76, 78, 81–82, 84, 177–78
    in kind, 54, 162
Taylor, Alfred B., 127
Taylor, Frederick, 187
Taylor, John, 138
Tefft, Thomas Alexander, 177
Thailand, *41*, 194, 224
Thompson, Elihu, 118, 186
Timor-Leste, *41*, 53, 226, 246n52
Tocqueville, Alexis de, 9–10
Togo, *41*, 225, 246n52
Tonga, *41*, 227
Totten, C. A. L., 139–41, 143
Toynbee, Arnold, 74
Tralles, Johann Georg, 91, 92
Trinidad and Tobago, *41*, 50, 227
Tunisia, *40*, 224, 246n52
Turkey, 5, 25, *40*, 49, *57*, 223
Turkmenistan, *41*, 55, 225
Tuvalu, *41*, 227
twenty-foot equivalent unit (TEU), 66

Uganda, *41*, 50, 226, 248n85
Ukraine, *41*, 225
United Arab Emirates, *41*, 226
United Kingdom, *41*, 42, 214, 226, 234, 235
United Kingdom of the Netherlands, *40*, 52, 230

United States Congress, 84–91, 144, 155, 218, 299, 230, 254n83, 262n38
    metric legislation, 93, 95–99, 103–5, 111, 131, 134, 143–44, 185, 191–216
United States Customs Service, 184
United States Department of the Treasury, 14–15, 91–93, 97, 99, 175, 180, 230, 231–32
    and counterfeiting, 175, 231
    Secret Service, 175
United States Mint, 91, 97, 175
Urey, Harold, 17
Uruguay, *40*, 56, *57*, 223
US Metric Board, 85, 105, 112, 218, 235
USSR, *41*, 47, 55
Uzbekistan, *41*, 225

Vanuatu, *41*, 227
Venezuela, *40*, 56, *57*, 223
Vestal, Albert, 99
Victor Emmanuel II, 48
Vietnam, *41*, 54, 246n52
Voltaire, 59

Washington, George, 86, 87, 208, 228
Watt, James, 118, 126, 195–96

Weber, Max, 164, 165
weights and measures
    adapted to metric units, 36–37
    divinely inspired, 148
    history of, 3, 7, 142, 207, 218, 253n61, 276n115
Wells, William C., 198
Western Sahara, *41*, 225, 246n52
Westinghouse, George, 268
Wigglesworth, Edward, 110
Wilberforce Mann, William, 127
William I of the Netherlands, 75
Wilson, Woodrow, 197
Wolfe, Tom, 116
World Calendar Association, 135
World Trade Club of San Francisco, 192

yard
    international yard, 103, 234
    world yard, 160, 233,
yardstick, 2, 3, 36, 159, 160, 161
Yemen, *41*, 227

Zambia, *41*, 227
Zelizer, Viviana, 175
Zimbabwe, *41*, 226, 248n85

www.ingramcontent.com/pod-product-compliance
Lightning Source LLC
Chambersburg PA
CBHW021848230426
43671CB00006B/315